Sustainable Development Goals Series

World leaders adopted Sustainable Development Goals (SDGs) as part of the 2030 Agenda for Sustainable Development. Providing in-depth knowledge, this series fosters comprehensive research on these global targets to end poverty, fight inequality and injustice, and tackle climate change.

The sustainability of our planet is currently a major concern for the global community and has been a central theme for a number of major global initiatives in recent years. Perceiving a dire need for concrete benchmarks toward sustainable development, the United Nations and world leaders formulated the targets that make up the seventeen goals. The SDGs call for action by all countries to promote prosperity while protecting Earth and its life support systems. This series on the Sustainable Development Goals aims to provide a comprehensive platform for scientific, teaching and research communities working on various global issues in the field of geography, earth sciences, environmental science, social sciences, engineering, policy, planning, and human geosciences in order to contribute knowledge towards achieving the current 17 Sustainable Development Goals.

This Series is organized into eighteen subseries: one based around each of the seventeen Sustainable Development Goals, and an eighteenth subseries, "Connecting the Goals," which serves as a home for volumes addressing multiple goals or studying the SDGs as a whole. Each subseries is guided by an expert Subseries Advisor.

Contributions are welcome from scientists, policy makers and researchers working in fields related to any of the SDGs. If you are interested in contributing to the series, please contact the Publisher: Zachary Romano [Zachary.Romano@springer.com].

More information about this series at http://www.springer.com/series/15486

Godwell Nhamo · David Chikodzi

Cyclones in Southern Africa

Volume 1: Interfacing
the Catastrophic Impact of Cyclone
Idai with SDGs in Zimbabwe

Godwell Nhamo
Institute for Corporate Citizenship
University of South Africa
Pretoria, South Africa

David Chikodzi
Institute for Corporate Citizenship
University of South Africa
Pretoria, South Africa

ISSN 2523-3084 ISSN 2523-3092 (electronic)
Sustainable Development Goals Series
ISBN 978-3-030-72395-8 ISBN 978-3-030-72393-4 (eBook)
https://doi.org/10.1007/978-3-030-72393-4

This Springer imprint is published by the registered company Springer Nature Switzerland AG
The registered company address is: Gewerbestrasse 11, 6330 Cham, Switzerland

Preface

Never did the Southern African region experience a devastating tropical cyclone prior to Cyclone Idai that made landfall in mid-March 2019. Tropical Cyclone Idai resulted in catastrophic impacts, mainly in Malawi, Mozambique and Zimbabwe. Prior to, and after Tropical Cyclone Idai's disastrous landfall, Southern Africa had experienced several tropical cyclones including Cyclones Kenneth (2019), Dineo (2017), Favio (2007) and Eline (2000). However, the pathway for Tropical Cyclone Idai was so different and unique and resulted in challenges in Early Warning Systems as there was no benchmark from the history books. The United Nations Office for the Coordination of Humanitarian Affairs (OCHA) reported 602 deaths, 1600 injuries, 1.85 million people in need and an estimated US$285 million emergency relief funding support as a result of this event. With these figures, Tropical Cyclone Idai became the most severe tropical cyclone to hit Southern Africa in living memory. Immediately after Tropical Cyclone Idai, there was Tropical Cyclone Kenneth that left 45 people dead, 374,000 more displaced and US$104 million in emergency relief funding needed. Based on the Rapid Impacts Needs Assessment report, the Zimbabwe Cabinet Committee on Environment, Disaster Prevention and Management put the total damages in Zimbabwe alone between US$548 million and US$622 million, with over US$800 million required for recovery if communities are to build back better. The agriculture, housing and transportation sectors were identified as the hardest hit. Given the foregoing, this book is dedicated to document and analyse the catastrophic impacts of Tropical Cyclone Idai in Zimbabwe and also add to the huge gap of knowledge in the fields of disaster risk reduction and management. The book further interfaces the findings with the 2030 Agenda for Sustainable Development that embeds 17 interlinked Sustainable Development Goals.

Pretoria, South Africa
Godwell Nhamo
David Chikodzi

Acknowledgements

We, the authors, Prof Godwell Nhamo and Dr David Chikodzi, wish to thank all the blind peer reviewers for their invaluable inputs during the writing and publishing process of this book. We thank Springer for taking on board this book project and for the quality control processes. We also wish to thank our families for their ongoing support of our work. The book project was coordinated through the Exxaro Chair in Business and Climate Change at the University of South Africa (UNISA). The Exxaro Chair is a research Chair funded by the Exxaro Resources (Pty) Ltd. Chairman's Fund and hosted by the Institute for Corporate Citizenship at UNISA. The Exxaro Chair was established in 2008 and is now in its fourth term running (2018–2022).

The authors wish to thank the Government of Zimbabwe and traditional leadership structures for granting permissions to undertake the research. Among some of the government ministries and agencies that granted permissions are the Ministry of Local Government, Public Works and National Housing; Ministry of Primary and Secondary Education; Ministry of Health and Child Care; Meteorological Services Department of Zimbabwe; Forestry Commission; The Postal and Telecommunications Regulatory Authority of Zimbabwe (POTRAZ); Manicaland Provincial Administrator's Office; Chimanimani District Administrator's Office; Chimanimani Rural District Council and the Chiefs and other traditional leadership structures in Chimanimani.

The authors also wish to acknowledge the hard work put by UNISA editing team that include the following: Nkeke Thosago, Dr Malvin Vergie, Louis Kotzé, Jane Franz, Shirley Steenekamp, Debbie Rodrigues, Lynette Posthumus, Willem Pretorius, Alexa Barnby, Elmarie Williams, Dolly Mathabatha and Alfred Nethanani. These individuals worked throughout the COVID-19 lockdowns to make sure that the quality of the book meets international language standards.

The authors further thank the following people that took park during the fieldwork, Joshua Chibvuma, Crecentia Gandidzanwa, Cowen Dziva, Knowledge Mwonzora, Itai Kabonga, Ernest Mando, Smart Mhembwe, Donald Chikoto, Emmanuel Maziti, Prof Soul Shava, Prof Amos Saurombe, Dr Talkmore Saurombe, Charity Denhere, Simukai Mukoyi and Matthew

Mare. We also thank all those that gave us responses, be it from the focus group discussions, household questionnaire survey, interviews and cooperated with us during field observations. Last but not least, the authors thank Exxaro Chair in Business and Climate Change Junior Researchers, Hlengiwe Precious Kunene and Nthabiseng Mashula for the work they performed as part of our team with internal chapter reviews.

This book is blind peer-reviewed. Apart from being the international norm, this double-blind peer review process is mandatory for South African-based authors in order to fulfil the requirements of the Department of Higher Education and Training's (DHET) policy for recognised research outputs for subsidy purposes. The authors invested their time to incorporate observations from the blind peer review process, an aspect that enhanced the quality of the product.

Contents

About the Authors

Godwell Nhamo, PhD Lead Author and Accounting Officer, is a Full Professor and Exxaro Chair in Business and Climate Change at the University of South Africa (UNISA). He is a National Research Foundation (NRF) C-Rated researcher undertaking research in the fields of climate change and governance, sustainable tourism, green economy and sustainable development. Prof Nhamo has conceptualised and completed 11 book projects (3 co-authored). The most recent being: *Counting the Cost of COVID-19 on the Global Tourism Industry* (2020); *Scaling up SDGs Implementation: Emerging cases from state, development and private sectors published by Springer* (2020); *SDGs and Institutions of Higher Education published by Springer* (2020) and *SDG 7: Ensure Access to Affordable, Reliable, Sustainable and Morden Energy by Emerald* (2020). Prof Nhamo has also published over 90 journal articles. Since 2013, Prof Nhamo has graduated 11 PhDs and hosted 10 postdoctoral fellows. Prof Nhamo's recently completed mega research projects include one on Cyclones, Tornados and Floods in the era of SDGs in Southern Africa and another on SDGs for Society. Professor Nhamo sits in a number of both international and national boards and has received several awards and recognitions for his outstanding work both locally and internationally. Finally, Prof Nhamo was one of the four-member African Union High-Level Panel that drafted the Green Innovation Framework for the continent. Email: nhamog@unisa.ac.za

David Chikodzi, PhD is a postdoctoral fellow with the Exxaro Chair in Business and Climate Change at the University of South Africa. His research interests are in climate change, water resources management, tourism, sustainable development and application of Earth Observation technologies for societal benefit. He has worked for over 10 years in academia at Great Zimbabwe University (Zimbabwe) and has published over 30 journal articles, one co-authored book titled *Counting the Cost of COVID-19 on the Global Tourism Industry* (2020) and several book chapters. Dr Chikodzi has also taken part in several local and international funded research projects across Southern Africa and has previously worked as a Research Scientist at the Scientific and Industrial Research and Development Centre in Zimbabwe. Dr Chikodzi is a former member of the ISIbalo Africa Young Statisticians and also the Zimbabwe Young Academy of Sciences. Email: chikod@unisa.ac.za

Part I

Introduction and Background

The Catastrophic Impact of Tropical Cyclone Idai in Southern Africa

1

Abstract

No words can describe the nature of the landfall and the resultant negative impacts from Tropical Cyclone Idai. The darkness from the storm of 15 March 2019 was described by one eyewitness as twofold the normal, the rainfall equated to having been pouring, and the sound of rolling stone boulders as if the world was ending abruptly. The cries of the dying and the injured could be vividly remembered by tormented survivors, and officials were left in shock as everybody was caught unprepared. This holding chapter profiles the catastrophic landfall of Tropical Cyclone Idai, drawing data and information from official documents and the fieldwork undertaken in some parts of Zimbabwe, including a household questionnaire survey, key informant interviews and focus group discussions. Data was also generated using remote sensing and Geographical Information Systems (GIS). The results proved beyond any shadow of doubt that Tropical Cyclone Idai was incomparable in terms of the damage and the manner in which the event took Zimbabwe many years back in development. To this end, the attainment of the 2030 Agenda for Sustainable Development by Zimbabwe became more remote. It was indeed, a catastrophic landfall!

Keywords

Cyclone Idai · Hurricanes · Typhoons · Landfall · Malawi · Mozambique · Zimbabwe

1.1 Introduction and Background

Catastrophic! Terrifying! Horrific! Unbearable! Disastrous! Calamity! Tragedy! Heartless! Adjectives run out when it comes to describing the nature and experience from Tropical Cyclone Idai landfall. Tropical Cyclone Idai was likened to the 2004 Indian Ocean tsunami by those who experienced it first hand, and the global citizenry took notice. Never before had Malawi, Mozambique and Zimbabwe experienced such due to a natural hazard. On the other hand, the warnings on increasing hazards were there as the Capacity for Disaster Reduction Initiative (CADRI) concluded that Zimbabwe was prone to climate-induced hazards that included floods, tropical cyclones and droughts (CADRI 2017). These hazards were also reported as regularly triggering food insecurity, disease and pest outbreaks.

The catastrophic landfall of tropical cyclones (also known as hurricanes in the Americas and typhoons in South East Asia) result in social, political, and economic destruction in societies, as well as physical damages across the natural environment and biodiversity. In the context of the 2030 Agenda for Sustainable Development (AfSD) and the aligned 17 Sustainable Development Goals (SDGs) (United Nations 2015), it becomes inevitable to relate how progress towards 2030 can be derailed by tropical cyclones. The disastrous landing of tropical cyclones may lead to increased poverty levels (SDG 2), thereby bringing about inequality (SDG 10) and gender-based violence (SDG 5). Tropical cyclones may also result in the destruction of the agricultural sector (SDG 2) and deteriorating health and wellbeing conditions (SDG 3) as health facilities are destroyed. The quality of education delivery (SDG 4), as well as water and sanitation provision (SDG 6) are also impacted negatively and at times take a long time to rehabilitate.

Other impacts from catastrophic tropical cyclones landfalls that are linked to climate change (SDG 13) include damage to energy (SDG 7), human settlements (SDG 11), as well as industry, trade and transportation infrastructure (SDG 9) (United Nations 2015). Communities are robbed of sustainable job opportunities (SDG 8), while massive internal displacements take place leading to insecurity (SDG 16). On the environmental front, forests and biodiversity (SDG 15) and the ocean economy (SDG 14) all get disturbed with huge negative economic impacts. In addition, when such events like Tropical Cyclone Idai happen, they demand genuine cooperation and partnership (SDG 17) among and between key stakeholders. Lastly, in the interest of the call under the AfSD to let no one be left behind (Nhamo et al. 2019), every effort should be made to ensure that disadvantaged communities—which include indigenous peoples, people with disabilities, women, the youth and children—are involved in all disaster risk reduction (DRR) and management stages (UNDRR 2015).

There are also cross-cutting matters that emerge from the SDGs that need addressing after tropical cyclone landfalls such as information and communication technology (ICT) and tourism (Dube et al. 2020; Nhamo et al. 2020). Other global development agendas that also get negatively impacted in terms of their implementation from tropical cyclones damage include the Sendai Framework on Disaster Risk Reduction (UNDRR 2015), the New Urban Agenda (United Nations Habitat 2016) and the Paris Agreement (UNFCCC 2015). These and other policies of interest are the subjects of discussion in Chap. 2, which deals with building and building back better (BBB) in the context of all global disasters.

Prior to, and after Tropical Cyclone Idai's disastrous landfall in March 2019, southern Africa had experienced several tropical cyclones. These tropical cyclones include Kenneth (2019), Dineo (2017), Favio (2007) and Eline (2000). Two other tropical cyclones were also experienced shortly after Tropical Cyclone Idai during the 2020/2021 rain season and these included tropical cyclones Chalane (2020) and Eloise (2021). Their pathways are depicted in Fig. 1.1. What is clear is that the pathway for Tropical Cyclone Idai was so different and unique. This resulted in some challenges in Early Warning Systems (EWS) as there was no benchmark from the history books. What emerges is that countries such as Mozambique, Seychelles, Comoros, Madagascar, Reunion and Mauritius remain on the tropical cyclone pathways.

Tropical Cyclone Idai made landfall in Mozambique and other above-mentioned countries in March 2019. The United Nations Office for the Coordination of Humanitarian Affairs (OCHA) reported 602 deaths, 1600 injuries, 1.85 million people in need and an estimated US$285 million emergency relief funding support as a result of this event (OCHA 2020). With these figures, Tropical Cyclone Idai became the most severe tropical cyclone to hit southern Africa in living memory. Immediately after Tropical Cyclone Idai there was Tropical Cyclone Kenneth that left 45 people dead, 374,000 more displaced and US$104 million in emergency relief funding needed. Based on the Rapid Impacts Needs Assessment (RINA) report, the Zimbabwe Cabinet Committee on Environment, Disaster Prevention and Management (CCEDPM) put the total damages in Zimbabwe alone between US$548 million and US$622 million (CCEDPM

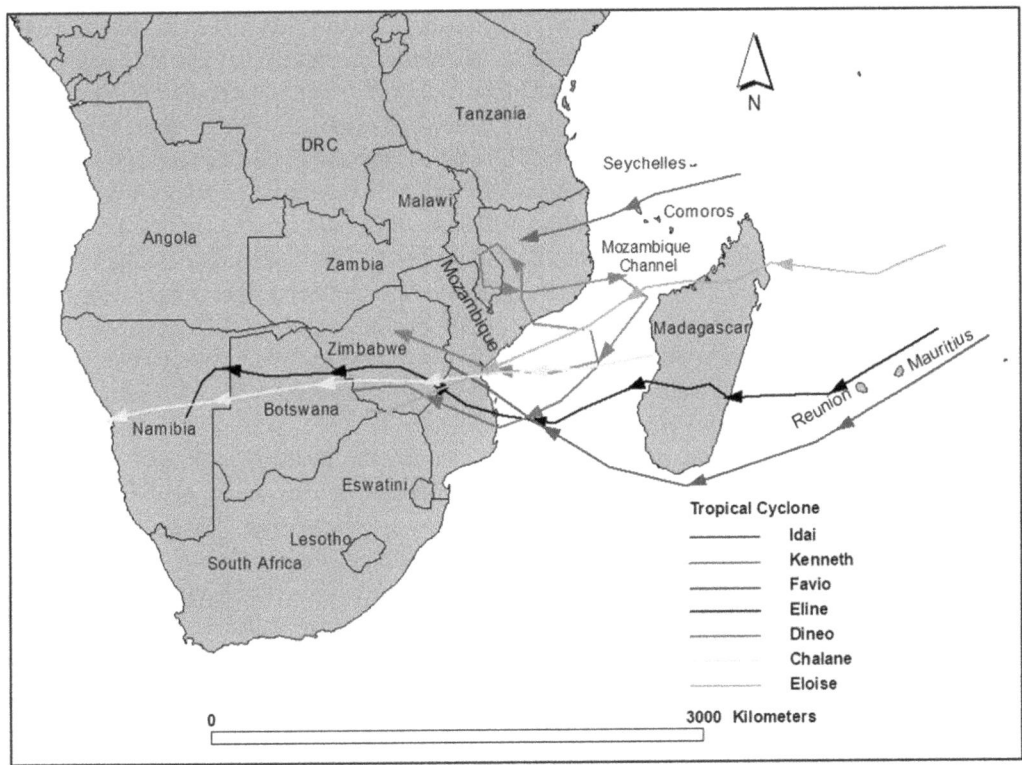

Fig. 1.1 Pathways of some tropical cyclones in southern Africa. Source: Nhamo et al. (2020)

2019), with over US$800 million required for recovery if communities are to build back better (BBB). The agriculture, housing and transportation sectors were identified as the hardest hit. For those with knowledge of the goings-on in the DRR and management spaces, 2019 became a year of massive damage in southern Africa. It was a tragedy! Tropical Cyclone Idai was classified by the Joint Typhoon Warning Centre (JTWC) as a Category 3 cyclone (ZMSD 2019) and sustained winds of up to 167 km/h.

Huge human life and economic losses from tropical cyclone landfalls are common. Hurricanes Michael (2018), Florence (2018), Maria (2017), Irma (2017), and Harvey (2017) collectively led to 3269 direct and indirect deaths (Shi et al. 2020). The hurricanes also resulted in an estimated US$325 billion in economic loss. They usually deposit huge amounts of rainfall accompanied by strong winds that leave a trail of destruction on many sectors including agriculture (livestock, forestry, crop and fisheries) (Winer

et al. 2020; Frame et al. 2020). The energy sector, including hydropower, is also vulnerable to tropical cyclones. In Mozambique, for example the Cahora Bassa hydropower plant and its distribution network usually gets hit (Uamusse et al. 2020), leaving southern Africa with blackouts like what happened during Tropical Cyclone Idai. South Africa had to send recovery teams to restore power as transmission lines and poles, transformers and other infrastructure went down. Paerl et al. (2019) observe a disturbing trend of increasing catastrophic tropical cyclone flooding in coastal North Carolina in the USA. The findings were based on continuous rainfall records of 120 years that revealed a trend of increasingly high precipitation associated with tropical cyclones. Extreme flooding was also reported during Hurricane Florence (Callaghan 2020). Ultimately, such trends have significant destruction across all economic sectors when the tropical cyclones make landfall and also move to other countries inland.

This chapter presents an overview of the catastrophic impacts of Tropical Cyclone Idai in Zimbabwe. The chapter is presented as the landing chapter for the entire book, bringing insights on the nature of damages caused by the tropical cyclone to selected sectors. The chapter raises an objective to determine the nature of damage that Tropical Cyclone Idai brought to Zimbabwe, with special reference and emphasis on the Manicaland Province that was hardest hit. The intention is to provide readers with a quick overview of what happened and kindle the reader's interest on this and other books addressing the impacts of Tropical Cyclone Idai beyond Zimbabwe.

1.2   Summary of the Methodological Orientation

The book uses a mixed method approach. Since each chapter details the methodology applied, this section will only summarise the methods without giving further details. As is the normal protocol for research, ethics clearances were obtained from both the University of South Africa (South Africa) and Chinhoyi University of Technology (Zimbabwe), which was a formal research partner. Furthermore, ministerial clearances and research permissions were also sought from Zimbabwe and those granted were from the Ministry of Local Government, Public Works and National Housing; Ministry of Health and Child Care; Ministry of Primary and Secondary Education; and the Postal and Telecommunications Regulatory Authority of Zimbabwe (POTRAZ). Research permission was also granted by the traditional chiefs in Chimanimani. These permissions and ethics clearances were enabling platforms for the research team of 17 individuals to get into the field, with some coming out of the country for three weeks.

Since the research was also taking place in a disaster area, all those involved undertook three days of training on how to conduct research in disaster areas. Included in the facilitators were two qualified psychologists, as well as practising

clinical psychologists. The training included hands-on scenarios likely to be encountered during fieldwork such as encountering grieving and remorseful respondents. A suitable timing had to be agreed upon to get to Chimanimani in terms of the desire to minimise the harm of respondents. Hence, seven months after Tropical Cyclone Idai was agreed as the appropriate timing leading to fieldwork end of September and in October 2019. Additional and ongoing follow-up, up to February 2021, were made through various platforms that included social media, calls, WhatsApp calls and emails.

Among the methods used was a household questionnaire survey ($n = 219$) that involved piloting in Chimanimani. The questionnaire survey was administrated offline on the QuestionPro platform, with researchers uploading it during the evenings following daily briefings. There were nine focus group discussions undertaken as granted by the Higherlife Foundation, Forestry Commission, ZMSD, at Kopa, Chimanimani District Medical Office, Chimanimani Ministry of Education, at Dombera Farm, with business representatives at Chimanimani Centre, and at Postal and Telecommunication Services in Chimanimani. There were 64 additional key informant open-ended interviews that included three chiefs and the Chimanimani East Member of Parliament. What was also interesting is the fact that the research team was officially introduced to the impacted communities, especially those in Kopa and Ngangu, by the Acting Chimanimani District Development Coordinator and his other operating government arms. The research team was also introduced to the Manicaland Provincial Development coordinator and his functionaries. All these protocols enabled the team to work late into the night interviewing community members without fear of crime. It was clear that the community gave their responses freely, as they had seen the teams introduced during the first two days of stay in Chimanimani.

There were many documents gathered during the fieldwork and after. Among such documents were official Tropical Cyclone Idai monthly status reports supplied both by the Department of Civil Protection and the Manicaland Provincial

Development Coordinator's Office. Other documents include those supplied by POTRAZ and the ZMSD. Other team members had a chance to attend to the United Nations Economic Commission for Africa (UNECA) on "Building back better: planning workshop for climate-resilient investment in reconstruction and development in cyclone-affected regions of Malawi, Mozambique and Zimbabwe". The workshop took place in Harare from 23 to 25 October 2019. Many documents presented during the workshop were used for the work and are available online at https://www.uneca.org/building-back-better. There was also the tracing of key Tropical Cyclone Idai media reports. The use of GIS and remote sensing was extensively utilised given the focus of the book. On the ground capturing of scenes through photographs was done extensively with thousands of photos being taken with permission. In addition, every effort was made not to have people or any recognisable faces in such photos. Lastly, the data analysis was done using document analysis protocols that borrowed some perspectives from grounded theory. Both the MS Excel and QuestionPro were also used to plot graphs and other descriptive statistics.

1.3   Catastrophic Impact of Tropical Cyclone Idai in Zimbabwe

This main section of the chapter comes in several subsections. These include the following: an overview of; what victims and residents heard and saw; rescue, death, injury, the missing, displacements and psycho-social support; impact on transportation and telecommunications infrastructure; as well as the health, water, sanitation and hygiene sectors. The chapter also has subsections on Tropical Cyclone Idai impacts on the agricultural sector; impacts on housing, business and related infrastructure; impacts on educational facilities; the impact on natural resources, wildlife and tourism attractions; challenges in early warning and other disaster risk reduction (DRR) management systems; and the (un)sung heroes involved in search and rescue as well as

missions. Each of these sections is discussed in turn in the next paragraphs.

1.3.1   An Overview

The nature, extent and depth of the devastation from Tropical Cyclone Idai is probably best summarised in the Manicaland Province Administrator's Office[1] report of 8 July 2019. The document gives the following encounter with the tropical cyclone:

> We have experienced tropical cyclones before notably Eline, Japhet, Dineo among others. They did not come with the level of destruction the province experienced from Cyclone Idai. The magnitude of Cyclone Idai was never anticipated. We used to encourage people to move to higher ground to avert flooding related disasters. However, Cyclone Idai was more devastating on high ground, especially sloping areas, than ever before. This was further compounded by the misty weather, wind and persistent rainfall for about four consecutive days, making it difficult for helicopters to fly to the epicentre of the disaster (Manicaland Province Administrator's Office 2019, p. 7).

Tropical Cyclone Idai hit Zimbabwe on 15 March 2019. Based on the information supplied by the Manicaland Provincial Administrator's Office during the fieldwork in 2019, the districts in the province were affected differently. Based on fieldwork, the most affected district and the epicentre was Chimanimani. This was followed closely by Chipinge district. Details regarding other districts are shown in Fig. 1.2

A summary concerning the impact of Tropical Cyclone Idai on several economic and social sectors from a household survey undertaken with mainly the victims from Chimanimani District is presented in Figs. 1.3 and 1.4. The "n" in the graphs represent the number of respondents per that category. What comes out clearly from the household survey is that there was extensive damage to many sectors. Further details will be

---

[1] The office has the mandate to coordinate DRR at provincial level and there was a name change transition to have it known as the Provincial Development Coordinator's Office.

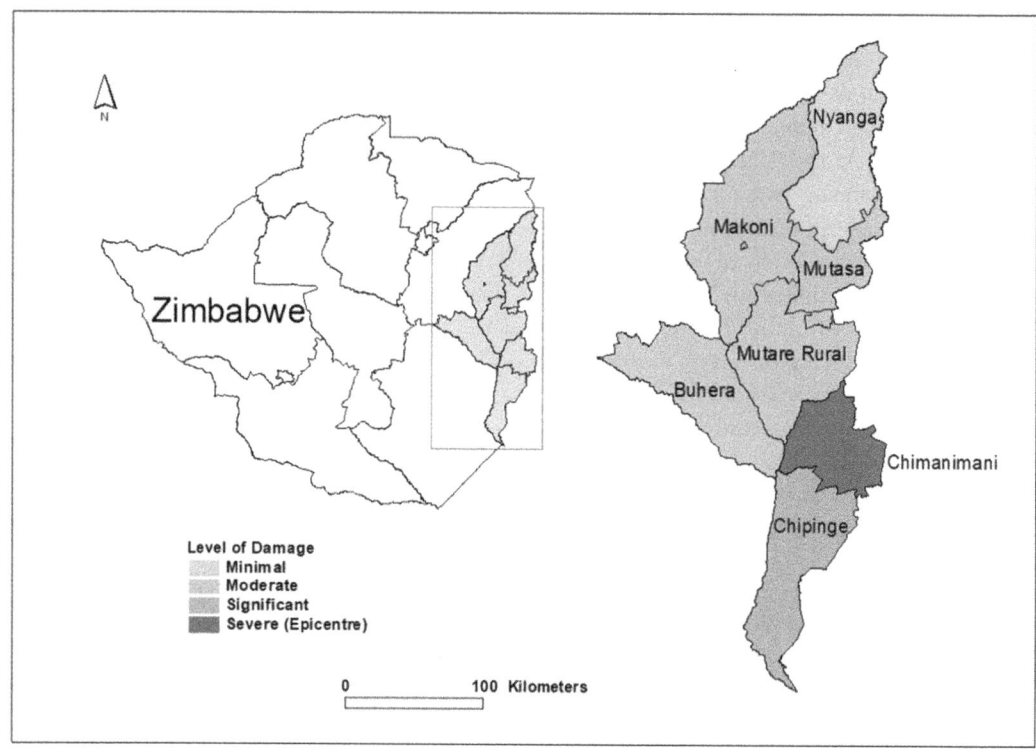

Fig. 1.2 Tropical Cyclone Idai Damage in Zimbabwe. Source: Authors

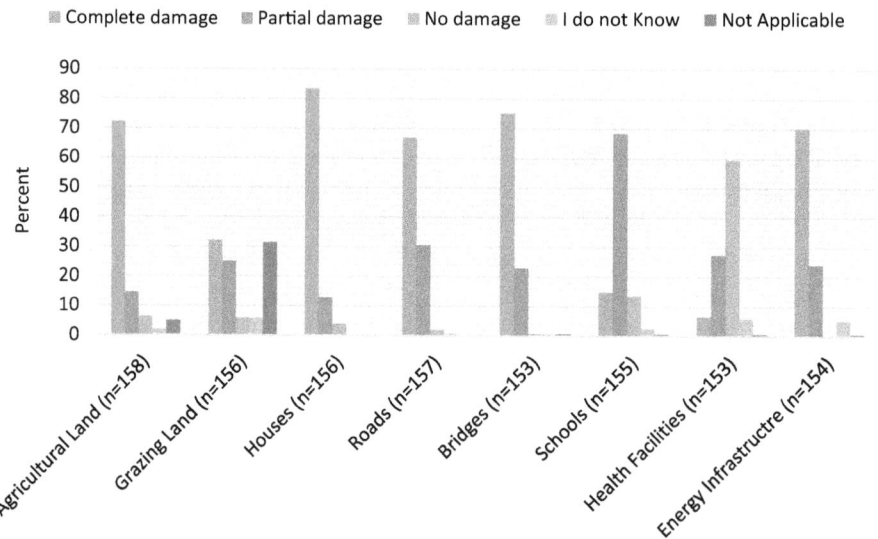

Fig. 1.3 Impact of Tropical Cyclone Idai on selected sectors (Part A). Source: Authors, Household Survey (2019)

presented in the relevant sections and/or in full chapters in the book.

A summary of Tropical Cyclone Idai impact is also summarised by Zimbabwe's Environmental Management Agency (EMA). The EMA (2019) highlights that in Chimanimani, 15 administrative wards were affected, with wards 15 and 21 being the most affected. In Chiping 10

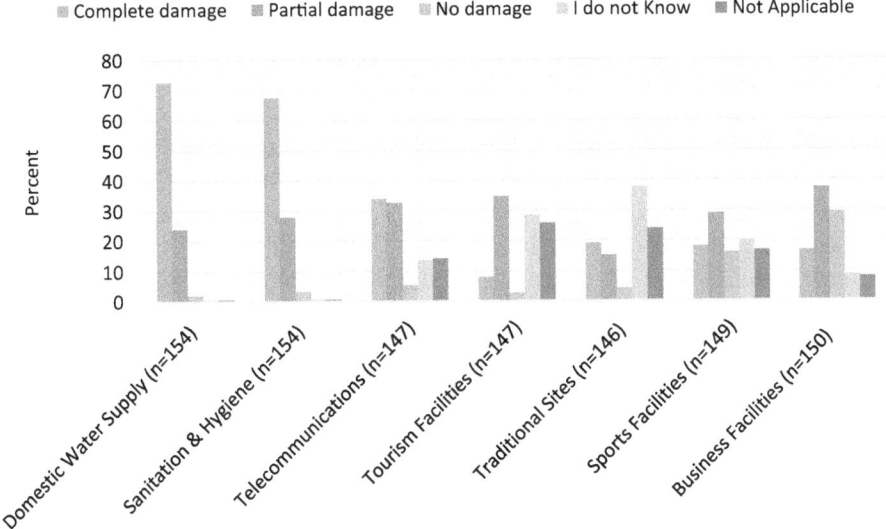

Fig. 1.4 Impact of Tropical Cyclone Idai on selected sectors (Part B). Source: Authors, Household Survey (2019)

administrative wards were affected. Further details concerning the damage are discussed later in the chapter.

1.3.2   What Victims and Residents Heard and Saw

In one of the interviews with a key informant (Respondent 1), it was clear that this respondent followed the news on Tropical Cyclone Idai closely five days before its arrival. Two to three days before the cyclone made landfall, the respondent indicated that there were indications that the cyclone was going to go through the northern parts of Chimanimani and over the Nyanga District side. Then a day just before landfall, the respondent highlighted that he got from an internet site that the cyclone was going to hit Chimanimani. This account confirms the unique pathway of Tropical Cyclone Idai presented earlier. Another internet site reported that the Chimanimani area would receive up to 750 mm of rain in 24 h. With this information, the respondent went to their timber sawmill and told workers that Tropical Cyclone Eline in 2000 gave 300 mm of rain, and Cyclone Idai would be much more. With this information, the respondent indicated that they dug a drainage trench surrounding

their house. The respondent then described the unfolding events starting Thursday 14 March 2019.

The respondent continued with their testimony, indicating that on Friday 15 March 2019 they shut down the whole sawmill operation. The rainfall increased, and the respondent kept emptying the rain gauge from time to time. At some point, they were getting about 25 mm an hour. "That Friday night it was raining very hard. At about 8 pm there was a period where there was virtually no wind, which we thought was the centre of the cyclone coming through". Then things changed and the cyclone became more aggressive. The respondent went on indicating that:

> People say that it wasn't thunder and lightning, it was rocks coming down. It may have been the case but in the sky, it sounded like lightning. At about 9pm or 10pm, we experienced a very heavy downpour. The tempo of the rainfall was high, and I recorded 70mm in 40 minutes. That's when I think most of the damage was done. That's when all the rocks and/or mountains came down. (Respondent 1).

From the account, the respondent recorded 735 mm in 24 h, and they indicated that at Dombera Farm they had recorded 950 mm. "Then on Saturday morning, we heard there was trouble in Ngangu Township". The respondent highlighted that when they went to Ngangu, what they witnessed had not been seen in his 60 years of stay in Chimanimani. It was catastrophic!

From a focus group discussion conducted at Dombera Farm, it was also confirmed that the rains started on Thursday 14 March 2019 and grid electricity went off that same night. It was then pouring Friday night, 15 March 2019 around 7 pm, and there was such thick darkness. In addition, the excess flowing water was very cold, and this could have aggravated the number of deaths of those who had survived the initial impact. The Chimanimani East Member of Parliament also confirmed the heavy rains. Respondent 3 from the National Association of Non-governmental Organisations (NANGO) presented their experience with Tropical Cyclone Idai: "Cyclone Idai came to Manicaland Province as a shocker with many devastating effects, which we did not foresee. We thought it was going to be one of those natural disasters, but not going to affect us as Idai did", indicated the respondent. In another description regarding the amount of rainfall, the Chimanimani East Member of Parliament had this to say:

> We got to the Chimanimani Mountains where the mountain range is solid granite rock, so all this rain was now pouring off the mountain and we came across this massive body of water coming from the mountain like waterfalls. It was a very dramatic sight for us, unfortunately, we didn't take any pictures. It was very scary. … So, all our major rivers were in flood, i.e., Umvumvumvu, Nyanyadzi, Nyahode, Haroni, and Rusitu.

Respondent 8 from Chimanimani Hospital compared Tropical Cyclone Eline of 2000 with Tropical Cyclone Idai. The main message was simple; these were two very different events. The respondent went further indicating that "Idai was totally different. We have never seen these stones, mud, trees falling and rolling in from the previous cyclones and no one anticipated that this cyclone was going to be so powerful". From the scenario being painted by eyewitnesses, there were deaths, injuries, displacements and other traumatic encounters. Some people were swept away in their sleep, and some were completely buried in mudslides in their houses, yet in other instances houses collapsed killing many. It was tragic!

1.3.3   Rescue, Death, Injury, the Missing, Displacements and Psycho-social Support

Data and information provided by the Manicaland Provincial Administrator's Office revealed that there were 341 deaths and 344 people unaccounted for. In addition, there were 43,883 households affected, 26,683 moderately affected and 2272 internally displaced persons (IDP). As for injuries, there was a total of 184 reported of which 164 were from Chimanimani District. From these numbers, the majority were from Chimanimani where 171 people died with 325 reported missing. Almost half of those households affected came from the same area, with almost the entire IDP residing in Chimanimani. A significant number of IDPs were accommodated in four camps (Fig. 1.5).

Given that the Chimanimani District borders Mozambique and that many rivers empty into that country, many bodies were washed away across the border into Mozambique. From the fieldwork, it emerged that 158 bodies were discovered in Rusitu River on the other side of the border and had to be buried there. The officials from the Chimanimani District Administrator's Office visited Mozambique and 75 samples were taken for DNA testing. Of these bodies, 51 were from Kopa and 24 from Ngangu, the two main points that were most impacted by the disaster.

As indicated earlier, there were four shelter camps set up for the IDPs. It emerges that from the original figure of 238 people at the Arboretum, there were now 309 people as of April 2020 (IOM Displacement Tracking Matrix 2020). However, those from Nyamatanda had reduced from 119 people to 113. For those at Kopa, the figure given in households remained the same at 59. The largest reduction in the number of households was from Garikai where the 83 were still remaining in April 2020 from the original figure of 185 households. The IOM Displacement Tracking Matix (2020) reported 309 people in Garikai in April

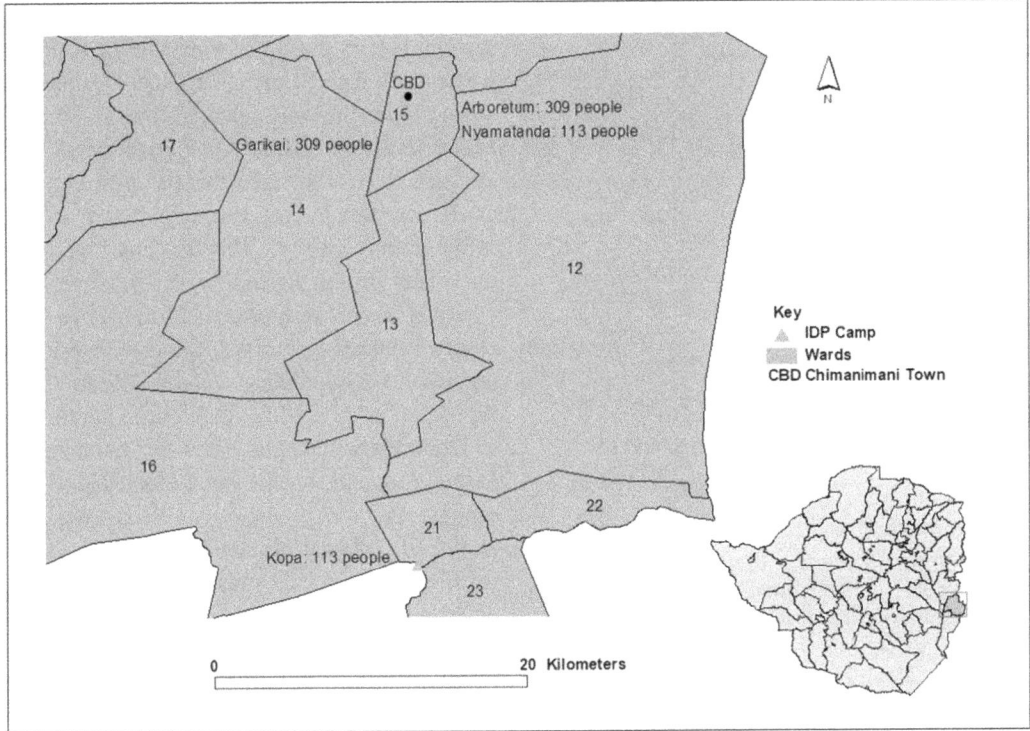

Fig. 1.5   Location of camps for IDPs (as of April 2020) in Chimanimani. Source: Authors

2020. The numbers as of April 2020 stood at 224 households and 859 individuals that remained in camps, one full year post-Tropical Cyclone Idai. Other camps that existed earlier and were closed shortly after the peak of search and rescue from the information provided included Ngangu Primary School (which had 195 people), Chimanimani Secondary School (260 people), Chimanimani Club (211 people), Chimanimani Methodist Church (120), Chimanimani Catholic Church (450 people), Chimanimani Hotel (±500 people) and Tongogara Refugee Camp in Chipinge (168). Altogether, 1654 people in these temporary shelters were later cleared.

From the field surveys, the following came up as places that experienced severe deaths in Chimanimani: Kopa, Ngangu, Rathmore Farm and Peacock Business Centre. There was also a bus from Harare that was swept away in a mudslide. The Chimanimani East Member of Parliament revealed that at Ngangu, there were many lives lost and bodies had been submerged in the mudslides and washed away into the rivers.

The soldiers went with the community and started the process of burying the deceased. At one point they buried about 45 bodies in one day and the numbers of people who were buried at the Chimanimani Heroes' Acre were in excess of about 60.

As for psycho-social support, the information from the Manicaland Provincial Administrator's Office revealed that this support was mainly provided to victims from Chimanimani and Chipinge. The details are summarised in Table 1.1. In addition, Chimanimani was designated a priority area for the re-issuance of identity documents. However, one focus group discussion at Kopa revealed that there were many more people who had not received their re-issued identity cards. Additional documents required included passports, driver's licences and educational certificates.

Generally, as of 8 July 2019, about 25 development partners were on the ground assisting with various disaster relief initiatives. In addition, there were also several United Nations

Table 1.1 Psycho-social support services as of 8 July 2019

Nature of Support	Total Numbers
Children reached with services	42,316
Alternative care arrangements, including foster care and places of safety for children	620
Unaccompanied minors and separated children assisted	470
Initial counselling and psycho-social support	106,573 out of 209,712 targeted
Households benefiting from food assistance	41,890
Old persons assisted	9120
People with disability assisted	2504 (152 children)
Child-friendly centres created (mothers and children met daily for psycho-social support)	8

Source: Authors, Data from Manicaland Province Development Coordinator's Office (2019)

agencies including United Nations Population Fund (UNFPA), United Nations Children's Fund (UNICEF), World Health Organisation (WHO), World Food Programme (WFP), United Nations Development Programme (UNDP) and the United Nations High Commissioner for Refugees (UNHCR).

However, with all these measures in place, the fieldwork revealed that by October 2019, there were still many people that had not received psycho-social support services. From a focus group discussion held at Dombera Farm and a key informant interview from Respondent 2, the Rathmore Farm where 25 people died had not been visited by any organisation, including government. Out of the 25 deaths, only five bodies were recovered and the rest remained unaccounted for. At some point, 60 families that lost their homes had to be accommodated at the farm manager's place, which has 12 bedrooms and a guest house. Overall, there were other psycho-social support issues teased out from the household survey. A concern was raised; whether the households witnessed an increase in certain matters that would require psycho-social support. The results are shown in Fig. 1.6.

The household survey revealed that there were several matters that were worrying the residents, particularly the victims of Tropical Cyclone Idai who were exclusively sampled to respond to the raised questions. Ninety-five percent of those sampled were of the view that the most traumatic experience was increased general fear of a repeat of the event (anxiety). This was confirmed from the focus group discussion at Dombera Farm where workers were said to form groups when clouds started gathering. Children were also reported to rush indoors each time it started raining, with the shout "Idai, Idai" heard in reference to Tropical Cyclone Idai. From this number, 81% strongly agreed that it was a challenge. Strong sentiments were also expressed regarding increased financial problems, mainly due to lost livelihoods, increased stress levels and increased prostitution. Alcohol and drug abuse were other challenges that emerged from the household survey as requiring attention from a psycho-social perspective.

1.3.4 Impact on Transportation and Telecommunications Infrastructure

As reflected in the household survey earlier, the transportation infrastructure—especially roads and bridges—was extensively damaged. This is the reason why emergency relief could only be delivered first by air about 5 days later. The Zimbabwe National Army (ZNA), 3 Brigade, started travelling to Chimanimani village on 17 March 2019 by road and encountered damaged bridges and landslides. At some point the soldiers had to walk long distances to the disaster zones. Figure 1.7 shows the road map with selected damage, not all of which is reflected on the map. An estimated 1500 km of road network was severely damaged (Chatiza 2020) and many parts remained in the same state for many months. The Member of Parliament for Chimanimani East gave a description of the power of the Nyahode River and the way it uprooted the bridge at Kopa. For close to 3 months, there were still some areas

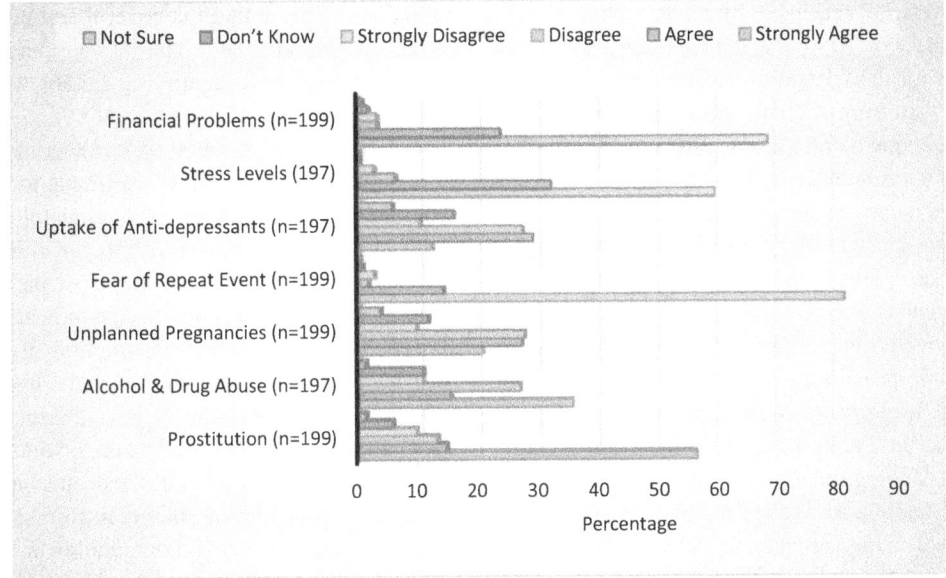

Fig. 1.6 Perceptions from household on issue requiring psycho-social interventions. Source: Authors, Household Survey (2019)

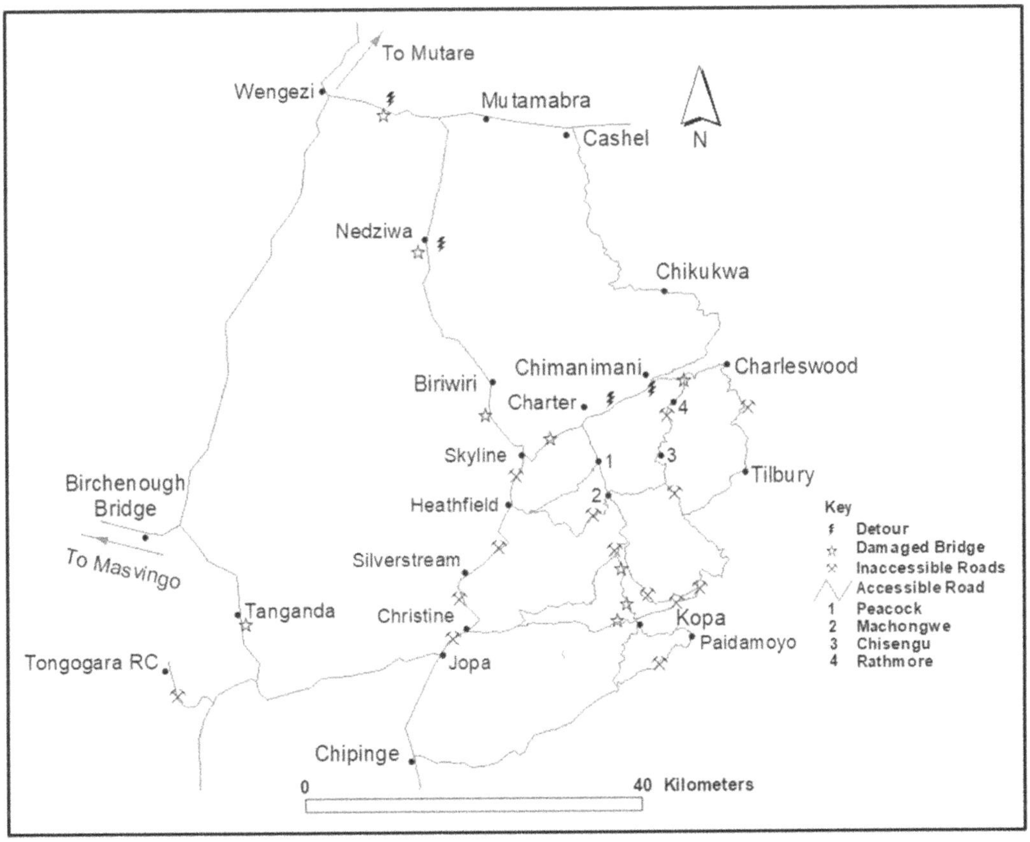

Fig. 1.7 Impact on selected roads and bridges networks. Source: Authors

that were not reachable by roads such as Vimba, Muchadziya, Ruwedza and Chikukwa.

As for the telecommunications infrastructure, there were similar challenges. Once grid electricity went down on Thursday night 14 March 2019, some mobile telecommunication base stations started facing challenges. This was aggravated by the fact that all of them ran out of diesel a few days later. This became a mega challenge as diesel supplies were completely cut-off due to damaged roads and bridges. All the major mobile network providers—including Econet, NetOne and Telecel—were down and one had not recovered at all by the time of fieldwork in October 2019. Underground and overhead cables for the fixed landline network, TelOne, were also extensively damaged at Ngangu and Kopa as reported in one of the focus group discussions. Data supplied by the Manicaland Province Administrator's Office revealed that 19.5 km of TelOne access roads were damaged, mainly with gravel washed away and gullied. As of July 2019, the damages had not been attended to due to funding challenges. Overall, the loss in the telecommunications sector stood at an estimated US$2.433 million (Manicaland Province Administrator's Office 2019). A whole chapter on the impact of the cyclone on telecommunications is presented elsewhere in the book series and interested readers can follow-up on it.

1.3.5   Impacts on the Health, Water, Sanitation and Hygiene Sectors

This section will focus more on health and sanitation matters as there is a whole chapter dedicated to water supply in the book. Health issues came to the fore during the Tropical Cyclone Idai. With so many injured and dead, one interview after the other bemoaned the lack of a government district hospital in Chimanimani. Respondent 5 summarised it well when they indicated that if the government was to hear their cry and wish, then the district hospital would be of major consolation post-Tropical Cyclone Idai. However, from the interview with the Member of Parliament, the

politician indicated that the government had this project on the cards as it, "has already budgeted for one and was now identifying suitable land to put up a district hospital".

From the household survey undertaken in Chimanimani East, an effort was made to determine the disease prevalence situation following Tropical Cyclone Idai. The results are shown in Fig. 1.8. Diarrhoea became the most prevalent disease, and this corresponds to the indication by households that there was destruction of many permanent water, sanitation and hygienic (WASH) facilities after the cyclone. During fieldwork, Respondents 6 and 7 from Nyamatanda Camp confirmed that a health inspector had just visited the camp after an outbreak to take samples of the piped water. Contamination was suspected from the spring/weir where monkeys and baboons could have defecated at the source, with there being limited water treatment in the tank before reaching consumers. Overall, 21,952 patients (14,115 female and 7837 male) had been attended to as of 16 May 2019 (Manicaland Province Administrator's Office 2019). The patients were given a range of treatments and vaccinations that included measles, rubella, vitamin A, HPV 1st dose and HPV 2nd dose.

Evidence of the occurrence of diarrhoea cases are further supported by responses from the household survey that showed 70.5% of the respondents indicating that domestic water quality was reduced after the cyclone, with the remaining percentage maintaining that there was no change. Water supply infrastructure also faced severe strain as grid electricity went off for about a month in some areas. Up to 96% of households surveyed indicated that they had a blackout after Tropical Cyclone Idai, with only 4% indicating otherwise. A focus group discussion with the District Medical Office staff revealed that the cyclone washed away most water supply and sanitation facilities in the affected areas, with the systems at Kopa and Ngangu having been completely destroyed. The low prevalence of cholera was also confirmed by Respondent 8 from Chimanimani District Council Hospital who noted that as of October 2019, they had not encountered any cases. In addition, the district

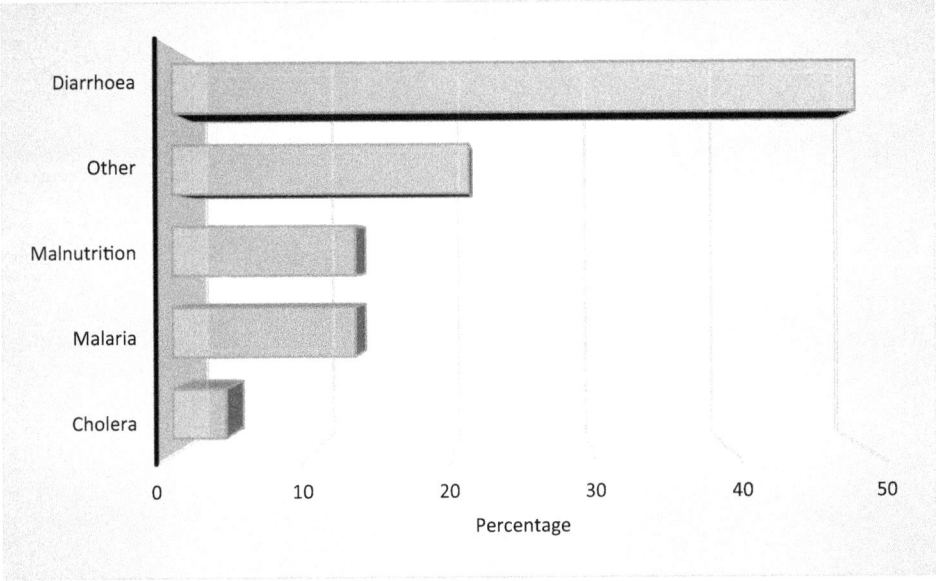

Fig. 1.8   Most prevalent diseases after Tropical Cyclone Idai ($n = 184$). Source: Authors, Household Survey (2019)

seems to have been ready as they had just come out of a cholera period since they border Mozambique where the incidences were ad hoc. Hence the doses of medication for cholera were adequate. There were also health issues concerning the donated food stuffs; this issue is considered in depth in the chapter dealing with ethics in disasters.

A summary of reported damages to health facilities is presented in Table 1.2. However, across the province, 2194 community water points that include boreholes, springs, piped water schemes and shallow wells were affected. In addition, a total of 9849 household sanitation facilities and related infrastructure that included BVIPs, uBVIPs, pit latrines and water closets were damaged. This resulted in about 256,461 people being exposed to risk.

Concerning the injured, Respondent 8 revealed that Chimanimani District Council Hospital had attended to 362 cases. The respondent indicated that the hospital did not have the capacity and that was also not prepared for Tropical Cyclone Idai. In fact, in terms of the recognised status of a hospital in Zimbabwe, Chimanimani District Hospital is at clinic level. Therefore, the magnitude of the disaster and the treatment and care

needs were too great for the hospital, and staff worked around the clock. One of the key issues that emerged was that there was a bit of complacency in the entire administration system and across the line ministries based on the experience of Tropical Cyclone Eline, which had manageable disturbances. Hence, one could always reflect and say some of the deaths might have been avoided with timely interventions. The respondent had this to say in terms of readiness: "We were not fully informed. We were only told on Saturday morning that the epicentre had been discovered in Rusitu (Kopa)".

1.3.6   Impact on the Agricultural Sector

One of the sectors that was hit hard was the agriculture sector, with its forestry, livestock and crop (including horticulture) subsectors. Common crops affected included maize, beans, tomatoes, seed maize, onions, cotton and sugarcane. Data and information from the Manicaland Province Administrator's Office revealed that there was damage to irrigation schemes across the province estimated at US$8.655 million. A

Table 1.2 Summary of damage to health facilities

District	Name of facility	Damaged area
Chipinge	Kopera	Water pump washed away together with concrete slab
	Paidamoyo	Kitchen for expectant mothers' shelter was destroyed
	Gumira	Part of roof blown away
	Ngaone	Part of roof blown away
	Nyunga	Extent of damage to be established
Chimanimani	Muchadziya	1 staff house roof blown off, roof of 2 rooms also blown off and health records destroyed
	Mutsvangwa	1 staff house roof blown off (3 rooms), staff toilet destroyed
	Biriiri	Water source damaged
		Staff houses leaking
	Rusitu	• Water supply system damaged
		• Sister-In-Charge office, dining hall and kitchen roof ridges were blown off
	Ngorima	Water source washed away
	Nyahode	• Three deep wells were destroyed
		• Eight BVIPs collapsed in population serviced by Nyahode Clinic
		• Roof of expectant mothers' shelter blown away
		• No electricity power supply

Source: Manicaland Province Development Coordinator's Office (2019, pp. 69–70)

total of 9315 hectares (16%) out of the planted 58,269 hectares of maize were destroyed in Chipinge, Chimanimani, Mutare and Nyanga districts. There were also 489 barns destroyed in Makoni, Mutare and Mutasa districts at an estimated cost of US$244,500. As for livestock, infrastructure information from Chatiza (2020) showed that 13 offices and 86 dip tanks were destroyed across the province and a total of 348 cattle, 17,000 chickens, as well as 222 goats and sheep were swept away.

Chimanimani is known for its vast plantations and as the key area supplying timber in the country and beyond its borders. To this end, damages in some of the farms like Rathmore Farm included the second largest timber mill in Zimbabwe that was swept away. Two tractors, one vehicle and other machinery, were also swept away. The value of the plant and equipment that was lost came to about US$1.86 million. Details supplied by the farm manager indicated that the farm is about 600 hectares in size of which 45 hectares were under avocado trees and 80 hectares under macadamia trees. An estimated 12 hectares due for harvest were lost to the cyclone for the macadamias and two hectares for avocados. Other losses from large industries and Rathmore Farm are presented in Table 1.3.

The ground visit and inspection of Rathmore Farm revealed the tragedy as some workers' houses were swept away from the married quarters' section. The picture before and after is shown in Fig. 1.9. Across the road to the right side was where the timber mill was located. From three separate rain gauge stations, the farm reported rainfall averaging 1963 mm in 2 days.

There were also other farms close by, such as Dombera, with similar land usage that were also hit hard. The story remained the same across the agriculture sector in Chimanimani and Chipinge, apart from tea and coffee plantations. These had minimal damage.

The destruction in the agriculture sector also came up in a key informant interview with the Chimanimani East Member of Parliament. It

Table 1.3 Impact in the forestry sector

District	Company name	Infrastructure, equipment and/or stock loss incurred
Chimanimani	Allied Timbers	Damages inflicted on water reticulation, roads, bridges, culverts, planted forests, fire towers
	Wattle Company	Loss of forestry resources; 700 ha affected in Chimanimani and Chipinge
		Lost production at both Silverstream factory and Vumba Timbers treatment plant
		Infrastructure damage (roads, bridges and buildings)
	Rathmore Farm	58 ha were lost (34 of pines and 24 of blue gums) out of a total of 224 ha (of which 110 was pine and the remainder blue gums)
Chipinge	Makande Estates	50 tons macadamia nuts washed away
		20 tons avocadoes destroyed
		350 ha maize destroyed

Source: Authors, Based on Manicaland Province Administrator's Office (2019, p. 123) and Fieldwork 2019

Fig. 1.9 Rathmore Farm before (13 March 2014) and after (25 March 2019). Source: Authors

emerged that the "traumatised community had lost their means of production; their fields, their livelihoods and livestock. In fact, every resident from Chimanimani was affected in one way or the other". The issue surrounding riverine agricultural activities came up twice from the Member of Parliament and Respondent 4 from a Harare-based non-governmental organisation (NGO) that operates in Chimanimani. While the Member of Parliament emphasised that livelihoods depended on alluvial fields that were swept clean by Tropical Cyclone Idai, Respondent 4 was of the view that streambank cultivation had got out of control in the area. With these divergent views, the Member of Parliament indicated that the government was now focusing on supporting the subsistence farmers to be resettled by putting up new irrigation schemes, with the alternative of solar powered boreholes. With households still in camps more than 18 months after the cyclone, the proposal from the Member of Parliament seems far-fetched indeed.

1.3.7 Impact on Housing, Business and Related Infrastructure

There was untold destruction of human settlements and business premises, especially at Kopa, Ngangu, Peacock and Rathmore Farm. Describing the situation at Kopa, Chimanimani East Member of Parliament summarised it as follows:

> The devastation was just out of this world. What happened at Kopa is that, Nyahode river comes down from the North, the Chipita river coming from the other side (Ndiyadzo) going into the Rusitu river. This confluence has three rivers meeting; the Chipita being not as big as the other two. We had a dam up on the Nyahode river at Nyabamba, a hydropower dam, which burst due to the flooding of Nyahode river carrying massive boulders down to Kopa, literally 30–40 tonne boulders, massive rocks that I believe were ripped off the river bed itself.

At Kopa, it was not only private houses that were swept away. There was also a police camp, institutional houses for agricultural extension workers and about 26 houses for Agricultural and Rural Development Authority (ARDA) as per

Respondent 9. The police camp had about 18 solar panels too. According to another eyewitness who was also rescuing those marooned, up to 60 people were pulled out using a rope tied to tree and electricity pole (Respondent 10). However, the rescue mission was cut short when a lorry was swept away by the flood and hit the tree, cutting off the rope. The additional tragedy meant all those who were using the rope to escape were likewise swept away.

What was also interesting was the description that the boulders that are visible in Fig. 1.10 were not yet there during the rescue. According to Respondent 10, the boulders only came later. This narrative seems to support the Chimanimani East Member of Parliament's account that the huge rocks were dislodged from upstream of Nyhode River and transported downstream. The Member of Parliament went further, indicating that they had traced the locations upstream from a helicopter. The power of the Nyahode River also came under the spotlight from Respondent 11 who testified that they do not know where the river gets its power from, as four times since Tropical Cyclone Eline in 2000, the Nyahode Bridge at Kopa had been swept away. Judging from the slope along the Nyahode River channel, there is a real chance that the huge rocks picked up velocity and rolled along the channel.

It is believed that the rocks contributed to the massive destruction that followed and took away almost all the houses after the episode when the rescue rope had broken. This event also followed the uprooting of the Nyahode Bridge, which Respondent 11 indicated had trapped a lot of water for a while. In fact, there was a story of one agricultural extension worker who was constantly monitoring the water rise till he went to sleep; the bridge that had been blocked by logs acted as a temporary dam wall. It was around 11 pm that night that the horrible and unthinkable happened as the powerfully released flood waters and huge rocks uprooted the entire bridge and deposited it about half a kilometre downstream. To add to the disaster, many people who had been woken up and warned to flee the flood refused, citing that in 2000, it was the same. The waters of Nyahone would not pass certain points/markers because it

Fig. 1.10 Some of the rocks deployed at Kopa from the Nyahode River flood waters. Source: Authors, Fieldwork (2019)

had never done so before. What a terrible end to lives that could have been saved. Another story of a lady who had US$25,000 and ZAR200,000[2], who went back after being rescued and got swept away, was narrated by Respondent 11 and supported by Respondent 10 and others who were listening to the story. This story is picked up in the chapter dealing with religion and disasters. The condition of the settlements at four selected key impact locations is shown in Figs. 1.11, 1.12, and 1.13.

The government, in liaison with the Chimanimani District Council, identified several areas for resettling the IDP. However, the main area was identified close to Nhedziwa Growth Point. The greatest challenge has been in resettling the IDP better. By the time of fieldwork in October 2019, a survey had been done at the main proposed location according to the Member of Parliament. The biggest worry was that the government did not want to relocate people to an area that is prone to tropical cyclones again. Hence, the process was taking long. Needless to say, almost a year since the discussions with the Member of Parliament, the IDPs were still in camps as of September 2020.

There was also significant damage to business premises. Figure 1.14 presents one business centre that was completely swept away during Tropical Cyclone Idai. The fieldwork also revealed that other business centres and outlets were swept away or partially damaged at Rathmore Farm, Ngangu and Kopa. There was also an entire Mawenje Lodge that was swept away. All these left communities without livelihoods as well as the loss of lives.

The Manicaland Province Administrator's Office supplied details of completely damaged and partially damaged private family houses and huts. The total for those destroyed stood at 17,012 across the province, with the bulk being in Chimanimani (8805) and Chipinge (6940). The total for those partially damaged came to 26,720 (11,217 in Chimanimani and 11,390 in Chipinge). The rest was shared by other districts, apart from Nyanga. From the two epicentres, 210 out of 1481 houses were destroyed at Ngangu, and 147 houses swept away at Kopa. Lastly, at Peacock Business Centre all the buildings were swept away by the floodwaters. However, Chatiza (2020) provides a good observation on some of the challenges that led to the massive destruction of buildings. He presents a case of violation of building codes, an aspect that was also visible during on the ground visits by this research team.

---

[2] The exchange rate was about US$1 = ZAR16.66 as of 14 September 2020.

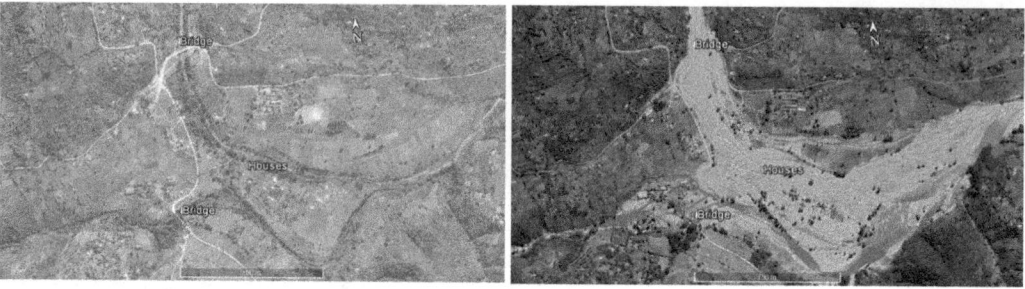

Fig. 1.11 Settlements at Kopa before (29 July 2016) and after (25 March 2019). Source: Authors

Fig. 1.12 Settlements at Ngangu before (21 November 2018) and after (25 March 2019). Source: Authors

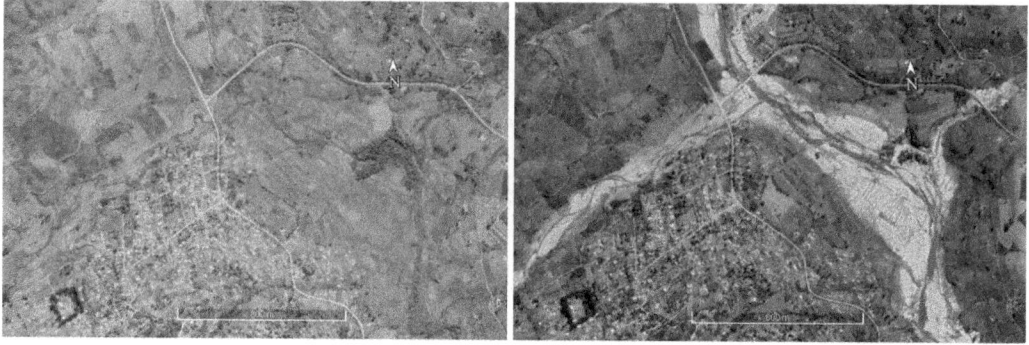

Fig. 1.13 Settlements at Machongwe before (8 April 2018) and after (6 April 2019). Source: Authors

To this end, Chatiza recommends that local authorities comply with appropriate building codes, even in rural set-ups.

1.3.8   Damaged Educational Facilities

Based on a focus group discussion with the representatives from the Ministry of Primary and Secondary Education, a total of 58 schools were reported to have been affected by the tropical cyclone. These schools were in Chimanimani (20), Chipinge (19), and Buhera (19). Sanitation facilities, particularly ablutions facilities, were part of the infrastructure hardest hit as these structures collapsed. Many classroom blocks had their roofs blown off and some walls collapsing. Some teachers' houses were also damaged. At St Charles Luangwa (Fig. 1.15), rock boulders

Fig. 1.14 Peacock Business Centre before (8 April 2018) and after (6 April 2019). Source: Authors

Fig. 1.15 St Charles Lwanga Secondary School before (8 April 2018) and after (6 April 2019). Source: Authors

rolled over after landslides into a dormitory resulting in two deaths. Respondent 20 from the school indicated that there were talks of relocating the school, with the land having been identified at Ruwaka. Overall, 43 schools (33 primary and 10 secondary) were shut in Chimanimani District as from 26 March 2019 and later opened 7 May 2019.

1.3.9 Impact on Natural Resources, Wildlife and Tourism Attractions

The Environmental Management Agency (EMA) reports that the climate of Chimanimani and Chipinge is characterised by high rainfalls and this comes in excess of 1200 mm annually (EMA 2019). To this end, the region mainly falls within the Agro-ecological Regions 1 and 2, although the other Agro-ecological Regions 3–5 are also scattered throughout the districts. EMA also notes two major causes of damage from the cyclone, namely flooding along with natural watercourses, and landslides and erosion on steep slopes ranging between 70° and 85° gradient. Much of the degraded areas was characteristic of second-generation vegetation of shrubs and grasslands. EMA (2019, p. 4) observed that, "there was headward erosion of all major watercourses hence volumes of water running in each channel was so intense that it flooded beyond naturally defined highest flood levels and all the affected settlements were along waterways, either constructed on a watercourse or flood plain". Given the massive volumes of floodwaters, river channel widths were widened, some more than three times. Watercourses changed in less than 24 h, especially at Kopa, Rathmore Farm and Haroni junction at Mawenje Lodge as the authors also witnessed in October 2019 during fieldwork. EMA further noted that the exposed soil profile shows large stones and rock boulders underneath, and it appears the underlying geology is mainly sedimentary rocks with flood plains characterised by deep fertile loam soils. A summary of selected land degradation by Tropical Cyclone Idai is presented in Fig. 1.16.

Tropical Cyclone Idai had a massive negative impact on natural resources, especially the short- to medium-term aesthetics. As huge boulders and trees were uprooted, gullies and other unwanted scenes were created. Figure 1.17 shows how several riverbeds were widened, and aquatic life was also swept along into the Indian Ocean where the Chimanimani River system drains. Some riverine vegetation has known value as producing edible fruits, storing water and providing traditional sources of medicines. All this was cleared and may take many years to recover. From the illustration there is also additional insight regarding the main road linking Chimanimani Town to the Chimanimani Mountains and also key tourism resorts there that included Outward Bound and many more as will be detailed in some chapters in this and other book series focusing on Tropical Cyclone Idai. The tourism sector was not spared either. Many attractions were damaged, yet a few

others were improved. One such tourism attraction in Chimanimani is the Golf Club and golf course. There was severe damage to the golf course, and this is shown in Fig. 1.18.

Apart from the golf course being damaged, wildlife was also swept away in the floods. Respondent 17 who grazes his sheep, a flock of about 50, indicated that there used to be three pythons in three spots along the stream. All these were swept away. The impact on wildlife was also confirmed by Respondent 18 from Rathmore Farm who indicated that they used to have problems with three monkey troops. Yet after the flood, they were dealing with only one troop. Another confirmation was that wild pigs and other small wildlife were also swept away (Respondent 19). Details concerning the potential of new tourism attractions, especially dark tourism, are the subject of discussion in a full chapter in the book.

1.3.10 Challenges in Early Warning and other DRR Management systems

Information provided by the Zimbabwe Meteorological Services Department (ZMSD)

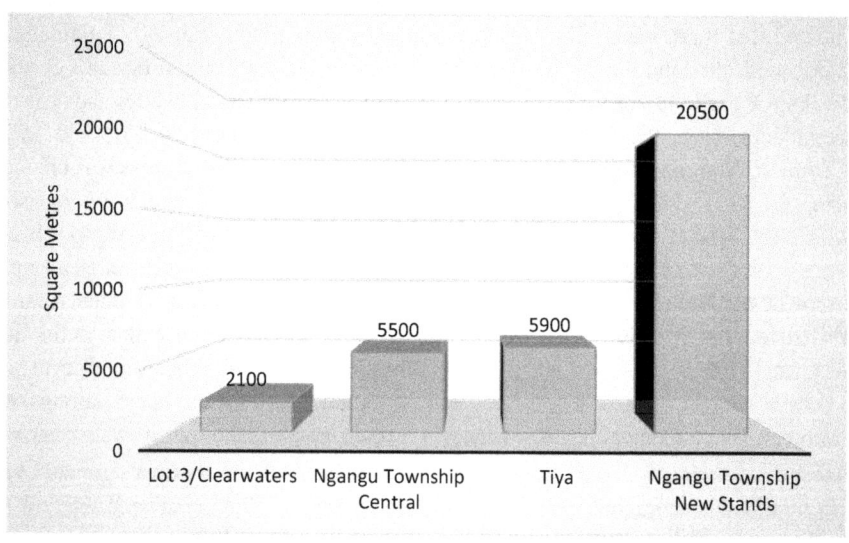

Fig. 1.16 Summary of land degraded in selected areas in Chimanimani. Source: Authors, Data from EMA (2019, pp. 7–8)

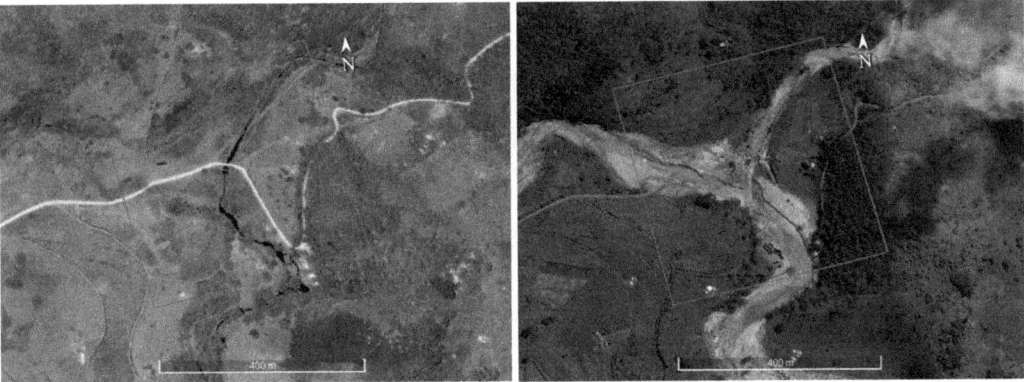

Fig. 1.17   Haroni River and tributary before (11 November 2018) and after (28 March 2019). Source: Authors

Fig. 1.18   Chimanimani Golf Course before (14 May 2006) and after (25 March 2019). Source: Authors

confirmed that EWS is indeed a challenge. Both the focus group discussion and the official written report reviewing the ZMSD's capacity to observe, forecast and respond to Tropical Cyclone Idai confirmed this (ZMSD 2019). It emerged that with the equipment, models and human capacity available, the ZMSD could only predict a maximum of 100 mm of rainfall in 24 h from Tropical Cyclone Idai (ZMSD 2019). To this end, the department indicated that without radar and other related equipment and technical capacity, it will remain a challenge to present accurate early weather-related warnings in the country. Several figures in terms of realised rainfall come up, and these include the 600 mm+ reported by the EMA (2019), close to 2000 mm in 48 h reported at Rathmore Farm and about 900 mm reported from Dembera Farm in 24 h as per the fieldwork done

by authors in October 2019. Whatever the case could have been, there was excessive rainfall from Tropical Cyclone Idai. The ZMSD revealed that during the peak of the cyclone, there was no electricity at its Belvedere Head Office as the generator did not have diesel and there is no solar power. This meant delays in the issuing of weather bulletins. In addition, the office did not operate around the clock and this resulted in information gaps in terms of EWS, and particularly during disaster times such as that of Tropical Cyclone Idai. The ZMSD also highlighted challenges with limited functional weather stations across the country as there were only 48 such, and from the list, some were poorly equipped and manned. During the United Nations Economic Commission for Africa (UNECA) workshop in October 2019, the ZMSD presented its wish list on equipment, offices and other necessities that revealed a scenario of an organisation in need and under capacitated in many aspects. A summary of resources needed to have a functioning EWS by the ZMSD in Manicaland Provincial Office was also presented in a Tropical Cyclone Idai report and is shown in Table 1.4.

From one of the Tropical Cyclone Idai Cabinet reports of July 2019, a DRR and management institutional framework for Zimbabwe emerged (Fig. 1.19). As noted earlier, on paper, there are institutional frameworks at all administration levels, with civil protection committees (CPCs) being the main platform (CADRI 2017). The CPCs exist at national, provincial and district levels. These multi-stakeholder CPCs, however, lack the capacity to effectively and efficiently respond to DRR and management matters. "Human, logistical and material resources required to undertake effective emergency preparedness, early action or immediate response activities are seriously lacking" (CADRI 2017, p. 49). CADRI further observed that the dual DRR and management response coordination—split between the National Civil Protection Committee chaired by Civil Protection Department (DCP) for flood and other sudden-onset disaster response, as well as the National

Food and Nutrition Council chaired by Office of the President and Cabinet (OPC) for droughts—presents institutional cohesiveness challenges. The permanent secretaries are also in the picture in terms of the everyday running of ministries; so are provincial administrators and district administrators. The arrangement further presents the efficient use of resources.

Given the persistent challenges of large-scale disasters in the country, international development partners established the humanitarian coordination structure. This structure has at the top level, the Humanitarian Country Team made up of representatives of NGOs, the Red Cross Movement, development partners and United Nations agencies operating in the country. At the sector level, the United Nations Resident Coordinator's Office chairs the meetings. However, the international humanitarian coordination structure is also limited in terms of resources, and this hinders robust preparedness and response coordination in support to the government. One of the key gaps identified in terms of relief efforts was that there were no warehousing facilities at all across all administrative levels.

In response to Tropical Cyclone Idai, Chatiza (2020) observe gaps in technical, financial and

Table 1.4 Resources needed in Manicaland Province meteorological stations

Instruments	Buildings	Staff	ICT and office equipment
Weather radars, seven automated weather stations (AWS), digital thermometers, digital barometers, standard rain gauges, sunshine recorders, thermo-hydrographs, and dynes	All offices need renovations. At least two family houses at each station. Office repairs needed at Chipinge, Chisumbanje, Nyanga, Buhera, and Rusape. New offices required at Mutare, Mukandi, Chisengu and Chimanimani weather stations	Provincial office needs 2 technical, 1 messenger and 1 general hand staff members, while all other offices require at least two technical and one general hand at each station	Provincial office needed a high-speed broad band connection, 4 desktops, 2 laptops, AWS digital display, and one land line. All other offices needed 1 desktop at each station, 1 ADSL connection at each station, and 1 printer at each station. New office furniture was also needed for all offices
Other essentials listed for the provincial office	4 × 4 off road vehicle, 120 L fuel per month, TV and fridge—provincial office		

Source: Authors, Based on ZMSD (2019, p. 23)

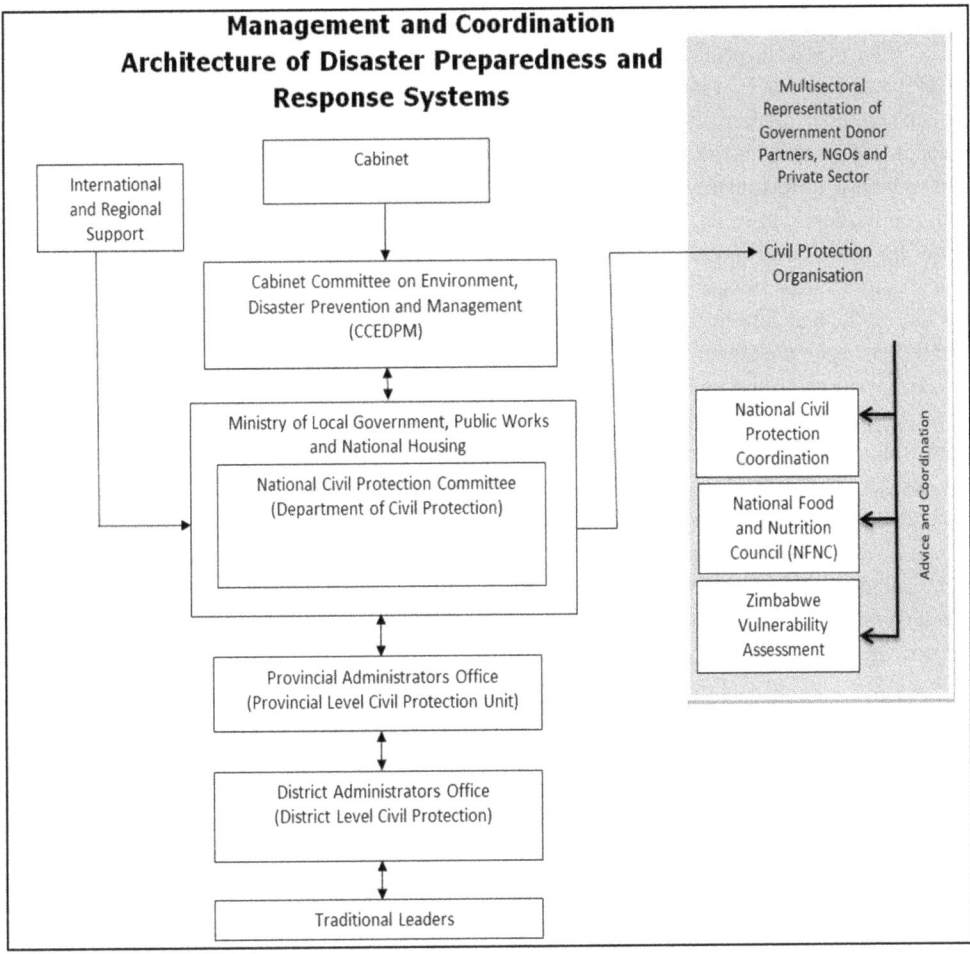

Fig. 1.19 DRR and management framework in Zimbabwe. Source: CCEDPM (2019, p. 13)

logistical capacities. Overall, a reactive, as opposed to a proactive, approach is used. This effectively means there is a lack of preparedness in institutions. A key informant from the National Association of Non-governmental Organisations (NANGO) was of the view that both the state and non-state actors were not adequately prepared to respond to Tropical Cyclone Idai (Respondent 3). This lack of readiness did not only include the government, especially the Civil Protection Department, but also civil societal organisations, which include the NGO operating both in Manicaland Province and even those in the rest of the country. In the respondent's view, there are a lot of things that went wrong and could have been done right in terms of EWS. This view comes from the fact that many stakeholders in positions

of authority knew that Tropical Cyclone Idai was coming 2 weeks prior to its arrival. The lack of readiness in DRR was also highlighted by the Member of Parliament for Chimanimani East. In his view, "Cyclone Idai caught us quite unprepared because we thought it would blow over like the others. Unfortunately, it was a Cyclone in its own class".

There were no adequate preparations for search and rescue, or relief (Respondent 3). An example is given where the Manicaland Provincial Civil Protection Committee started meetings only 3 days before the cyclone. The respondent painted a picture of what could have gone wrong in terms of DRR, particularly preparedness and response. There was an allegation that there was neither a Manicaland provincial nor national policy on internally displaced persons (IDP) in place. Information dissemination about Tropical Cyclone Idai is said to have taken place late. Another challenge was that the government asked people to run to higher ground, and this is where, unfortunately, Tropical Cyclone Idai hit the hardest. However, although this was the case, on the ground interviews revealed that the impacts of 15 March were mainly in the middle of the night, and due to the darkness, many people did not have a chance to escape to higher places. Responding to the challenges for search and rescue, the Chimanimani Member of Parliament had this to say: "Our first point of call was to try access the area using air through our airforce; helicopters were mobilised but due to the cloud cover, it was not practical to get in by air. Unfortunately, that was the situation for the next 3–4 days".

Regarding relief, Respondent 3 was of the view that there was no standing Disaster Relief Fund. This was aggravated by the fact that the budget prior to Tropical Cyclone Idai did not make provision for such or other disasters. Furthermore, CPCs, especially at district level, were not functional prior to Tropical Cyclone Idai. This brought about serious administration challenges including some indications of lack of on the ground command-and-control as some administrators in Chimanimani were alleged to have been off stations and giving wrong information regarding the state of the disasters. Furthermore, the need for permanent safe shelters or safe shelter sites was highlighted by Respondent 3, indicating that prior to Tropical Cyclone Idai, there were none in Chimanimani. Other matters concerning disaster relief came to the fore that one would not consider randomly and worth documenting herein. In the respondent's judgement (which the authors agree to):

> For example, women were not given pads in time. And when it was time to give them, you know what the government and NGO did wrong again? The pads were not accompanied by pants. So, we were giving people a new pad with an old pant. That was wrongly done in terms of the human rights focus.

Matters concerning the exercise of authority also came under the spotlight. Respondent 3

highlighted the great political support and will by citing that President Emerson Dambudzo Mnangagwa had to cut short his United Arab Emirates trip to attend to the disaster. The support by the president was also highlighted by the Chimanimani East Member of Parliament who revealed that three times, the president visited Chimanimani after Tropical Cyclone Idai. However, more could be done in terms of devolution of powers as time passed while waiting for the president to appoint ministers to visit the disaster areas (Respondent 3). The NGOs also had to wait for a long time to get approval to enter the disaster areas. There were also aspects of suspicion from the government and the respondent sums it as follows:

> When the international NGO came in, initially it was not so welcomed because of the politics maybe. So, aid was there but people wanted to know where they were coming from and who had given them aid, how come they had mobilised more resources than the government within a few days. I will give an example of Miracle Missions; … because Miracle Missions had two choppers within 3 days. They also had 11 trucks and resources had been mobilised by just a mere church organisation within 4-5 days after Idai. When the government was still in the boardroom, NGOs had already begun mobilising.

Another hurdle was faced by trucks ferrying disaster aid. According to Respondent 3, there were instances when trucks with food and non-food items would go to Chimanimani and then be told to go back to the provincial offices to declare what they have first and perhaps be given to officers to assist in distributing it. Yet victims on the ground needed those resources. There were a lot of politics, with everything having to be passed by the government offices. Hence, in one encounter, 15–20 trucks parked by the government buildings at the province awaiting declaration and approval. Then they would be told that there is a need of warehousing these items and they would be distributed slowly.

Drawing from a focus group discussion at Dombera Farm, it emerged that there was a historical record pointing out a similar event happening in 1942. This account is said to be confirmed by the fact that the pathways opened by the floodwaters revealed old channels.

Unfortunately, people had settled in some of these old pathways.

1.3.11 The (Un)sung Heroes of Tropical Cyclone Idai Rescues and Other Missions

The story of the Tropical Cyclone Idai may not be complete without reflecting briefly on selected cases of bravery, kindness and related missions. Preference in this section is given to narratives that are not covered in full chapters later in the book such as the Higherlife Foundation and partners, as well as the Chimanimani Hotel, management and staff. The first from the selected cases is that of two gentlemen interviewed at Kopa who were involved on the "mission impossible" rescue of those trapped on the now flooded island. The two gentlemen were among the group that tied the rescue ropes to an electricity pole and a tree, pulling out about 60 people as documented earlier. The second story is of a gentleman from Ngangu whose name was given as Steven Hamudi (Dzionere). Mr Hamudi is reported to have saved many lives from being swept away by the ravaging floodwaters, yet he lost his life at the very last moment. His home was safe. *The Herald* carried an article of the posthumous honour conferred by President Mnangagwa on Mr Hamudi on 13 August 2019 (Murwira 2019). Based on President Mnangagwa's proclamations of honours to individuals, corporate, religious organisations and diplomats that were involved in various missions pertaining to Tropical Cyclone Idai, particularly search and rescue, relief and initial recovery work, the following names also emerged: the United Nations family that was represented by Ambassador Bishow Parajuli, Dean of African Ambassadors accredited to Zimbabwe and Democratic Republic of Congo Ambassador, Mwawapanga Mwanananga; Bitumen World represented by Mr Andre Zietsman, Fossil Contracting represented by Mr Ronald Mashura, and Mr Itai Samhere. Both Mr Samhere and Hamudi were conferred with the Zimbabwe Bravery Award in the Gold Cross category posthumously as they died in community rescue mis-

sions. Other names that received the Jairos Jiri Humanitarian Award are listed in Box 1.1.

Based on the fieldwork, there were strong res-

Box 1.1 Jairos Jiri Humanitarian Award in Recognition of Those Heroes During Tropical Cyclone Idai

1. Dzingie Augastine	9. Chimupini Douglas
2. Dilloon Plaxedes	10. Murata Professor
3. Sithole Tichaona	11. Masaiti Farai
4. Chinamira Bongai	12. Masaiti Peter
5. Wonderful Zikhalo	13. Tasiyana Eliphas
6. Chirara Gloria	14. Strive Masiiwa
7. Joshua Sacco	15. Billy Reutenbach
8. Chitopo Tendai	16. Andre Zietsman

Source: Authors, Based on Pindula News (2019, online)

---

ervations from some community members regarding the official and final list that went to the president. Other names and associations that came up during fieldwork include the following: retired Colonel Patrick Kaneta from Kopa, Mr Shen Kid, the Chimanimani Tourism Association (CTA), Mr Doug Van Deruit, and the mountain guides, and Saurombe Logistics & Consulting. The CTA is said to have provided diesel for the mobile phone network boosters, food hampers, clothes and maintenance of victims' houses. Mr Kid provided the timber for many coffins to allow descent burials of the dead. Mr Van Deruit was one of the citizens who mobilised with friends to bring in a helicopter from Harare. Saurombe Logistics & Consulting transported donated emergency relief worth about ZAR1.131 million that was stuck in Midrand, South Africa, due to a lack of transportation. The lack of transport was discovered during the 3-day Tropical Cyclone Idai author's workshop organised by the University of South Africa under the Exxaro Chair in Business and Climate Change that conceptualised and was leading the entire research project. The donations were made by the Zimbabwe Catholic Community in South Africa, a faith based organisation (FBO). The FBO received overwhelming response and collects goods that included tents, blankets, clothes, toiletries, pharmaceutical and food. The goods were handed over to Caritas Mutare (Zimbabwe), a Non-Governmental Organisation under the Catholic Church that assisted in the distribution in Chimanimani. The total number of recorded beneficiaries were 249 households that translated to approximately 1245 individuals. In addition, about 417 students also received sanitary hampers.

The third story could be linked to the Zimbabwe and South African Defence Forces that all did wonderful work during search and rescue as well as relief and recovery. On the ground conversations in Chimanimani confirmed that the forces handled themselves well and were true to duty. Lastly, Chatiza (2020, p. 6) concurs that Tropical Cyclone Idai confirmed that "Zimbabweans across social, economic and political contexts have shown that they can unite in responding to disasters". The acts of kindness were of such a nature that some lost their lives in rescue missions; others sacrificed their businesses.

1.4   Book and Chapter Outline

This book comes in 6 parts and 15 chapters. The orientation of the chapters is in a manner that each chapter can be read as a full standalone piece, with key sections that include an abstract, methodology and key findings. The book comes in 5 parts. Part I presents the introduction and background and is made up of this single chapter looking at the catastrophic impacts of Tropical Cyclone Idai in southern Africa. Part II focuses on the conceptual and global framework for DRR and is composed of three chapters. Chapter 2 presents a new model on DRR, codenamed the B4 Model that brings together perspectives on building and BBB. This is followed by addressing the use of and contestations in earth observation technologies in DRR in Chap. 3. Chapter 4 is pitched to address the naming of tropical cyclones, narrowing down to Cyclone Idai in this case.

Part III is made up of four chapters and discusses the impact of Tropical Cyclone Idai on selected sectors and how the sectors incorporated harnessing elements from the BBB principle under the Sendai Framework on DRR. Chapter 5 documents the scaled-up illegal gold mining activities in Chimanimani post-Tropical Cyclone Idai, with Chap. 6 being dedicated to presenting perspectives on energy infrastructure and the BBB concept. Chapter 7 articulates matters regarding floods in the midst of a drought, as brought up by Tropical Cyclone Idai in southeastern Zimbabwe, while Chap. 8 analyses BBB domestic and irrigation water supply systems.

Part IV details Tropical Cyclone Idai acts of kindness. Chapter 9 presents a rare scenario of ethical philanthropy and social responsibility during natural disasters, from the Higherlife Foundation. Chapter 10 presents another unique act of kindness from the Chimanimani Hotel with Tropical Cyclone Idai bringing out the humanitarian deed that overtook business interests. In Part V the book looks ahead in order to act today and three chapters make up this part. Chapter 11 looks at ethical and human rights dilemmas during Tropical Cyclone Idai, while Chap. 12 brings up religious engagements during Tropical Cyclone Idai and implications on BBB. Chapter 13 then presents exploring the potential of dark tourism in the aftermath of Tropical Cyclone Idai in Chimanimani, while Chap. 14 looks at flood hazard modelling in the eastern parts of the Save Catchment drawing lessons from Tropical Cyclone Idai. The last part, Part VI, brings up conclusions and policy recommendations and is made up of another single chapter. This last chapter, Chapter 15, draws up lessons and recommendations to enhance the B4 Model.

References

CADRI (Capacity for Disaster Reduction Initiative). (2017). Capacity Assessment of the Disaster Risk Management System in Zimbabwe. Harare: CADRI.

Callaghan, J. (2020). Extreme rainfall and flooding from Hurricane Florence. *Tropical Cyclone Research and Review*. https://doi.org/10.1016/j.tcrr.2020.07.002.

CCEDPM (Cabinet Committee on Environment, Disaster Prevention and Management). (2019). Report to

Cabinet by the Cabinet Committee on Environment, Disaster Prevention and Management on the Prioritised Cyclone Idai and Drought Programmes and Projects. Harare: CCEDPM.

Chatiza, K. (2020). Cyclone Idai in Zimbabwe: An analysis of policy implications for post-disaster institutional development to strengthen disaster risk management. Harare: Oxfam International.

Dube, K., Nhamo, G., Chikodzi, D. (2020). COVID-19 cripples global restaurant and hospitality industry. Current Issues in Tourism. https://doi.org/10.1080/13683500.2020.1773416.

EMA (Environmental Management Agency). (2019). Cyclone Idai Assessment Report in Chimanimani and Chipinge Districts. Harare: EMA.

Frame, D.J., Wehner, M.F., Noy, I., Rosier, S.M. (2020). The economic costs of Hurricane Harvey attributable to climate change. Climatic Change, 160:271–281. https://doi.org/10.1007/s10584-020-02692-8.

IOM Displacement Tracking Matrix. (2020). Tropical Cyclone Idai Response Multi-Sectoral Location Assessment (MSLA)—Round 6: Chimanimani, Manicaland. Retrieved from https://displacement.iom.int/system/tdf/reports/Zimbabwe_DTM_Multi-Sectorial%20Location%20Assessemnt_April%202020%20rev.pdf?file=1&type=node&id=8979 (Accessed 24 August 2020).

Manicaland Province Development Coordinator's Office. (2019). Manicaland Province Cyclone Idai Disaster Consolidated Report as at 20 May 2019. Mutare: Manicaland Province Development Coordinator's Office.

Murwira, Z. (2019). Zimbabwe: President Honours Cyclone Idai Heroes. Retrieved from https://allafrica.com/stories/201908130033.html (Accessed 15 September 2020).

Nhamo, G., Nhemachena, C., Nhamo, S. (2019). Is 2030 too soon for Africa to achieve the water and sanitation sustainable development goal? Science of the Total Environment. https://doi.org/10.1016/j.scitotenv.2019.03.109.

Nhamo, G., Nhemachena, C., Nhamo, S. (2020). Using ICT indicators to measure readiness of countries to implement Industry 4.0 and the SDGs. Environmental Economics and Policy Studies, 22:315–337. https://doi.org/10.1007/s10018-019-00259-1.

OCHA (United Nations Office for the Coordination of Humanitarian Affairs). (2020). Cyclones Idai and Kenneth. Retrieved from https://www.unocha.org/southern-and-eastern-africa-rosea/cyclones-idai-and-kenneth (Accessed 9 September 2020).

Paerl, H.W., Hall, N.S., Hounshell, A.G., Luettich Jr., R.A., Rossignol, K.L., Osburn, C.L., Bales, J. (2019). Recent increase in catastrophic tropical cyclone flooding in coastal North Carolina, USA: Long-term observations suggest a regime shift. Scientific Reports, 9:10620. https://doi.org/10.1038/s41598-019-46928-9.

Pindula, (2019). President Mnangagwa Awards Cyclone Idai Responders. Retrieved from https://news.pindula.co.zw/2019/08/12/president-mnangagwa-awards-cyclone-idai-responders/ (Accessed 6 May 2021).

Shi, L., Olabarrieta, M., Nolan, D.S., Warner, J.C. (2020). Tropical cyclone rainbands can trigger meteotsunamis. Nature Communications, 11:678. https://doi.org/10.1038/s41467-020-14423-9.

Uamusse, M.M., Tussupova, K., Persson, K. (2020). Climate Change Effects on Hydropower in Mozambique. Applied Sciences, 10, 4842; https://doi.org/10.3390/app10144842.

United Nations Habitat. (2016). United Nations Habitat III: New Urban Agenda. New York: United Nations Habitat Secretariat.

UNDRR (United Nations Office for Disaster Risk Reduction). (2015). Sendai Framework on Disaster Risk Reduction (2015-2030). New York: UNDRR Secretariat.

UNFCCC (United Nations Framework Convention on Climate Change). (2015). Paris Agreement. Bonn: UNFCCC Secretariat.

United Nations. (2015). Transforming our World: The 2030 Agenda for Sustainable Development. New York: United Nations Secretariat.

ZMSD (Zimbabwe Meteorological Services Department). (2019). Review of the Meteorological Services Department's capacity to observe, forecast and respond to future extreme weather events: An assessment of Tropical Cyclone Idai. Harare: ZMSD.

Part II

Conceptual and Global Framework for Disaster Risk Reduction

The B4 Model (Building and Building Back Better) in Disaster Risk Reduction and Management

2

Abstract

This chapter draws on global literature addressing the growing and popular concept of building back better (BBB) after disasters. It then adds another "B" to the new proposal we are presenting code named B4 (building and building back better), implying that communities should not only BBB after disasters but also have building better as part of everyday DNA. Matters like the choice of building materials, where to build and how to build, as well as proper planning and the respect of planning laws by politicians and others in positions of influence should all be in the mix. The chapter also incorporates the issues surrounding climate resilient building and buildings. Hence, there is a need to disrupt the built environment profession and have resilient and adaptive societies. To strengthen arguments, examples are used from primary research undertaken on Tropical Cyclone reached Zimbabwe on 15 March 2019 and the floods that followed in April 2019 in South Africa. Ultimately, the chapter presents a strong case for harmonising planning, development agendas and politics. This includes disaster risk reduction (DRR), sustainable development goals (SDGs), Sendai Framework on Disaster Risk Reduction, New Urban Agenda, the Paris Agreement and regional visions such as Africa Agenda 2063. The chapter also brings the notion of "all-of-society engagement and partnership" during disasters.

Keywords

BBB · B4 · Disaster risk reduction · Sendai Framework · Paris Agreement · SDGs

2.1 Introduction and Background

The phenomenon of climate change that brings to the fore perspectives on resilience, adaptation, and building back better (BBB) remain central in the disaster risk reduction (DRR) and management spaces of engagement. Initiatives have been ongoing to raise awareness and call for action on climate change (UNFCCC 2015) and the 2030 Agenda for Sustainable Development (AfSD) that got ratified in September 2015 for implementation starting January 2016 (United Nations 2015). From the 2030 AfSD are 17 sustainable development goals (SDGs) that the world committed to attain (United Nations 2015; Nhamo et al. 2020) as guided by the BBB principle embedded in the United Nations Sendai Framework for Disaster Risk Reduction (UNDRR 2015). Other important landmarks include the United Nations Habitat's New Urban Agenda that comprehensively embrace SDG 11, focusing on sustainable human settlements (United Nations Habitat 2016). The New Urban Agenda sums up the relationship between the key concepts addressed in this chapter. Bullet 13(g) challenges the world to "Adopt and implement disaster risk reduction and management, reduce vulnerability, build resilience and responsiveness to natural and human-made hazards and foster mitigation of and adaptation to climate change" (United Nations Habitat 2016, p. 5). This drive is also supposed to be age and gender responsive. From the African continent, there is the Africa Agenda 2063 (African Union Commission 2014), and other similar regional policy set-ups exist across all other continents.

The 2030 AfSD is unequivocal when it comes to addressing matters of climate change. SDG 13 places global responsibility in taking "urgent action to combat climate change and its impacts" (United Nations 2015, p. 14). In addition, five targets are outlined, with many more indicators. The full responsibility to implement the climate SDG is given to the United Nations Framework Convention of Climate Change (UNFCCC). By default, the Paris Agreement coming out of the UNFCCC processes is the global instrument upon which matters of climate change are addressed (UNFCCC 2015). There are several incidences in which the 2030 AfSD mentions DRR. In bullet 14, the 2030 AfSD highlights global health threats and more frequent and intense natural disasters as areas of concern as these threaten to reverse the gains in development from the Millennium Development Goals (MDGs). Without getting into depth, bullet 33 refers to the need to strengthen resilience and promote DRR. Disasters are also alluded to under targets 1.5 (SDG 1), 2.4 (SDG 2), 11.5 (SDG 11), 11.b (SDG 11), and 13.1 (SDG 13). In addition, the express mentioning of the Sendai Framework for DRR is also made as follows under Target 11.b:

> By 2020, substantially increase the number of cities and human settlements adopting and implementing integrated policies and plans towards inclusion, resource efficiency, mitigation and adaptation to climate change, resilience to disasters, and develop and implement, in line with the Sendai Framework for Disaster Risk Reduction 2015-2030, holistic disaster risk management at all levels. (United Nations 2015, p. 22)

From Target 11.b, other matters subject to further discussions in this chapter include resilience and adaptation. Bullet 29 requests the global community to strengthen resilience in communities hosting refugees, especially those from the developing global south. Target 1.5 demands that resilience be built for the poor and residents in vulnerable situations. This has to be done so that their exposure and vulnerability to climate change-related extremes, as well as economic, social and environmental shocks and disasters, are minimised. With hindsight on how the COVID-19 pandemic has ravaged vulnerable communities, the observations from the 2030 AfSD cannot be more accurate. The need for cities to be resilient to disasters is also highlighted under Target 11.b.

Resilience is also aligned to climate change adaptive capacity discussed under Target 13.1. The target calls upon governments and their development partners to "Strengthen resilience and adaptive capacity to climate-related hazards and natural disasters in all countries" (United Nations 2015, p. 23). Under SDG 14 that deal with the ocean (blue economy), Target 14.2 stipulates the desire to "sustainably manage and protect marine and coastal ecosystems to avoid significant adverse impacts, including by strengthening their resilience" by 2020 (United Nations 2015), thereby informing the New Urban Agenda (United Nations Habitat 2016).

The Sendai Framework remains the overarching platform for managing disaster risk. The framework succeeded the "Hyogo Framework for Action (HFA) 2005–2015: Building the Resilience of Nations and Communities to Disasters" (UNDRR 2015). The Sendai Framework acknowledges that disasters are escalating, with over 1.5 billion people having been affected by such in the reporting period of the HFA. To this end, one of the guiding principles of the Sendai Framework is to allow the managing of disaster risk with the intention to protect "persons and their property, health, livelihoods and productive assets, as well as cultural and environmental assets, while promoting and protecting all human rights, including the right to development" (UNDRR 2015, p. 12). The UNDRR further outlines that effective DRR demands an "all-of-society engagement and partnership" (UNDRR 2015). The Sendai Framework then identifies preparedness (before), search and rescue (during), as well as relief and recovery (after) as the key elements in the DRR cycle, with the national and local spatial levels as critical spaces of engagement. The BBB concept (discussed in more detail later) is outlined as one of the priority areas. What is also of interest to note is the broadening of the scope of what constitutes disasters. The Sendai Framework broadens the scope to cover both natural and human-induced hazards, as well as related environmental, technological and biological hazards. This witnessed the promotion of health resilience across the document. This discovery challenges the world to re-think global frameworks. Had this warning been taken seriously, possibly the world could have been better prepared for the COVID-19 pandemic.

From the African Union Agenda 2063, climate change—especially adaptation—is prioritised in the context of sustainable development. Bullet 15 mentions that the continent "shall address the global challenge of climate change by prioritising adaptation in all our actions, drawing upon skills of diverse disciplines and with adequate support" (African Union Commission 2014, p. 6). If this action is to be realised, African leaders agreed that there needs to be adequate means of implementation. Such means of implementation would include affordable technology development and transfer, capacity building, financial and technical resources. Furthermore, seven aspirations are spelt out to include the following: (1) desire to have a prosperous Africa based on inclusive growth and sustainable development, (2) having an integrated continent that is politically united and based on the ideals of Pan Africanism, (3) call for good governance, democracy, respect for human rights, justice and the rule of law, (4) have a peaceful and secure continent, (5) a continent with a strong cultural identity, common heritage, values and ethics, (6) a place where development is people driven, recognising the role women and youth can play, and (7) a continent that is an influential global player and partner.

Given the foregoing, this chapter spells out an objective to elevate the building and building back better (B4) model. The B4 model embraces an additional "B" on the BBB principle in DRR. The chapter's methodology mainly utilises the document and critical discourse analysis (Mbatu 2020; Mendes et al. 2020). Additional data and insights are drawn from two sets of fieldwork: one from Chimanimani (Zimbabwe) where Tropical Cyclone Idai arrived in March 2019, and also another from Port St Johns in the Eastern Cape Province of South Africa where there were devastating floods in the same year in April. Limited data were also generated from household surveys. It is hoped that the B4 model will enhance both the national, regional and global preparation and engagements with any form of disasters apart from the common climate change hazards.

2.2   A Literature Survey: Towards the B4 Model

This literature survey comes in four sub-sections. The next sub-section looks at climate change as the dominant global disaster. This is followed by a focus on resilience, another conceptual framework that has and still is shaping DRR and management work. Next in the sequence is a deliberation on adaptation and adaptive capacity, and how such relate to resilience, with the fourth sub-section focusing on the BBB principle. Each of these sub-sections is presented in the following paragraphs.

2.2.1   Climate Change as the Dominant Global Disaster

Climate change remains the dominant global disaster, with the scaling up of efforts since the ratification of the Paris Agreement in December 2015 (UNFCCC 2015). This is the reason why the world maintained calls to continue focused and address it amid the COVID-19 pandemic (Quéré et al. 2020). The Sendai Framework reminds us that many disasters associated with climate change continue to increase in frequency and intensity and continue to eat into the gains that have been made towards sustainable development (UNDRR 2015). The Sendai Framework then makes reference to the UNFCCC processes, upon which climate change matters are debated and finalised. From the New Urban Agenda (United Nations Habitat 2016), climate change (SDG 13) is acknowledged and linked to culture, as well as sustainable consumption and production (SDG 12) (United Nations 2015).

The New Urban Agenda (United Nations Habitat 2016) goes further to acknowledge the Paris Agreement concluded in 2015 under the UNFCCC, and indicates that human settlements should seek to reduce greenhouse gas (GHG) emissions with the view to maintain such below the two degrees Celsius above pre-industrial levels. With the help of other stakeholders and United Nations agencies, every effort was to be channelled towards limiting the temperature rise to 1.5°C above pre-industrial levels. The reason is that GHG emissions (also interchanged with carbon emissions) remain the main cause of global warming that leads to climate change, and ultimately the extreme weather events including floods, tornados, tropical cyclones, droughts and many more. This space brings up the notion of climate change mitigation, an act of reducing carbon emissions. To this end, the United Nations Habitat (2016) encourages the climate-effective design of spaces, buildings and construction, services and infrastructure, as well as resilience building.

2.2.2   Drawing from the Concept of Resilience

There is growing global awareness of resilience as a cornerstone to overcoming disasters, and probably the best starting point to consider matters on resilience is the Sendai Framework. In 63 instances, the Sendai Framework mentions the word "resilience" in contrast to "adaptation" being mentioned only once. In fact, out of the four priority areas, Priority 3 is dedicated to

investing in DRR for resilience. The integration of DRR and the building of resilience into national development agendas and policies is encouraged (UNDRR 2015). Drawing from its earlier 2009 work, the UNDRR defines resilience as follows: "The ability of a system, community or society exposed to hazards to resist, absorb, accommodate to and recover from the effects of a hazard in a timely and efficient manner, including through the preservation and restoration of its essential basic structures and functions" (UNDRR 2020a online). This definition expands upon the UNFCCC 2007 definition that indicates resilience as "The ability of a social or ecological system to absorb disturbances while retaining the same basic structure and ways of functioning, the capacity for self-organization and the capacity to adapt to stress and change" (UNFCCC 2020a online). From the Sendai Framework, governments were called upon to invest in the economic, social, health, cultural and educational resilience of persons, communities and countries and the environment. Effectively, resilience had to be built across the economic, social, health and environmental sectors. Building and investing in resilience means saving lives, preventing and/or minimising losses, as well as ensuring efficient and effective recovery and rehabilitation. The desire to embrace the concept of, and build resilience in health systems from a national, right up to the WHO level is also highlighted under bullet 3(i) as follows:

> To enhance the resilience of national health systems, including by integrating disaster risk management into primary, secondary and tertiary health care, especially at the local level; developing the capacity of health workers in understanding disaster risk and applying and implementing disaster risk reduction approaches in health work; promoting and enhancing the training capacities in the field of disaster medicine; and supporting and training community health groups in disaster risk reduction approaches in health programmes, in collaboration with other sectors, as well as in the implementation of the International Health Regulations (2005) of the World Health Organization (UNDRR 2015, p. 19).

The business sector is not left out either. The Sendai Framework envisages the sector increasing its resilience to protect livelihoods and assets

(UNDRR 2015). For this to happen quicker, partnerships, as now expressly highlighted in SDG 17 (United Nations 2015), need to be built. The United Nations Global Compact was allocated the role to work with businesses to build resilience in the context of sustainable development. Resilience in new and existing infrastructure is also promoted—an element that brings up perspectives from this chapter that seeks to elevate the fourth "B" in the B4 model. Among the infrastructures mentioned are water, transportation and telecommunications, educational facilities, hospitals and other health facilities.

The New Urban Agenda also brings up the concept of resilient urban development that is supported by environmentally sustainable growth. The same sentiments are raised in the Paris Agreement in Article 7(9)(e), which stipulates that "building the resilience of socioeconomic and ecological systems, including through economic diversification and sustainable management of natural resources" needs to take place (UNFCCC 2015, p. 11). The resilience of communities and livelihoods is also considered in the Paris Agreement. From an urban perspective, the resilience of settlements is discussed in line with disasters, particularly climate change-induced hazards such as floods, droughts and heat waves (United Nations Habitat 2016). Building resilience in the context of the New Urban Agenda embraces several aspects and perspectives. Among such is included being more proactive as opposed to being reactive to disasters. It also includes all hazards and all-of-society approaches. These approaches focus on raising public awareness as well as harnessing timely and effective local responses to address disaster risk.

There is also growing attention from academia on resilience. Forrest and Milliken (2018) focus on building resilience to disasters. In their view, governments the world over were grappling with increased risks emanating from disasters that included climate change, social inequality, urbanisation and aging infrastructure. The Resilient America Programme (RAP) was born out of the National Academies report on Disaster Resilience that presented a national imperative in 2012. The RAP sought to bring science into resilience-

> **Box 2.1 The Seven Themes of Resilience Are as Follows:**
>
> 1. The distinction between resilience as a system trait, process, or outcome;
> 2. The importance of resilience as a strategy for dealing with uncertainty;
> 3. A shift from understanding resilience to active resilience building;
> 4. The incorporation of transformation into resilience;
> 5. The increasingly normative interpretation of resilience;
> 6. the growing emphasis on measuring and evaluating resilience; and
> 7. The mounting critiques of the resilience agenda demanding attention.
>
> Moser et al. (2019, p. 1).

The many disciplines spoken about lead to a situation where other authors speak of resilience in the context of tourist destinations and natural disasters (Filimonau and de Coteau 2019). The authors found that while local tourism stakeholders in Grenada were aware of potential destruction from natural disasters on their destinations, they failed to put in place effective measures to build destination-wide and organisational resilience. Bešinović (2020) discusses resilience in the context of railway transport systems. Resilience in the railway transport systems is then described to incorporate aspects of vulnerability, survivability, response, recovery, mitigation and preparedness. Croese et al. (2020) use the lens of urban resilience in localising the SDGs, while Anholt and Sinatti (2020) paint a picture where the framing of resilience and its execution depends on the local contexts and vested interests of key stakeholders and power play. The authors use the example of the European Union and its agenda to build resilience to insecurity. Hence, resilient building became a refugee containment strategy.

Logan and Guikema (2020) find that there were shortfalls in the dominant approaches to assessing resilience as it mainly focuses on evaluating community characteristics or infrastructure functionality. The focus on communities is also confirmed by Sheehan and Fox (2020) when they looked at building resilience in the context of the COVID-19 pandemic. Zhao et al. (2020) add another limitation, thus, the dominance and focus on urban areas, compared to poor and rural villages. This results in shortfalls in the approach's ability to provide actionable insights. Hence, Logan and Guikema (2020) argue that communities require access to other services usually excluded inform the resilience debate such as food, education, health care and cultural amenities. This is in addition to the conventional focus on water, power, sanitation and communications. From the authors' perspectives, a new conceptualisation of resilience should capture what they call essential services. Using two cases from Hurricanes Florence and Michael, the authors show how decision-makers and planners could apply the new framing on "equitable access to essentials approach to community resilience … that will enable communities not only to bounce back from a disruption, but to bound forward and improve the resilience and quality of life" (Logan and Guikema 2020, p. 1538). The proposed new framing of resilience makes sense. Ultimately, Parker (2020) is of the view that resilience must embrace adaptation, change and transformation.

2.2.3   The Concepts of Adaptation and Adaptive Capacity

Adaptation is another building block concept mentioned in the Sendai Framework, and this is done in the context of climate change (UNDRR 2015). The New Urban Agenda further stipulates that there should be medium- to long-term adaptation planning processes in societies (United Nations Habitat 2016). In addition, city-wide assessments of climate vulnerability and impacts should be undertaken in order to inform the policies (strategies, plans, programmes and actions). To this end, access to several multilateral disaster funds such as the Green Climate Fund, the Global

Environment Facility, the Adaptation Fund and the Climate Investment Funds needed to be accessed and made available for such purposes. With specific reference to its application in the context of climate change, the UNFCCC (2020b online) defines adaptation as follows:

> Adjustments in ecological, social, or economic systems in response to actual or expected climatic stimuli and their effects or impacts. It refers to changes in processes, practices, and structures to moderate potential damages or to benefit from opportunities associated with climate change.

To this end, the adaptation process embraces four main stages that include the need to (1) assess impacts, vulnerability and risks, (2) plan for adaptation, (3) implement adaptation measures, and (4) monitor and evaluate adaptation (UNFCCC 2020b). In line with the global need to verify actions, the verification perspective can be added into the last step. To add to the adaptation process, the UNFCCC identifies other enablers and/or means of implementation to the adaptation process that include raising awareness and ambition, having the necessary political spaces of engagements, information and knowledge sharing, technical and institutional capacities, financial resources and the engagement and active involvement of stakeholders. In the context of the SDGs (United Nations 2015), all the stakeholders should be involved. Since adaptation actions depend on the specific country, region, community, business or organisation, a "one-size-fits-all" approach is not practical (UNFCCC 2020b online).

The UNFCCC goes on to identify some of the adaptation actions that include "building flood defences, setting up early warning systems for cyclones and switching to drought-resistant crops, to redesigning communication systems, business operations and government policies" (UNFCCC 2020b online). From the highlighted key interventions, there is a clear overlap with resilience, which dominates the Sendai Framework (UNDRR 2015). This leads to unnecessary confusion, particularly given that the global community has warmed up and has just started embracing the adaptation agenda under the Paris Agreement. However, the main challenge is that adaptation in this context is restricted to climate change, a perspective that could exclude other key global disasters such as what was witnessed with the COVID-19 pandemic. Some who may wish to be critical of the global development agenda may allege that the increasing push for the resilience agenda may delay progress towards adaptation and the 2030 AfSD as citizens start re-aligning processes and thoughts towards the resilience space.

There are several other critical terms related to climate change adaptation including adaptive benefits, adaptation costs, adaptive capacity, maladaptation, climate change, climate variability and vulnerability. The definitions for these and other terms can be found online. However, for the purposes of this chapter, three other definitions drawn from the Intergovernmental Panel on Climate Change (IPCC) as endorsed by the UNFCCC are considered (Box 2.2).

Box 2.2 Additional Definitions in the Climate Change Adaptation Space

Adaptive capacity (in relation to climate change impacts): The ability of a system to adjust to climate change (including climate variability and extremes) in order to moderate potential damages, to take advantage of opportunities or to cope with the consequences (IPCC AR4 2007).

Maladaptation: Any changes in natural or human systems that inadvertently increase vulnerability to climatic stimuli; an adaptation that does not succeed in reducing vulnerability but increases it instead (IPCC Third Assessment Report 2001).

Vulnerability: The degree to which a system is susceptible to, or unable to cope with, adverse effects of climate change, including climate variability and extremes. Vulnerability is a function of the character, magnitude and rate of climate variation to which a system is exposed, its sensitivity and its adaptive capacity. Therefore adaptation would also include any efforts to address these components (IPCC AR4 2007).

Source: UNFCCC (2020b online)

Across all United Nations documents, the Paris Agreement remains the main framework document addressing climate change adaptation. However, as the world has witnessed under the COVID-19 pandemic, both the adaptation and resilience spaces need broadening to include all disasters. In 47 instances in the Paris Agreement, the word "adaptation" is mentioned (UNFCCC 2015). Concluded during the Conference of the Parties to the UNCCC (COP 21) in December 2015, the Paris Agreement had to embrace adaptation that had played second class status to the mitigation agenda (Nhamo and Nhamo 2016). Two main implementation instruments to the Paris Agreement were agreed upon, namely the Nationally Determined Contributions (mainly mitigation oriented) and the National Adaptation Plans (UNFCCC 2015). The entire Article 7 of the Paris Agreement is dedicated to addressing adaptation. Article 7(1) states that:

> Parties hereby establish the global goal on adaptation of enhancing adaptive capacity, strengthening resilience and reducing vulnerability to climate change, with a view to contribute to sustainable development and ensure an adequate adaptation response in the context of the temperature goal referred to in Article 2.

Some key points coming from Article 7(1) include the fact that adaptation is being linked to resilience, an element that dominates in the Sendai Framework, as well as contributing to the SDGs, particularly SDG 13. In Article 7(4), adaptation is linked to mitigation, acknowledging that mitigation measures will reduce the need for additional adaptation efforts in the future and so will be the reduction in costs. From Article 7(5), "adaptation action should follow a country driven, gender responsive, participatory and fully transparent approach, taking into consideration vulnerable groups, communities and ecosystems" (UNFCCC 2015, p. 9). A call is also made to have adaptation measures informed by the best available science, traditional, indigenous and local knowledge systems. Furthermore, adaptation had to be integrated into socioeconomic and environmental policies and actions as appropriate. The cooperation on adaptation by Parties should be aligned to the Cancun Adaptation

Framework. Part of the Cancun Adaptation Framework as enshrined in Article 7(8)(d&e) focuses on the following:

> Assisting developing country Parties in identifying effective adaptation practices, adaptation needs, priorities, the support provided and received for adaptation actions and efforts, and challenges and gaps, in a manner consistent with encouraging good practices; and Improving the effectiveness and durability of adaptation actions. (UNFCCC 2015, p. 9)

In Article 9(1), the developed country Parties to the UNFCCC are supposed to provide the means of implementation, including financial and technological resources (UNFCCC 2015). However, such resources have always been difficult to mobilise (Nhamo and Nhamo 2016). The biggest take away is that at least adaptation has been moved from being framed as a purely local problem to being a global challenge (Setzer et al. 2020). Another perspective that has found its way onto the global platform concerning climate change adaptation is traditional ecological knowledge (TEK). TEK is useful in addressing climate change adaptation in Sarawak in Malaysia (Hosen et al. 2020). Through TEK the communities have managed food security in spite of persistent droughts and wildfires.

Santos et al. (2020) raise practical adaptation perspectives from a case study on Portugal's Setubal municipality. With reference to the Sendai Framework, the authors find that the municipality has been able to withstand multiple disasters that include landslides, earthquakes, tsunamis, floods and the COVID-19 pandemic. What the municipality did to adapt was to instal equipment such as emergency signs and electronic panels that allow the dissemination of information to the communities in real time. However, Williams et al. (2020) have a different picture regarding local government capacity in Mauritius. The authors find that effective climate change adaptation is hindered by elements that include the lack of technical know-how, as well as financial and human resources.

The subject of maladaptation is also given considerable space in the literature. In Fiji, Piggott-McKellar et al. (2020) discovered that

seawalls meant to safeguard communities against coastal pressure have resulted in artificial and unanticipated dams. The walls trap the water along their landward sides due to ineffective designs and the manner in which the walls were constructed. Kundu et al. (2020) observe a major maladaptation gap in the agricultural space in coastal Bangladesh. To this end, the authors call upon stakeholders in the disaster-prone country to understand more on maladaptation in order to avoid mistakes in planning processes. In another study, Mendelsohn and Zheng (2020) view the building of very tall coastal walls to manage 100-year storm surges from tropical cyclones too costly for USA communities.

Neset et al. (2019) focus on maladaptation in the Nordic agriculture sector. Their findings reveal that several adaptation measures led to maladaptation. For example new technical equipment such as drying resulted in higher energy and investment costs and an increase in GHG emissions. Increased fertiliser use led to eutrophication, while increased pesticides use resulted in poor soil, water and food quality. Poor soil quality also resulted from increased tillage that was adopted as an adaptation measure as nutrients leaked. The installation of new drainage systems led to depleted wetlands, while new irrigation systems caused a shift in flooded areas. The shift to new crop types also led to the increased use of pesticides and fertilisers as well as GHG emissions. Within the settlements setup, poorly conceptualised adaptation strategies that include the erection of flood barriers may also result in the failure by communities to adapt and learn to live with the floods in certain instances.

2.2.4   The BBB Principle

The BBB concept is enshrined in the Sendai Framework as one of the four priority areas. The fourth priority focuses on "Enhancing disaster preparedness for effective response and to 'Build Back Better' in recovery, rehabilitation and reconstruction" (UNDRR 2015, p. 21). To BBB, the lessons from past disasters point to the fact that governments need to integrate DRR into national development plans. This also has a further advantage of making communities more resilient to disasters. Regional cooperation in DRR and management is encouraged, including undertaking common exercises and drills, as well as developing protocols that facilitate sharing of responses capacity and resources. Vulnerable countries are also identified in the Sendai Framework, and these include Small Island Developing States (SIDS), landlocked developing countries, African countries, and middle-income countries. The New Urban Agenda is also unequivocal about BBB, outlining it in bullet 78 that it must embrace "post-disaster recovery process to integrate resilience-building, environmental and spatial measures and lessons from past disasters, as well as awareness of new risks, into future planning" (United Nations Habitat 2016, p. 16). The resilience should also address urban habitats that seek ecosystems-based adaptation.

Drawing from experiences on COVID-19, the World Resources Institute (WRI) argues that BBB should be aligned to the SDGs and the Paris Agreement (WRI 2020). The WRI is of the view that COVID-19 provides lessons on how the climate and biodiversity crises will affect the world if not addressed earnestly. As such, building an inclusive, green and resilient recovery was now urgent, and all governments had to come to the party. Zhao et al. (2020) propose a framework for resilience development in poor villages following the 2008 Wenchuan Earthquake that had to integrate the BBB principle. They find that many years after the disaster areas—including social life systems, economic production systems, and natural ecosystems—still require improvement. The next section brings together the framing of resilience, adaptation, BBB and DRR, and management in a proposed new model code named the B4.

2.3 BBB After Cyclones and Floods in Southern Africa

This section brings in cases and examples on BBB from southern Africa, specifically drawing from Tropical Cyclone Idai that hit Zimbabwe and floods that hit Port St Johns in the Eastern Cape Province of South Africa in March and April 2019. From some primary fieldwork in Port St Johns (South Africa) and Chimanimani (Zimbabwe) that took place between September and December 2019, the elements of BBB could be witnessed.

The catastrophic arrival of Tropical Cyclone Idai left visible and untold damage to infrastructure, houses, roads, bridges and the social fibre. From the epicentre of the disaster in Kopa in Chimanimani, two key bridges were swept away—one from the Nyahode River and the other from the Rusitu River. The remains of the swept bridge from the Nyahode River could be located some half a kilometre or so away from the original site (Fig. 2.1). At the time of the fieldwork, work had already started on the second Barley bridge at the Nyahode River. Barley bridges are temporary, and the work was being done jointly by the South African and Zimbabwean defence forces; the first Barley bridge was put across the Rusitu River (Fig. 2.2). What emerged from one focus group discussion during fieldwork was that it was the fourth time the Nyahode Bridge had been swept away since Topical Cyclone Eline in February 2000. Clearly, the BBB scenario may not be playing out in this situation. This implies that a better design for a stronger and resilient bridge is needed at Nyahode River at Kopa.

At the other location that was also severely damaged by Tropical Cyclone Idai, which is the second hotspot, the remains of one teachers' house from Ngangu Primary School are presented in Fig. 2.3 and the replacement that was also BBB is seen in Fig. 2.4. The process of BBB on the house involved using more modern building material and moving it away from the flood zone significantly. However, while elements of BBB are evident from the Ngangu Primary rebuilt house, this was not the same elsewhere in the Chimanimani District as hundreds of households remained in emergency shelters awaiting their resettlement one and a half years after the disaster.

The BBB perspectives from tropical cyclones in Zimbabwe, specifically from Chimanimani,

Fig. 2.1 Remains of a bridge swept away by the flood waters from Cyclone Idai at Nyahode River. Source: Authors, Fieldwork 2019

Fig. 2.2 A temporary Barley bridge coming up across the Rusitu River. Source: Authors, Fieldwork 2019

remain in the infancy stages. For example the 2019 fieldwork revealed that there were still some government structures destroyed by Tropical Cyclone Eline in February 2000 that had not been repaired. In Fig. 2.5 is shown a school dormitory that was damaged in 2000 and still not repaired 20 years later. Tropical Cyclone Eline devastated Mozambique, Zimbabwe and South Africa between February and March 2000 (Reason and Keibel 2004). Similar scenarios

were already playing out during Tropical Cyclone Idai, especially if one considers the essential services approach to building resilience presented by Logan and Guikema (2020). From one focus group discussion with the displaced at Kopa, it was clear that there were a lot of non-infrastructure essentials that were needed to facilitate the BBB of social, religious and business aspects. Critical issues that could bring back livelihoods quicker included issuance of IDs, passports, drivers'

Fig. 2.3 House destroyed during the cyclone at Ngangu Primary School. Source: Authors, Fieldwork 2019

Fig. 2.4 Replacement house built back better. Source: Authors, Fieldwork 2019

licenses, birth certificates, death certificates, education certificates, hospital records and other documentation. Many women were making a living from cross border trading getting into Mozambique and South Africa that was cut short because passports were swept away or soiled. Of importance to note is the quicker arrangement of issuing national IDs done by the Registrar's office that included mobile posts.

One of the pressing issues in BBB is what to do with the problem faced from the disaster and also how to deal with families and households that would have been displaced temporarily or permanently. From the household survey ($n = 212$), the majority of the respondents were of the view that the challenge of flooding from Tropical Cyclone Idai could be resolved by relocating the affected households. An estimated 68.87% were of this view, while 20.76% thought geo-engineering and improved drainage could assist. The remaining percentage did not know or gave other reasons. One of the interviews

Fig. 2.5 School dormitory in Chimanimani damaged by Tropical Cyclone Eline in 2000. Source: Authors, Fieldwork 2019

with the Chimanimani District Residents Association Acting Chair indicted that they had been in discussions with the District Council to resettle those whose houses were damaged badly and/or washed away. The preference was for those in this bracket to get new stands close to where they originally were so that livelihoods would be minimally disturbed. However, the District Council in liaison with the Ministry of Local Government identified an area close to Bumba, which was indicated to be about 65km away from the affected areas. The Bumba area also falls in a much drier region (agro-ecological region 3) compared to the affected area that is mainly in agro-ecological regions 1 and 2. The affected areas have very high rainfall, cooler temperatures and spring water. And bananas can grow both in valleys and in mountains.

Within the Bumba area, the water is mainly underground from boreholes and wells, with bananas grown mainly along river banks. This scenario presents a changed livelihood, and in terms of BBB and adaptation, it will lead to some form of maladaptation. Hence, the 210 households that were targeted for relocation as per the Acting Chair would be disadvantaged. Given that many households rely on subsistence agriculture, especially banana planting, the Bumba soils were

indicated to be poor and would not sustain such activities. One of the local alternative relocation places was identified by the Acting Chair as Frog and Fern, right in Chimanimani Town. The other identified potential relocation area was Nhuta Farm. However, apart from the challenges associated with the relocation, there were undertones in the interview showing that Chimanimani had been outgrown by Chipinge Town that now had a magistrate court, district hospital and other services, yet it was established well after Chimanimani. All these services were lacking in Chimanimani, with residents having to go to Chipinge for such.

Linked to the BBB principle, Zimbabwe's largest cement manufacturing company, Pretoria Portland Cement (PPC) Zimbabwe, a subsidiary of PPC Pty Ltd based in South Africa, handed over 90 tonnes of cement for the Cyclone Idai Victims Housing Project (Chronicle 2020). The cement came in the form of 1800 bags of the SURECEM brand, with the donation made in July 2020. The donated cement would make it possible to have safe, sustainable and secure houses post Tropical Cyclone Idai. A total of 30 houses were to be constructed.

Another fieldwork exercise was undertaken in December 2019 to investigate elements of BBB after the April 2019 floods in Port St Johns Local

Municipality in the Eastern Cape Province of South Africa. On 22 April 2019, the O.R. Tambo District Municipality (2019), as the responsible authority for disaster risk, reported the incident that caused massive damage to property and displacements. A total of 882 households were affected with 226 families left homeless mainly from the Greens Farm. Eleven political wards were impacted. Asked to indicate how the issue of flooding could be resolved, 44% of the household responses (n=292) indicated the installation of a proper storm water drainage system would assist, 34% indicated a need to relocate affected households, while 11% thought improved waste collection could also assist. The remaining 12% did not know how. Cases of BBB after the Port St Johns floods emerged during the fieldwork in 2019. Figure 2.6 presents an old homestead that was ravaged by the floods. Following the interventions by government and other stakeholders, the homestead was relocated to a nearby location and built again in a much better way (Fig. 2.7). There were pole and mud round huts prior to the floods, and this scenario is common in the affected areas.

Port St Johns is also difficult to access during flood emergencies. This is due to the fact that it only has one access route, which passes through mountains that frequently produce landslides in the event of heavy rains. This is especially so on mountains close to Port St Johns River Lodge. These landslides block the roads and hamper evacuation efforts and emergency services that normally react from Umtata. In the aftermath of the April 2019 flood disaster, field observations and key informant interviews confirmed that the South African government initiated a programme to make the roads more resilient to mudslides and damage from heavy rains. This included stabilisation of slopes at vulnerable locations, and the construction of erosion and mass movement gabions. The BBB project also involves the surfacing and re-surfacing of frequently damaged roads using more resilient standards and materials, as well as putting proper drainage structures such as culverts in place. Plans are also in place to open up an alternate and old access route into the town in order to avoid the town getting cut off completely during emergencies. However, the town, which is located on a lagoon, barely a few

Fig. 2.6 Pole and mud, as well as other brick and motor structures before the floods. Source: Authors, Fieldwork 2019

Fig. 2.7 Much improved newly built structures following the floods. Source: Authors, Fieldwork 2019

metres above sea level and on the Umzimvubu River mouth, will always be vulnerable to flooding due to its location. There is therefore serious talk of relocating the town to safer areas with a number of suggestions being made including close to the airport. However, efforts towards possible relocation to this specific area are being frustrated by land claims being made by a nearby chieftainship.

2.4   Putting It All Together: The B4 Model

Drawing from the discussions herein, the B4 model is proposed and developed (Fig. 2.8). This concept acknowledges the excellent work that has been done in BBB societies following disasters, addressing climate change, matters of resilience and adaptation. Even the politicians have embraced the concept of BBB following the COVID-19 pandemic. However, there seems to be a gap in the BBB narrative as societies need to build better now, thus, the B4 concept as presented by adding an additional "B". In countries

like South Africa, and probably many more, infrastructure and housing projects have been carried out *ad hoc*, leaving such developments vulnerable to extreme weather events such as floods and storms. The Reconstruction and Development Programme (RDP) houses, built as a post-apartheid project to better the lives of poor black people, at times collapse on a large scale following severe storms, thus providing evidence that these would not have been built properly in the first place.

What emerges from the B4 concept are several elements that include the following: (1) that governments and their development partners need to build resilience now, (2) governments and other stakeholders need to BBB after disasters, (3) we need to embrace the notion of continuous improvement, (4) we need to take a global solidarity pathway, and (5) we need to take the essential services approach to BBB. No country is immune to disasters; some take the nature of pandemics such as COVID-19. We all need each other. The notion of continuous improvement emanates from the manner in which progressive industries and organisation seek to be on the look

Fig. 2.8 The B4 model
as proposed. Source:
Authors

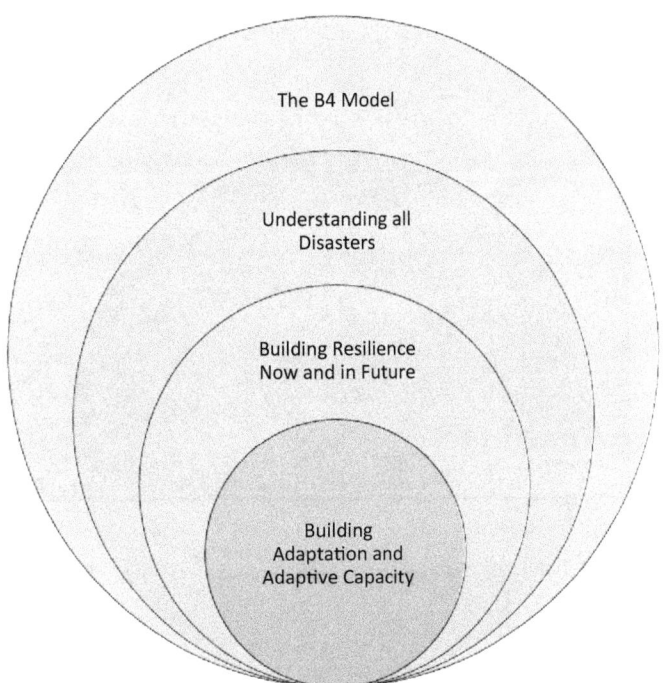

The B4 Model

Understanding all
Disasters

Building Resilience
Now and in Future

Building
Adaptation and
Adaptive Capacity

out to better themselves daily in order to survive competitors and other harsh operating environments.

A typical example of building better, and also building badly, is reflected from one of the studies undertaken in the Belvedere North and Epworth areas in Harare, Zimbabwe. In Fig. 2.9 is shown a wetland that was left open for years by the City of Harare that was later encroached into as pressure mounted for development space. In terms of building better, the original plans to preserve the wetlands were on point as such wetland have many functions that link directly to adaptation and resilience in human settlements, including flood management, water purification, biodiversity conservation and enhance city aesthetics. From Fig. 2.10, a new development to accommodate a shopping mall constructed by a Chinese company in 2013 is visible and substituted the long-term and probably acceptable urban agriculture use (Mutisi and Nhamo 2015). The other pronounced feature on the left in the image is the Zimbabwe National Sports Stadium, also built by the Chinese in the 1990s. Up to 42% of the wetland was disturbed in the Belvedere North wetland. Similar challenges of wetlands

degradation and invasion in the City of Harare were also observed in the Epworth Area with 63% of the wetland disturbed mainly for settlement construction. The encroachment into wetland areas surrounding many cities in Zimbabwe has been growing over the years since independence in April 1980. Some of the areas of concern include Parklands East in Bulawayo and the Zimre Park, Ruwa extensions that border Harare and Ruwa Township, and Monavale Vlei in Harare, which is also a Ramsar site. The real issue is that modern cities need to grow upwards rather than outwards. This is so because the supply of good land has run out and what remains in the peripheries and pockets of open land in the cities are usually wetlands and other fragile pieces of land that planners originally left and coded undevelopable. As indicated earlier, these portions of land acted as buffers to extreme weather events linked to climate change and other hazards.

Drawing from the work of Borie et al. (2019), stakeholders in DRR and management are encouraged to talk more of resilience thinking. This broader approach, married to systems thinking, can shape the manner in which the proposed

Fig. 2.9 Belvedere North wetland before settlement (Google Earth April 2011). Source: Mutisi and Nhamo (2015, p. 114)

Fig. 2.10 Belvedere North wetland after the mall and other developments (Google Earth, April 2014). Source: Mutisi and Nhamo (2015, p. 114)

B4 model should be implemented. It will not only be a model for DRR and management but also a way of life across disciplines and used by all stakeholders. In BBB, the perspectives on resilient, bouncing back quicker, adapt and taking along all stakeholders emerge (World Bank 2018). However, although there are overlaps, the distinction between resilience and adaptation remains (Bellinson and Chu 2019). In addition, it is also clear that the concepts of resilience and adaptation have their roots in DRR and climate change. Hence, as we move into the future, the B4 model demands that we continue learning by doing and adjusting processes and protocols on the go. From the USA, Kurth et al. (2019) proposed operationalising the resilience concept in the building industry by going beyond risk management. This approach resonates well with the B4 model. From the authors, "resilience, as applied to buildings, calls issues of building functionality, recovery and adaptation into question. Overall, these elements help address the issue of how to achieve continuity of service provision from a building or cluster of buildings following some stress" (Kurth et al. 2019, p. 481). The authors further acknowledge that resilience in building may also result in maladaptation. However, the existing investment routes and building regulations present a challenge to mainstreaming resilience in building. Drawing from the need to deconstruct resilience, Jordan (2019) proposes that there be adequate understanding of the intersection of vulnerabilities experienced by women due to sociocultural complexities, norms and practices. Hence, the B4 model should be gender and people with disability-responsive too.

2.5 Conclusion and Recommendations

The BBB principle, as well as the B4 model, are not easy frameworks to execute. From the literature reviewed and also what was emerging on the ground in Chimanimani (Zimbabwe) and Port St Johns (South Africa), there still remain major gaps to be filled now and in the future. Hence, the main reason why the B4 model is being proposed, as this would assist in taking care of future disaster risk on development conceptualisation and the design of infrastructure and essential services to be DRR-compliant, thereby building resilience. Apart from bouncing back quicker, efficiently and inclusively after disasters, the real deal is to build better now.

The B4 model embraces broader DRR and management perspectives. It goes beyond local, regional and global discourses surrounding climate change action to address this persistent global challenge. The B4 model embraces sociocultural, traditional and religious settings, presenting an opening for youths, people with disabilities, women and children. The model takes into consideration the broader resilience debate, going beyond characterising communities and organisations, and brings up the addition "B" for building better to the already popularised BBB principle from the Sendai Framework. The B4 embraces and harmonises the broader adaptation framing from DRR and climate change literature. This chapter, therefore, conceptualised the B4 model to include three main discourses on resilience, adaptation and the BBB principle. It also brings up the notion of learning by doing and learning from doing, as well as continuous improvement and elevating the roles of communities in DRR and management. The B4 model further promotes and elevates preparedness (readiness) in DRR as taking this proactive, as opposed to the reactive approach to save lives and assets, and assist in bouncing back quicker. Elevating readiness in DRR means doing things right the first time around. Hence, governments need to plan properly, allocate adequate resources and have the built environment live resilience from project conceptualisation to design, construction, operation and maintenance, to decommissioning.

The planners and politicians remain key players in organising the spatial and socioeconomic spaces of engagements in readiness for all kinds of disasters, and not only have a bias on climate change. The B4 model also sees a single world that needs to embrace global solidarity in dealing with disaster risks now and in the future. It also encourages the collapsing of disciplines when it comes to DRR and management. Finally, but not least, the B4 model goes beyond the 2030 AfSD and its intertwined 17 SDGs and their many targets. It also goes beyond the Africa Agenda 2063 or any other regional agendas. Hence the B4 model must be part of the global DNA as it is an agenda for safety, poverty eradication, environmental stewardship and prosperity. It embraces both the hardware and software perspectives that make life happen and brings up the "all-of-society engagement and partnership" approach to DRR. It is a model that can bring back smiles and happiness to the faces of many desperate global citizens living under a dark cloud of disaster risk and other hazards.

References

African Union Commission. (2014). Agenda 2063: The Africa We Want. Addis Ababa: African Union Commission Secretariat.

Anholt, R. & Sinatti, G. (2020). Under the guise of resilience: The EU approach to migration and forced displacement in Jordan and Lebanon. Contemporary Security Policy, 41(2), 311-335, https://doi.org/10.1080/13523260.2019.1698182.

Bellinson, R. & Chu, E. (2019). Learning pathways and the governance of innovations in urban climate change resilience and adaptation, Journal of Environmental Policy & Planning, 21(1), 76-89, https://doi.org/10.1080/1523908X.2018.1493916.

Bešinović, N. (2020). Resilience in railway transport systems: a literature review and research agenda. Transport Reviews, 40(4), 457-478, https://doi.org/10.1080/01441647.2020.1728419.

Borie, M., Pellinga, M., Ziervogel, G., and Hyams, K. (2019). Mapping narratives of urban resilience in the global south. Global Environmental Change, 54, 203-213. https://doi.org/10.1016/j.gloenvcha.2019.01.001.

Croese, S., Green, C., Morgan, G. (2020). Localizing the Sustainable Development Goals Through the Lens of Urban Resilience: Lessons and Learnings from 100 Resilient Cities and Cape Town. Sustainability, 12, 550; https://doi.org/10.3390/su12020550.

Chronicle. (2020). PPC Zimbabwe donates 90 tonnes cement to Cyclone Idai victims' housing project. Retrieved from https://www.chronicle.co.zw/ppc-zimbabwe-donates-90-tonnes-cement-to-cyclone-idai-victims-housing-project/ (Accessed 7 September 2020).

Filimonau, V. & De Coteau, D. (2019). Tourism resilience in the context of integrated destination and disaster management (DM2). International Journal of Tourism Research, 22, 202–222. https://doi.org/10.1002/jtr.2329.

Forrest, S. & Milliken, C. (2018). Building Resilience to Disaster: From Advice to Action. European Review, 27(1), 17–26. https://doi.org/10.1017/S1062798718000522.

Hosen, N., Nakamura, H., Hamzah, A. (2020). Adaptation to Climate Change: Does Traditional Ecological Knowledge Hold the Key? Sustainability, 12, 676; https://doi.org/10.3390/su12020676.

Jordan, J.C. (2019). Deconstructing resilience: why gender and power matter in responding to climate stress in Bangladesh. Climate and Development, 11(2), 167-179, https://doi.org/10.1080/17565529.2018.1442790.

Kundu, S., Kabir, M.E., Morgan, E.A., Davey, P., Hossain, M. (2020). Building Coastal Agricultural Resilience in Bangladesh: A Systematic Review of Progress, Gaps and Implications. Climate, 8, 98; https://doi.org/10.3390/cli8090098.

Kurth, M.H., Keenan, J.M., Sasani, M., Linkov, I. (2019). Defining resilience for the US building industry, Building Research & Information, 47(4), 480-492, https://doi.org/10.1080/09613218.2018.1452489.

Logan, T.M. & Guikema, S.D. (2020). Reframing Resilience: Equitable Access to Essential Services. Risk Analysis, 40(8), https://doi.org/10.1111/risa.13492.

Mbatu, R.S. (2020). Discourses of FLEGT and REDD + Regimes in Cameroon: A Nongovernmental Organization and International Development Agency Perspectives. Forests, 11, 166. https://doi.org/10.3390/f11020166.

Moser, S., Meerow, S., Arnott, J., Jack-Scott, E. (2019). The turbulent world of resilience: interpretations and themes for transdisciplinary dialogue. Climatic Change, https://doi.org/10.1007/s10584-018-2358-0.

Mendelsohn, R. & Zheng, L. (2020). Coastal Resilience Against Storm Surge from Tropical Cyclones. Atmosphere, 11, 725; https://doi.org/10.3390/atmos11070725.

Mendes, R., Fidélis, T., Roebeling, P., Teles, F. (2020). The Institutionalisation of Nature-Based Solutions—A Discourse Analysis of Emergent Literature. Resources, 9, 6. https://doi.org/10.3390/resources9010006.

Mutisi, L. & Nhamo, G. (2015). Blue in the green economy: Land use change and wetland shrinkage in Belvedere North and Epworth localities, Zimbabwe. Journal of Public Administration (Special Issue on Green Economy and Local Government). Vol 50 (1): 108-124.

Neset, T.S., Wiréhn, L., Klein, N., Käyhkö, J., Juhola, S. (2019). Maladaptation in Nordic agriculture. Climate Risk Management, 23: 78-87. https://doi.org/10.1016/j.crm.2018.12.003.

Nhamo, G., Nhemachena, C., Nhamo, S. (2020). Using ICT indicators to measure readiness of countries to implement Industry 4.0 and the SDGs. Environmental Economics and Policy Studies, 22, 315-337.

Nhamo, G. & Nhamo, S. (2016). Paris (COP 21) Agreement: Loss and Damage, Adaptation and Climate Finance issues. International Journal of African Renaissance Studies (IJARS), 11(2), 118-138, https://doi.org/10.1080/18186874.2016.1212479.

O.R. Tambo District Municipality. (2019). Floods that affected Port St Johns Local Municipality. Port St Johns: O.R. Tambo District Municipality Disaster Management Centre.

Parker, D.J. (2020). Disaster resilience: A challenged science. Environmental Hazards, 19(1): 1-9, https://doi.org/10.1080/17477891.2019.1694857.

Piggott-McKellar, A.E., Nunn, P.D., McNamara, K.E., Sekinini, S.T. (2020) Dam(n) Seawalls: A Case of Climate Change Maladaptation in Fiji. In. W. Leal Filho (ed.), Managing Climate Change Adaptation in the Pacific Region, Climate Change Management, pp. 69-84. https://doi.org/10.1007/978-3-030-40552-6_4

Quéré, C. L., Jackson, R.B., Jones, M.W., Smith, A.J.P. et al. (2020). Temporary reduction in daily global CO_2 emissions during the COVID-19 forced confinement. Nature Climate change, https://doi.org/10.1038/s41558-020-0797-x.

Reason, C.J.C. & Keibel, A. (2004). Tropical Cyclone Eline and its unusual penetration and impacts over the southern African mainland. Weather Forecast, 19, 789–805.

Santos, A., Sousa, N., Kremers, H., Bucho, J.L. (2020). Building Resilient Urban Communities: The Case Study of Setubal Municipality, Portugal. Geosciences, 10, 243; https://doi.org/10.3390/geosciences10060243.

Setzer, J., de Murieta, E.S., Galarraga, I., Rei, F., Pinho, M.M.L. (2020). Transnationalization of climate adaptation by regional governments and the Regions Adapt initiative. Global Sustainability, 3, e10, 1–10. https://doi.org/10.1017/sus.2020.6.

Sheehan, M.C. & Fox, M.A. (2020). Early Warnings: The Lessons of COVID-19 for Public Health Climate Preparedness. International Journal of Health, 50(3), 264–270. https://doi.org/10.1177/0020731420928971.

United Nations Habitat. (2016). United Nations Habitat III: New Urban Agenda. New York: United Nations Habitat Secretariat.

UNDRR (United Nations Office for Disaster Risk Reduction). (2015). Sendai Framework on Disaster Risk Reduction (2015-2030). New York: UNDRR Secretariat.

UNFCCC (United Nations Framework Convention on Climate Change). (2020a). Glossary of key terms. Retrieved from https://www4.unfccc.int/sites/NAPC/Pages/glossary.aspx (Accessed 5 September 2020).

UNFCCC (United Nations Framework Convention on Climate Change). (2020b). What do adaptation to climate change and climate resilience mean? Retrieved from https://unfccc.int/topics/adaptation-and-resilience/the-big-picture/what-do-adaptation-to-climate-change-and-climate-resilience-mean (Accessed 5 September 2020).

UNFCCC (United Nations Framework Convention on Climate Change). (2015). Paris Agreement. Bonn: UNFCCC Secretariat.

United Nations. (2015). Transforming our World: The 2030 Agenda for Sustainable Development. New York: United Nations Secretariat.

Williams, D.S., Rosendo, S., Sadasing, O. and Celliers, L. (2020). Identifying local governance capacity needs for implementing climate change adaptation in Mauritius. Climate Policy, 20(5): 548-562, https://doi.org/10.1080/14693062.2020.1745743.

World Bank. (2018). Building Back Better: Achieving resilience through stronger, faster and more inclusive post-disaster reconstruction. Washington DC: World Bank.

WRI (World Resources Institute). (2020). Global Dialogue on Responding to the COVID-19 Pandemic and Economic Crisis: Building Back Better Aligned to the SDGs and the Paris Agreement. New York: WRI.

Zhao, L., He, F., Zhao, C. (2020). A Framework of Resilience Development for Poor Villages after the Wenchuan Earthquake Based on the Principle of "Build Back Better". Sustainability, 12, 4979; https://doi.org/10.3390/su12124979.

Abstract

The frequency and impact of disasters have
been on the increase worldwide. As disasters
continue to strike with a heavy toll on communities, there has also been an increase in the
number of digital tools and systems available
to both the emergency responders and
impacted societies. There has been an
observed widespread utilisation of earth
observation (EO) technologies in disaster risk
reduction (DRR), which has led to calls for
more ethical scrutiny in all their applications.
Through the use of available literature, the
objective of this chapter is to examine the use
and contestations of EO technologies in
DRR. The findings show that there are frameworks that guide the practice of EO at global
level. The identified ethical dilemmas in the
use of EO in DRR include the protection of
privacy and security, handling and utilisation
of sensitive crowd-sourced data, communication of EO products to non-experts and respect
for individuals and communities during data
creation and distribution. The review of EO
methods and outputs through a standing
review board and core invention of methods
such as through participatory techniques are
some of the recommendations for more effective satisfaction of community needs. Users
and consumers of EO should continue to
observe ethics matters into the future.

Keywords

Ethics · Earth observation · Disaster risk
reduction · Satellite imagery · Unmanned
aerial vehicle

3.1 Introduction and Background

When there is a risk of disaster occurrence and/or
when disaster strikes, a cocktail of time-critical
data gathering and analysis efforts is activated.
These efforts, as observed by Lambert (2016),
draw from a diverse range of data sources and
platforms. Disaster risk reduction (DRR) methods, say Bunker and Sleigh (2018), critically
require spatial systems that connect location,
communities, persons and their activities in order
to develop accurate situational awareness. Earth
observation (EO) technologies have increased in
efficacy for timely data collection during disasters the world over. This is because disasters usually occur as discrete, widespread and high-impact
incidences. This means that rapid and systematic
field-based evaluations of the situation will be
difficult, if not altogether impossible, to do (Yuan
and Liu 2018). Ortiz (2020) posits that during
disasters, EO can be used for rapid assessment,
presentation of information on land use, determi-
nation of the extent of infrastructure damage,

facility mapping and monitoring and evaluation. Furthermore, EO can be used for hazard simulation and impact modelling as well as the identification of hotspots and the at-risk populations. Lambert (2016) sees EO as a key spatial decision support system for achieving high effectiveness in decision-making on resource allocation and optimisation.

The growth and evolution of EO technologies and their applications in DRR have been rapid and transformational (Scull et al. 2016). The changes have had profound effects on DRR as every new incident has been seen to trigger some new innovation in the EO realm in as much as dealing with the disaster concerned (Haworth 2018). For example during the Haiti earthquake in 2010, there was widespread use of crowd-sourced maps, created by a web of volunteers. The maps were invaluable during search and rescue operations (Hunt et al. 2016). Application of EO in DRR is therefore a rapidly evolving domain, which also brings with it the potential for contestations not only due to unethical use of EO products but the pressing need for ethical scrutiny of EO datasets during disasters (Neslen 2017). In particular, the improved access to near real-time and higher resolution EO products that can be distributed rapidly via telemetry or online databases has raised the debate on the permissible and ethical utilisation of EO datasets (Wetherholt and Rundquist 2010). Ethics has been defined by Harris (2013) as the "study of what makes actions right or wrong". From a utilitarian point of view, ethics assesses behaviour as being just or not when results of the entire chain of events are more beneficial to society or individuals when compared to the negative effects (IEP 2011). In the context of EO, Harris (2013) highlights that actions are ethically correct when their outcome brings benefits to more people compared to those disadvantaged. In other words, Harte and North (2004) posit ethical behaviour as targeting the greatest good for the greatest number.

All disasters raise ethical issues, questions and contestations. DRR stakeholders are now starting to take notice of the ethical issues and contestations brought about by disaster responses that

disseminate EO datasets (Haworth 2018). As a developing field, disaster ethics tries to gain an in-depth understanding of handling ethical issues during disaster periods. The process involves and encompasses numerous specialised professions in DRR. This leads to the need for studies on the best ethical approaches to take during disaster prevention ("preventive ethics"), disaster response ("ethics of response") and disaster recovery ("post-disaster ethics") (Moatty 2017). It has been observed that those responding to disasters grapple with diverse ethical dilemmas, but limited training or guidelines, have been developed to address this situation (Moatty 2017). However, a key step towards addressing ethical issues in DRR is for practitioners to have an appreciation of ethical dilemmas that can potentially occur during specific forms of disasters (Mitrović et al. 2019). A methodical examination of literature to identify key ethical issues in the use of EO during disaster situations is therefore key in devising relevant guidelines, training needs and further research.

Palen and Liu (2007) highlight that in the past, disaster management was mainly conducted for and not with the affected community. To this end, it followed a civil defence model, which is mostly a top-down approach. Furthermore, the process of emergency management was mainly reactive and largely geared towards operation during the actual disaster event (Haworth 2018). Approaches to disaster management have since evolved to place at the core communities and their vulnerabilities, as well as moving from emergency response to risk reduction with increased public participation (Manyena et al. 2011). This paradigm shift occurred as a result of the promulgation of the 2005–2015 Hyogo Framework for Action. This is a key United Nations strategic plan of action for the management of disasters that was agreed on by 168 member states. The Hyogo Framework was the predecessor to the 2015–2030 Sendai Framework for DRR, which shifted the emphasis from disaster management to community involvement and DRR (UNISDR 2015). Under the DRR model, there was a drift from the hardcore top-down approach to a more dynamic and collaborative approach that facilitates multi-organisational, intersectoral and inter-governmental collaboration (Haworth 2018). As these changes were taking place, the need was emerging for striking an ethical balance between the needs of intervening organisations such as United Nations humanitarian agencies, NGOs, national emergency management organisations, local government authorities, community service organisations and the needs of local communities (Lambert 2016).

Neslen (2017) is of the opinion that during disaster times, ethical behaviour can be conceived and understood in different ways by the different players involved in the DRR process. These different players may include the victims, volunteers, politicians, local community leaders, rescue operation experts, humanitarian elements, the media and other professionals. Ethical issues and contestations arise due to the diversity of players involved during the different DRR phases. The diverse players usually have diversified tasks that they engage in during the DRR process. These players also bring with them a wide range of professional conduct and operation codes, which in the end may create confusion and disharmony. Contestations arise when the interests of the impacted communities conflict with those responding to the emergency. This led to Wallemacq (2018) asking questions such as how ethical differences can be reconciled or if each disaster type needs separate ethical codes, and even how the codes can be applied. Although EO is now an indispensable component of DRR, its development and utilisation in DRR has outpaced the capacity of communities to produce related legislation, policies and standard procedure manuals that oversee its utilisation (Wetherholt and Rundquist 2010).

Although different "codes of conduct" have been developed by various institutions to try and guide the behaviour of EO professionals, in most cases these are not legally binding. This chapter aims at unpacking the utilisation and contestations in the adoption of EO technologies in DRR. The importance is underscored of ethical scrutiny of EO products not only during their application but from the design of the products, as well as their initial consideration for use.

The materials and methods that were used to collect and analyse data are discussed in the following.

3.2 Materials and Methods

The chapter is mainly literature-based. Snyder (2019) argues that reviewing literature lays a strong foundation for and gives context to any form of study. Literature-based studies assist in information advancement, produce procedures for strategy and practice, deliver evidence of an impact, can stimulate novel ways of thinking and guide the assimilation of new ideas for specific areas of study. If well executed, they can be used as the foundation of future studies. The literature-based review was found to be useful for this study because it integrates results and viewpoints reached from various research topics, which enables the research to address study objectives to the depth that is difficult to reach in one study.

Data was gathered by searching relevant literature from electronic databases. The key words used to search electronic databases included phrases such as EO + ethics, remote sensing + ethics, GIS + ethics, EO + DRR + ethics, EO + humanitarian services + ethics. All these phrases yielded data that was very relevant to the study. The literature obtained was first grouped into categories of relevant, somewhat relevant and not relevant. The relevant literature was further studied with the aim of extracting key issues concerning contestations and ethical issues in the utilisation of EO in DRR. The key issues extracted were classed into themes, followed by a systematic write-up of the identified themes. The global frameworks addressing the acquisition and utilisation of EO datasets were also analysed for their ethical soundness using obtained literature as well as the extent to which ethical issues are taught during the training of EO practitioners.

The next section presents the findings of the research. In particular, global frameworks guiding the ethical use of EO, ethical issues in the utilisation of EO in DRR and possible ways of dealing with the ethical dilemmas are explained and discussed.

3.3   Presentation of Data and Discussion of Findings

This part consists of a discussion of the global frameworks on ethical EO and a focus on identified ethical issues. The sub-section on identified ethical issues is further sub-divided into privacy and security concerns, method of map generation, respect for individuals and communities, handling volunteered geographic information (VGI), communicating EO data and addressing ethical issues in EO use during DRR.

3.3.1   Global Frameworks on Ethical EO

At global level, initiatives have been put in place to guide the ethical use of EO products in general, and specifically during DRR as manifested through extreme events which include cyclones, flooding, droughts, wildfires and strong winds. These initiatives include the United Nations Principles Relating to Remote Sensing of the Earth from Space, the Space and Major Disasters Charter, Global Earth Observation Systems of Systems (GEOSS) and the Sendai Framework of Disaster Risk Reduction.

To regulate the use of EO technologies at global level, the United Nations in 1986 formulated and approved several principles. The principles were meant to guide EO players in matters to do with the "remote sensing of the earth from outer space". These principles were deliberated on between 1970 and 1986 and eventually agreed on in 1986. Some of the agreed principles were enshrined in ethical statements (United Nations 1986a). Aspects covered in the principles include that remote sensing must promote the protection of humankind from disasters in accordance with international legislation, as well as advancing country-to-country cooperation (Wetherholt and Rundquist 2010). The principles also incorporate ecological ethics promoting the adoption of EO technologies in the conservation and management of the natural environment (Harris 2013). The most contentious principles are mainly Principles IV and XII, which focus on balancing

the rights of all nations to utilise "the outer space in accordance with the Outer Space Treaty" and the rights of all countries' sovereignty over who observes what is happening over their territories (Hobe et al. 2009).

From an ethical perceptive, the overall benefits of using EO are far higher when compared to the perceived negative consequences related to sovereignty. This makes it ethical to collect data from other countries from outer space as long it is meant for the general good, such as improvement of environmental monitoring and management as well as DRR. Since 1986, more countries have developed their own EO capability, which has seen some countries changing their stance on sovereignty access to EO data. Nigeria in the past was not happy with the commercialisation of EO data and protested satellite fly-passes through its territory (United Nations 1986b). However, after it launched its own satellite in 2003, the position of the country on the matter shifted, and Nigeria has already commercialised its EO assets that are collecting data from other nations even without their explicit consent. This shows that the benefits that communities stand to derive from the use of EO outweigh the negative effects, which makes it ethical to use it even sometimes without the express consent of the territories through which satellites make repeated fly-passes.

The Space and Major Disasters Charter is a global initiative to provide a unified response to the provision of satellite data to those affected by disasters for use in monitoring and response activities (International Charter 2020). The Charter is composed of "space agencies and space system operators from around the globe". During periods of disaster, the Charter supplies EO data to at-risk countries or societies. It also provides EO datasets to those that have already been hit by disasters in order to assist in search and rescue as well as recovery operations. The rationale of the Charter is the realisation that EO data provides information that is key in supporting decision-making during DRR (International Charter 2020). In the Charter, appeal for EO data can only be done by authorised users (AUs) on behalf of the affected communities. To qualify as an AU, one has to be a government or quasi-government official with a mandate of civil protection, DRR authorities and security institutions from countries that are formally registered by the Charter (Bessis et al. 2004). However, activation of the Charter can be done on behalf of countries that do not have an AU. An ethical aspect of the Charter is the fact that AUs take delivery of critical EO products at no cost to them, regardless of the fact that some of the products are commercialised (Ito 2005). The Charter is also a unified and reputable response which provides high-quality EO datasets that reduce the risk of misinformation during disaster periods. The provision of EO data during the DRR process after Charter activation is in direct fulfilment of Principle XI of the United Nations principles relating to remote sensing of the earth from outer space. Principle XI specifically calls on states to use EO to promote the protection of humankind from disasters. Figure 3.1 shows the operational layout of the Charter.

The Charter is activated by an AU through a call to a round-the-clock on-duty operator. The operator authenticates the caller and checks if all the information regarding the disaster has been provided in the user form. This information includes the scope, location and type of hazard. If all the information is found to be correct, the Charter is then escalated by the on-duty operator, who within an hour must send the disaster information to an emergency on-call officer, who in most cases is an engineer from one of the Charter members. The officer then analyses the nature and scope of the hazard together with the AU in order to determine the space assets best suited to support DRR activities, delivery methods, as well as time frames involved. Once finalised, requests for the most appropriate EO products are sent to Charter members for immediate actioning. Once the Charter is up and running, a project manager takes over from the emergency on-call officer. The manager is chosen from among the space operators to oversee all the chain of events until the conclusion of the process (Bessis et al. 2004).

The Group on Earth Observations (GEO) is an intergovernmental organisation with 111 member countries, the European Commission and around 129 participating organisations (GEO 2020). The group aims to improve and coordinate EO systems and to provide with speed open source EO datasets for the purpose of answering society's need for informed decision-making (Harris 2013). GEO developed the Global Earth Observation Systems of Systems (GEOSS), which aims at improving access to and application of EO technologies in nine areas of societal benefit. The areas are DRR, agriculture monitoring, water resources management, health, energy,

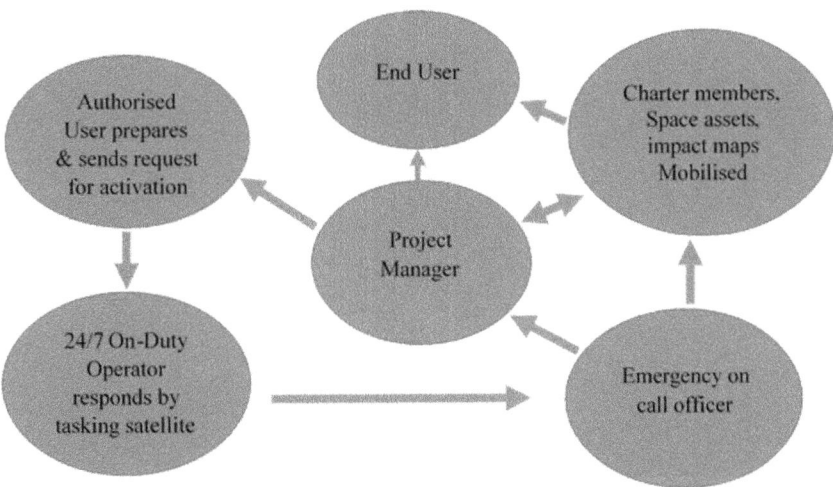

Fig. 3.1 Flow of communication and action in the Charter. Source: Authors, adapted from The Space and Major Disasters Charter (International Charter 2020)

climate, observation of weather, ecological systems and biological diversity (GEO 2020). The intention of GEOSS is to have a single platform through which there is acquisition and efficiently coordinated distribution of EO products so as to maximise their utility. Members of GEOSS participate on the platform because of their endeavours or ethical duty regarding EO utilisation for the common good of societies. This is also one of the highlighted principles under the United Nations Outer Space Treaty (Hobe et al. 2009). GEO also partners with United Nations agencies and its member states to help them improve their implementation and monitoring of global initiatives such as the Sustainable Development Goal Agenda 2030, the Paris Agreement on Climate and the Sendai Framework for DRR. The GEO open access data and sharing principles are guided by the need for minimal time lag to acquire EO datasets, minimal costs and sometimes provision of free-of-charge EO data for the benefit of society. Most of the products from this platform are therefore provided in near real-time and are open access. This enables near real-time monitoring of key disasters such as flooding, wildfires, cyclones and drought, thus enabling effective early warning for at-risk communities. The GEONETCast system of systems provides the backbone for EO data provision under GEO principles. Figure 3.2 provides the layout of the GEONETCast telecommunication satellite-based data distribution system.

The use of EO is also encouraged in the Sendai Framework of DRR under Priority 1 which is meant to improve the understanding of disaster risk at national and local levels. The Framework highlights that strategies for DRR must be informed by an in-depth understanding of hazard risk in all its dimensions. It further encourages the "development, periodic update and dissemination of location-based disaster risk information such as hazard risk maps to decision makers, the general public and the at risk communities in an appropriate format" (UNISDR 2015, p. 15). The Sendai Framework advocates for the promotion of and access to real-time reliable data from outer space and ground-based observations such as GIS, ICT as well as social media in DRR. The above are ethical components also contained within the GEO, the Charter as well as the United Nations Outer Space Treaty. Furthermore, the Sendai Framework promotes the enhancement of technological transfer, access, distribution and utilisation of non-sensitive EO products, as well as maintaining and strengthening in situ and remotely sensed earth and climate observations for purposes of DRR (UNISDR 2015). The Sendai Framework therefore encourages the ethical use of EO technologies in DRR.

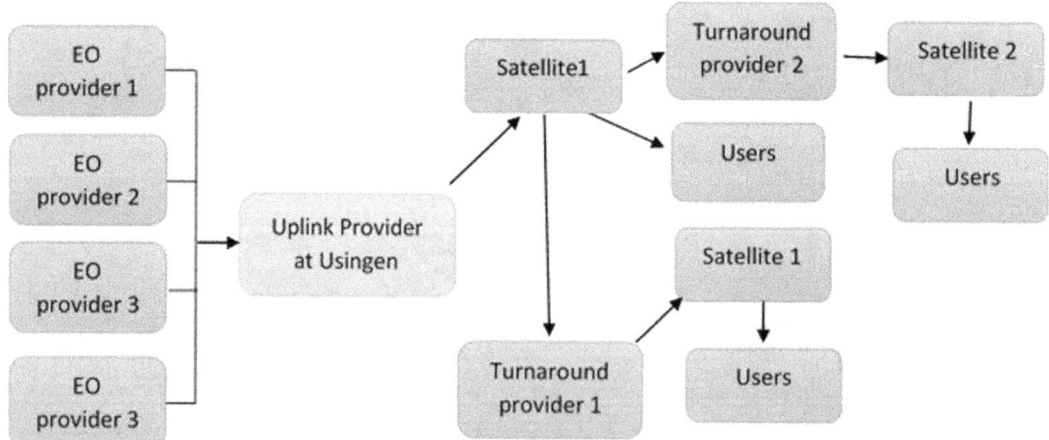

Fig. 3.2 Layout of the GEONETCast satellite-based data distribution system. Source: Authors, adapted from GEO (2020)

The next subsection considers the identified ethical issues in EO.

3.3.2 Identified Ethical Issues

3.3.2.1 Privacy and Security Concerns

Wetherholt and Rundquist (2010) highlight privacy and security concerns as being at the core of EO utilisation. The big data revolution, the Fourth Industrial Revolution (4IR) and the launch of commercial satellites with sub-metre spatial resolution have heightened these fears. Slonecker et al. (1998) observe that in the past, EO data accessible in the public domain was in course spatial resolution. As such, access and use of such data did not raise contestations around ethical use and personal privacy. The data had limited utility and could not be applied in the observation of sensitive areas such as private property. Furthermore, access to EO data was constrained by its exorbitant costs, limited computer processing power, limited availability of essential peripherals, as well as EO software. It was also easier to give oversight to the EO industry because most of the acquisitions were done and processed through government affiliates, collected at low temporal and spectral resolution, and their use was not very widespread (Hunt et al. 2016). As a result, EO data was not seen as threating confidentiality or national security. The use of EO data was effectively bottlenecked to public agencies and well-financed universities. However, in the past 30 years or so, the situation has changed rapidly with civilian and military intelligence access to EO datasets almost at par, raising ethical issues and contestations related to protection of privacy and security (Rao and Murthi 2006).

Wetherholt and Rundquist (2010) highlight that the restructuring of EO infrastructure from the control of a few national governments to commercial and more international governments has heightened security fears as well as created changes in EO data distribution and access platforms. This has meant that EO data is no longer accessible to only well-funded institutes and those with access to high-speed computers and expensive software. Advances in computing speed and the widespread availability and affordability of personal computers have also stimulated the growth in EO data utilisation. These changes in EO infrastructure have spurred debates around their impact on general security and individual privacy. However, applying the principle of the "greatest good for the greatest number", the use of high-resolution data during disaster periods remains in most cases ethical regardless of the disadvantages. Figure 3.3 shows

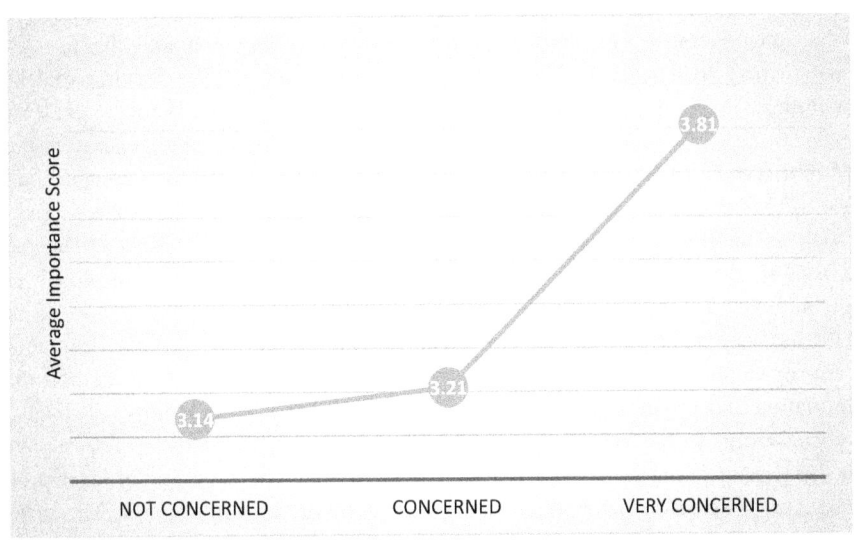

Fig. 3.3 Concern about privacy in the use of EO. Source: Authors, data from Scull et al. (2016)

the extent of privacy concerns due to the rapid spread in the use and availability of EO data from a study done by Scull et al. (2016). More people are concerned about privacy issues in the use of EO compared to those that are not.

Ashford (2015) is of the opinion that EO data collected for humanitarian purposes may cause safety fears if adversaries try to use such data to single out specific societies for punitive measures because they will be sharing undesirable information. Geopositioning enhances the ethical dilemma of connecting persons to a specific location. This tends to increase the risk of such information being used to target certain communities if data is traceable. There have been ongoing debates on whether it is proper for crisis maps to be made open source in order to support rapid and effective decision-making or allowing limited access to reduce safety risks (Schneier 2015). The creation of and adherence to robust EO data security protocols are increasingly becoming standard practice for those in DRR, given that even in non-conflict situations, such data can be abused or weaponised. Good examples are contact tracing mobile applications and maps of deadly or stigmatised infectious diseases such as Ebola and recently COVID-19 (Dube et al. 2020). This could compromise anonymity and expose individuals to harm and increased surveillance by the government. Maps sometimes show operational data which may be used to recognise areas seen to be uncooperative to humanitarian workers, which may lead to collective punishment from authorities (Ashford 2015).

3.3.2.2 Method of Map Generation

The process of generating maps for DRR from EO data such as satellite and unmanned aerial vehicle (UAV) imagery and GIS is becoming increasingly automated with limited human intervention. This has been made possible through the development of sophisticated computer algorithms, simulation and prognostic modelling by means of virtual reality technology to mimic conditions on the ground (Hunt et al. 2016). These maps can be used in selecting hotspots, planning human resources and equipment required, including human resources selection and development,

preparing field missions, as well as in advocacy endeavours. As a result, DRR maps have shifted from being static geographic representations to include near real-time information. Searle (2019) observes that governments and international NGOs use EO in creating maps that only fully trained mapping experts can use. These maps are therefore not useful for the at-risk community because they require an exclusionary level of know-how to understand.

The accuracy and ethical grounding of DRR maps produced without the involvement and ownership of the local communities have been questioned (Bendor 2014). Some like Canevari-Luzardo et al. (2017) have advocated for using participatory approaches to knowledge production as a way of facilitating the collection of data and crafting effective DRR interventions. Participatory community mapping is observed to be an important component of this approach due to the attention it gives to the local context of vulnerability and the respect it gives local knowledge. The maps derived using this methodology can be more accurate due to the unique way in which they combine physical, societal and economic factors. These can assist in creating a common appreciation of susceptibility issues and can therefore provide the basis for crafting effective DRR strategies (Preston et al. 2011).

Canevari-Luzardo et al. (2017) envisage participatory GIS (PGIS) as a "highly multidisciplinary mapping technique that uses EO to elicit, represent and validate local spatial knowledge" in order to co-produce knowledge. It has emerged as one of the many map generation methods that have evolved over time to counter the possibility of marginalising the local society, given the complexity of most GIS methods. The method does away with the top-down approach inherent in conventional mapping operations, as well as reducing the reliance on purely quantitative data (Elwood et al. 2012). As a method that seeks to stimulate the acceptability of map generation through the participation of local actors, PGIS encourages the advance of user-produced and demand-driven information. This has the effect of raising utilisation of maps as spatial decision support systems to make effective decisions during DRR because they will be grounded in information co-production with local communities (Preston et al. 2011).

According to Canevari-Luzardo et al. (2017), hazard risk assessments are usually performed with the aim of assisting intervention agencies in making informed choices but never really encourage the direct participation of the populations at risk. PGIS can therefore help encourage the at-risk population manage risks and reduce their exposure. Elwood et al. (2012) observe a weakness in such approaches to mapping. This is because the credibility of outputs from PGIS can be questionable due to the influence of individual and community bias as well as the adoption of inexpert methods during data collection and analysis. This elevates the levels of uncertainty about the locational accuracy of the maps. Furthermore, some argue that in spite of being categorised as participatory, PGIS is still not the best for use by non-educated and relatively poor communities because of the advanced levels of technicality involved, together with the high costs related to the acquisition, learning and utilisation of remote sensing and GIS software.

3.3.2.3 Demonstrating Respect for Individuals and Communities

The use of EO imagery for advocacy, fundraising and raising awareness in DRR has raised concerns and questions about whether such images are not harmful to affected individuals or communities and show respect for them (Madianou et al. 2015). These concerns are further heightened by the sense of uncertainty and enhanced vulnerability that is felt by people and societies in times of disasters. The gathering and circulation of digital data during disasters brings to the fore the question of ethics. Issues of informed consent, rights to the data and if the victims are aware of and consent to the purposes to which the information gathered about them will be used and the security of the data and where it will be kept need to be carefully considered. Dissemination of sharp resolution imagery captured from UAVs, for example, requires express permission from identifiable people or property titleholders to avoid breaching their rights to privacy. Skipping the informed consent process constitutes significant harm to the dignity of affected parties and also demeans their freedom (Madianou et al. 2015). It is therefore important to inform the affected societies on the mapping exercises to be carried out during disaster periods and to elicit their consent and concerns. Affected communities need to be informed on how the resulting maps will be used and how the use will benefit them, ownership of the final product needs to be discussed and how the victims can decide on the possible uses of the maps or imagery when the emergency phase is over (Mitrović et al. 2019).

3.3.2.4 Handling Volunteered Geographic Information (VGI)

VGI is the use of crowd-sourcing, citizen science and/or voluntary recruitment of large numbers of private citizens in the formulation of EO products (Wang et al. 2016). According to Bruce et al. (2014), the development of VGI has been facilitated by the 4IR, which has seen the rapid development of disruptive technologies such as social media, location-based technologies, graphics for visualisation, broadband communication, cloud storage and smartphones. VGI is characterised by a diversity of practices such as contributions made in response to disaster events and global crowd-sourced mapping efforts (Haworth 2018). Wang et al. (2016) observe that social media serves as the VGI platform through which public wisdom is expressed when disasters occur. It signifies unparalleled changes in the nature of information, methods of spatial information creation, as well as its distribution and utilisation.

Despite the importance of VGI in the context of DRR, its use raises several ethical questions related to the reliability of maps created by non-experts as a tool in supporting decision-making during disasters. Haworth (2018) highlights some of the biggest challenges of VGI to be associated with matters of the credibility of source, quality of the data used, validation of the end product, management of the data and perceived legal worries related to privacy and liability. Klonner et al. (2016) observe the stampede for contributions by both researchers and volunteers to be concentrated only during the response to the disaster instead of the entire DRR cycle, especially during mitigation or preparedness stages. The under-representation of particular groups and individuals in the process of VGI is also of concern to McCall et al. (2015) due to the digital divide. Haworth (2018) explains that trust in VGI is a function of data quality, source credibility and contributor reputation. McCall et al. (2015), however, argue that in DRR it is safer to use reputable local community members who have proven expertise in VGI operations and participatory mapping compared to a network that permits everyone to volunteer contributions. This can improve trust in VGI products and their usefulness in DRR. Haworth (2018) portrays the extent to which VGI can be useful in DRR to rely on whether the network architecture is designed to capitalise on the increasing popularity of citizen science. If DRR organisations are not able to adapt to the changes in data formulation practices and growing citizen participation, then the opportunities for using VGI will become lost prospects. An important area of the prospective use of VGI is that it can be a social practice, where prominence is given to the possibility for improved social cohesion and the participatory nature of people cooperatively contributing to EO data collection, rather than individual observations.

3.3.2.5 Communicating EO Data

Communicating and packaging technical products from EO to non-technical but important stakeholders such as at-risk populations and DRR managers has always been a difficult task. Efficient communication was identified by Mitrović et al. (2019) as essential when disasters strike. For information provided during disasters to be ethical, it needs to be timely, truthful and relevant to the situation. Experts who produce key EO products have a responsibility to communicate them in an ethically accepted manner to both the at-risk/affected populations and the incident managers. Effective communication should be clear, simplified and elaborating on jargon. Good communication and simplified presentation of EO products prepare at-risk commu-

nities for hazards which may significantly reduce the destruction and casualties (Mitrović et al. 2019). Respectable communication must also show concern and sympathy for the affected communities.

Communication breakdowns, on the other hand, can potentially have grave implications that may lead to unnecessary destruction and losses, as well as human suffering. Poor communication results in confusion and conflict among responders. Communication needs to address more than just informational needs, but must also include weighing the risks of false positives and false negatives due to inconclusive EO data (Gaitonde and Gopichandran 2016). The level of uncertainty in the EO algorithms and models used to generate outputs to support decision-making also need to be communicated and known. Gaitonde and Gopichandran (2016) argue that during disasters decisions are made using inconclusive data. This brings about an ethical responsibility to make DRR plans that are responsive to the situation on the ground as more data becomes available. When gaps in available information are not well communicated, contestations emerge and the safety and health of the victims as well as of the responders can potentially be harmed (Mitrović et al. 2019). This raises the broad theme of social justice as an area of ethical concern in the use of EO in DRR.

Since timely interventions are very important during disaster periods in order to save lives and property, balancing the time taken between the stages of striving to meet the necessary ethical procedures before any information is given out and the subsequent communication processes becomes very important. Critical information regarding the possibility of a hazard occurring should be released with enough time lag and packaged in the appropriate language to allow the potential victims and those positioned to help them take appropriate action. The coordination of all parties involved in DRR during an incident is also key to the success of the process, and it becomes more ethical if there is involvement of the affected persons or society.

3.3.2.6 Addressing Ethical Issues in EO Use During DRR

Searle (2019) acknowledges four policy balances that need to be struck when using technologies such as EO in DRR operations: (1) balancing DRR uses of EO technologies with other sources of information; (2) balancing the needs of disaster responders and the needs of those the disaster affected when exploring uses of EO technologies; (3) balancing the short-, medium- and long-term interests of those affected by disasters and; (4) balancing EO capacities to both centralise decision-making and facilitate individual and community autonomy during disasters. Emery (2016) observes that in the use of new EO methodologies in DRR, it needs to be authoritatively demonstrated that the innovation benefits both the responders and affected communities. In many cases not much research is done on the impact of these methodologies except assumptions that the innovation will benefit the communities. Searle (2019) highlights that too often, the benefits of a new innovation to responders outweigh those to the affected community, and preferences of those being assisted are usually minimised.

Lafferranderie (2001) suggested that one way of effectively promoting the ethical use of EO data was to impart ethics in remote sensing modules at universities. This was, however, seen to be challenging to implement as there was very limited discussion of ethics in top EO textbooks normally used in EO syllabus development (Lafferranderie 2001). Wetherholt and Rundquist (2010) observe that the most detailed ethics in EO were produced by professional societies that have designed "codes of ethics such as the American Society for Photogrammetry and Remote Sensing". To this end, Scull et al. (2016) conducted a study to determine the degree to which ethics was being taught in EO modules at universities and colleges around the United States. The study also assessed the degree to which lecturers perceived that ethics needed to be taught, and barriers to the addition of ethical use in EO modules. They found that most instructors saw the importance of including ethics in EO modules. Most of these instructors already discussed ethical uses, but there was a wide variation on the time dedicated to the topic, which varied from as little as 5 min to as much as 2 h. In addition, more than 52% of the participants either agreed or strongly agreed that ethics in EO education was an important factor in reducing undesired uses (Scull et al. 2016). Significant barriers to the inclusion of ethics in the EO curriculum included the lack of information and limited time to cover ethics (Scull et al. 2016). This therefore creates the need for ethical uses in EO to be given more prominence in leading remote sensing textbooks and other sources commonly used for curriculum development. Due to the geospatial revolution, which is trending together with 4IR, lessons on EO should go far beyond the technical aspects of the technology to how it relates to society. A stand-alone module on EO and society will therefore go a long way in better preparing scholars for ethical use of EO.

3.4 Conclusions

EO remains a very important source of data in DRR due to its unique capabilities to provide comprehensive, synoptic and multitemporal coverage of disaster-impacted areas at regular intervals and with quick turnaround times. However, the widespread use of EO during disaster periods has been met with an increased demand for a review of ethical issues in its utilisation, given the contestations that are emerging. The chapter gave a comprehensive review of the ethical considerations that need to be applied when using EO technologies in DRR. At global level, frameworks that guide the ethical utilisation of EO include United Nations Principles Relating to Remote Sensing of the Earth from Space, the Space and Major Disasters Charter, Global Earth Observation Systems of Systems and the Sendai Framework of Disaster Risk Reduction. The guidelines provided by the global initiatives feed into national or individual organisations' ethical principles in the practice of EO.

Identified ethical issues in the use of EO in DRR include the protection of privacy, security and dignity of affected communities, and proper handling and utilisation of volunteered geo-graphic information which sometimes trends faster than official communication in the era of social media. Other issues relate to the communi-cation of EO products in a manner that is under-standable to both the decision makers and the affected communities, as well as demonstrating respect for individuals and communities during the creation and distribution of EO products for DRR. Ethical issues in EO utilisation in DRR therefore involve pursuing morally sound deci-sions. Achieving this in an era of crowd-sourced and open-source data is very complex, but must involve good information, sound values, engage-ment of appropriate stakeholders and the ability to make balanced decisions.

The limited ability to make ethically sound decisions in EO data can potentially create another disaster within a disaster not only for vic-tims but also for those labouring to assist them. Possible ways of mitigating ethical violations and contestations in utilising EO during DRR may include regularly reviewing EO methods and outputs through a standing review board as is usually done for research which involves human subjects or which is likely to impact on them. Other ways include the localisations of the cre-ation of EO products through core invention methods such as participatory mapping, which may result in more effective satisfaction of com-munity needs. Above all, universities and col-leges need to do more to impact ethical reasoning in the use of EO as they train and develop EO experts.

References

Lambert, A. (2016). Disaster Data Assemblages: Five Perspectives on Social Media and Communities in Response and Recovery. 2016 49th Hawaii International Conference on System Sciences. https://doi.org/10.1109/HICSS.2016.280.

Ashford, W. (2015). Internal threat among biggest cyber security challenges, says former FBI investiga-tor. Computer Weekly 29 June 2015. http://www.computerweekly.com   /news/4500248908/Internal-threat-amongbiggest-cyber-security-challenges-says-former-FBI-investigator.(Accessed 22 May 2020).

Bendor, A.P. (2014). In West Africa's Ebola Crisis, a Mobile Phone-Based Hero for Health Workers. Chapel Hill: IntraHealth International. http://www.intrahealth.org/blog/west-africa%E2%80%99sebola-crisis-mobile-phone-based-hero-health   workers#. VyprSmQrJcx. (Accessed 23 May 2020).

Bessis, J. L., Béquignon, J. and Mahmood, A. (2004). Three typical examples of activation of the interna-tional Charter "Space and Major Disasters", Advances in Space Research, 33, pp 244-248.

Bruce, E., Albright, L., Sheehan, S. and Blewitt, M. (2014). Distribution patterns of migrating humpback whales (*Megaptera novaeangliae*) in Jervis Bay, Australia: A spatial analysis using geographical citi-zen science data. Applied Geography 54:83–95.

Bunker, D and Sleigh, A. (2018). The Future of Spatial Systems for Disaster Management, Geospatial and Temporal Information Capture, Management, and Analytics in Support of Disaster Decision Making. Proceedings of ISCRAM Asia Pacific, Sydney.

Canevari-Luzardo, L., Bastide, J., Choutet, I. and Liverman, D. (2017). Using partial participatory GIS in vulnerability and disaster risk reduction in Grenada, Climate and Development, 9:2, 95-109, https://doi.org/10.1080/17565529.2015.1067593.

Dube, K., Nhamo, G. and Chikodzi, D. (2020). COVID-19 cripples global restaurant and hospitality industry. Current Issues in Tourism. https://doi.org/10.1080/13683500.2020.1773416.

Elwood, S., Goodchild, M. F. and Sui, D. Z. (2012). Researching volunteered geographic information: Spatial data, geographic research, and new social practice. Annals of the Association of American Geographers, 102(3), 571–590.

Emery, J.R. (2016). The possibilities and pitfalls of humanitarian drones. Ethics and International Affairs, 30(2) : 162-4.

Gaitonde, R., and Gopichandran, V. (2016) The Chennai floods of 2015 and the health system response. Ind J Med Ethics,1(2):71–75.

Group on Earth Observations (2020). Earth Observation for Impact. www.earthobservations.org. (Accessed 24 May 2020).

Harris, R. (2013). Reflections on the value of ethics in relation to earth observation, International Journal of Remote Sensing, 34:4, 1207-1219, https://doi.org/10.1080/01431161.2012.718466.

Harte, N. and North, J. (2004). The World of UCL. London: UCL Press.

Haworth, B.T. (2018). Implications of volunteered geo-graphic information for disaster management and GIScience: A more complex world of volunteered geography. Annals of the American Association of Geographers, 108:1, 226-240, https://doi.org/10.1080/24694452.2017.1321979.

Hobe, S., Schmidt-Tedd, B., Schrogl, K.U. and Goh, G. (eds). (2009). Cologne Commentary on SpaceLaw, volume 1. 256 pp. Cologne: Carl Heymanns Verlag.

Hunt, M., Pringle, J., Christen, M., Eckenwilerd, L., Schwartz, L. and Davé, A. Ethics of emergent information and communication technology applications in humanitarian medical assistance. Int Health 2016; 8: 239–245. https://doi.org/10.1093/inthealth/ihw028.

IEP. (2011). Internet Encyclopedia of Philosophy. www.iep.utm.edu/ethics/.(Accessed 20 May 2020).

International Charter (2020). Activating the Charter. www.disastercharter.org/how-the-charter-works. (Accessed 23 May 2020).

Ito, A. (2005). Issues in the implementation of the International Charter on Space and Major Disasters. Space Policy 21: 141–9.

Klonner, C., Marx, S., Uson, T., De Albuquerque, J.P. and Hofle, B. (2016). Volunteered geographic information in natural hazard analysis: A systematic literature review of current approaches with a focus on preparedness and mitigation. International Journal of Geo-Information 5:103.

Lafferranderie, G. (2001). How to "entrench" the regulation of human activities in space. Space Policy 17 (2): 77-80.

Madianou, M., Longboan, L. and Ong, J. (2015). Finding a voice through humanitarian technologies? Communication technologies and participation in disaster recovery. Int J Commun 2015;9:3020–38.

Manyena, S. B., O'Brien, G., O'Keefe, P. and Rose, J (2011). Disaster resilience: A bounce back or bounce forward ability? Local Environment 16 (5): 417–24.

McCall, M. K., Martinez, J. and Verplanke, J. (2015). Shifting boundaries of volunteered geographic information systems and modalities: Learning from PGIS. ACME: An International E-Journal for Critical Geographies 14 (3):791–826.

Mitrović, V.L., O'Mathúna, D.P. and Nola, I.A. (2019). Ethics and Floods: A Systematic Review. Disaster Medicine and Public Health Preparedness VOL. 13/NO. 4. https://doi.org/10.1017/dmp.2018.154.

Moatty, A. (2017). Post-flood recovery: an opportunity for disaster risk reduction? In: Vinet F, ed. Floods. Volume 2 – Risk Management. London: ISTE Press and Elsevier; 2017:349–363.

Neslen A. (2017). Flood disasters more than double across Europe in 35 years. The Guardian. https://www.theguardian.com/environment/2017/jan/19/flood-disasters-more-than. (Accessed 15 May 2020).

Ortiz, D. (2020). Geographic information systems (GIS) in humanitarian assistance: A meta-analysis. Pathways: A Journal of Humanistic and Social Inquiry, Vol.1, Iss. 2.

Palen, L., and Liu, S.B. (2007). Citizen communications in crisis: Anticipating a future of ICT-supported public participation. In Proceedings of the SIGCHI Conference on Human Factors in Computing Systems, ed. M. B. Rosson and D. Gilmore, 727–36. San Jose, CA: ACM.

Preston, B. L., Yuen, E. J. and Westaway, R. M. (2011). Putting vulnerability to climate change on the map: A review of approaches, benefits, and risks. Sustainability Science, 6(2), 177–202.

Rao, M. and Murthi, S. (2006). Keeping up with remote sensing and GI advances—Policy and legal perspectives. Space Policy 22 (4): 262–273.

Schneier, B. (2015). Data and Goliath: The Hidden Battles to Collect Your Data and Control Your World. New York: W.W. Norton & Company.

Scull, P., Burnett, A., Dolfi, E., Goldfarb, A. and Baum, P. (2016). Privacy and ethics in undergraduate GIS curricula. Journal of Geography, 115:1, 24-34, https://doi.org/10.1080/00221341.2015.1017517.

Searle, M. (2019). Striking a balance: Disaster responders' and affected communities' interests in new technologies. Singapore: Nanyang Technological University.

Slonecker, E. T., Shaw, D. M. and Lillesand, T. M. (1998). Emerging legal and ethical issues in advanced remote sensing technology. Photogrammetric Engineering and Remote Sensing 64: 589–95.

Snyder, H. (2019). Literature review as a research methodology: An overview and guidelines. Journal of Business Research, 104, pp 333–339. https://doi.org/10.1016/j.jbusres.2019.07.039.

UNISDR. (2015). United Nations. Sendai Framework for Disaster Risk Reduction 2015–2030. UNISDR. Geneva.

United Nations. (1986a). Principles Relating to Remote Sensing of the Earth from Space, UN General Assembly, A/RES/41/65, 95th Plenary Meeting, December 3, 1986. United Nations, New York.

United Nations. (1986b). Reports of the COPUOS Legal Sub-Committee, April 8, 1986 (A/AC.105/C.2/SR.440) and June 12, 1986 (A/AC.105/SR.289), United Nations, Vienna.

Wallemacq, P. (2018). Natural disasters in 2017: lower mortality, higher cost. Cred Crunch 50:1-2. Retrieved from: https://www.cred.be/publications. (Accessed 10 June 2020).

Wang, Z., Ye, X., Tsou, M.H. (2016). Spatial, temporal, and content analysis of Twitter for wildfire hazards. Nat Hazards 83 (2016) 523–540.

Wetherholt, W.A. and Rundquist, B.C. (2010). A survey of ethics content in college-level remote sensing courses in the United States. Journal of Geography, 109:2, 75-86, https://doi.org/10.1080/00221341.2010.482161.

Yuan, F. and Liu, R. (2018). Feasibility study of using crowdsourcing to identify critical affected areas for rapid damage assessment: Hurricane Matthew case study. International Journal of Disaster Risk Reduction 28, pp 758–767. https://doi.org/10.1016/j.ijdrr.2018.02.003.

Naming of Tropical Cyclones: The World Meteorological Organisation Calls It Idai, While Grassroots Know It as *Dutumupengo*

4

Abstract

It often happens that communities give names to things with meanings attached and by their very nature names carry connotations. Names can remind us of the history of a generation, an event, a religious belief and more. However, when it comes to cyclones (or hurricanes or typhoons in other regions) what remains puzzling to many is how the names are conceived. Contestations have emerged in that such names were traditionally feminine and without local meaning. This study investigates the way in which tropical cyclones are named with particular reference to Cyclone Idai. It also documents how the people in Chimanimani, Zimbabwe, who were affected by the cyclone felt about the name and its implications. Drawing mainly from a methodological orientation including document and critical discourse analysis, as well as interviews and a household survey, it emerged that the World Meteorological Organisation (WMO) named it Cyclone Idai, while the locals know it better as *Dutumupengo* (one hell of a troublesome storm). Some locals were not impressed by the name given by the WMO, although these names already exist in a register long before the cyclones happen. It was just a coincidence that Idai happened to be a Shona name and was also named by a Zimbabwean meteorologist. The work recommends that local communities and political heads be educated concerning the naming of tropical cyclones so that there is no confusion or frustration. Furthermore, there is a need for a compromise, with local communities being given a chance to maintain their local names and/or refine them so that they have meaning. Local and national registers with agreed national names may then be part of a country's heritage.

Keywords

Cyclone Idai · Naming · WMO · Contestations · Females

4.1 Introduction and Background

The United Nations' 2030 Agenda for Sustainable Development (AfSD) that sets a global agenda to eradicate poverty, protect the environment and transform our future (United Nations 2015) makes reference to issues of culture. These are matters that relate to names and the naming of tropical cyclones. The United Nations acknowledges that there is natural and cultural diversity, and these should be recognised as they are critical enablers of sustainable development. This notion is further embedded in Target 4.7 of SDG 4 that deals with the need to attain inclusive and equitable quality education by 2030. The cultural dimensions that are also closely linked to the concept of dark tourism which is associated with death and pain-related events such as Tropical Cyclone Idai are also incorporated within several SDGs, including Target 8.9 of SDG 8 dealing with sustainable and decent jobs. Cultural dimensions furthermore feature in Target 12.b which seeks to promote local culture and products from SDG 12, dedicated to the attainment of sustainable consumption and production. Hence, when debates on naming occur, they will also be drawing from the global goal to achieve sustainable development and inclusive under the established 2030 goals, targets and the indicators.

What then are tropical cyclones? Before explaining the concept, it is important to note that the word "cyclone", first used by Henry Piddington, the president of the Marine Courts of Inquiry at Calcutta (now Kolkata) during the British rule in 1848 (Asokan and Girija 2018), which depicts the coiling of a snake. However, the best possible response is one that comes from the WMO. The WMO (2020a) online description of a tropical cyclone highlights describes it as follows:

> A rapid rotating storm originating over tropical oceans from where it draws the energy to develop. It has a low-pressure centre and clouds spiralling towards the eyewall surrounding the "eye", the central part of the system where the weather is normally calm and free of clouds. Its diameter is typically around 200 to 500 km, but can reach 1000 km. A tropical cyclone brings very violent winds, torrential rain, high waves and, in some cases, very destructive storm surges and coastal flooding. The winds blow counter clockwise in the Northern Hemisphere and clockwise in the Southern Hemisphere. Tropical cyclones above a certain strength are given names in the interests of public safety.

Notably, different terminology is used to depict such weather phenomena in different regions of the world (WMO 2020a). They arise in ten tropical cyclone basins and are referred to as the hurricane in the Caribbean Sea, the Gulf of Mexico, the North Atlantic Ocean and the eastern and central North Pacific Ocean. Typhoons are associated with the western North Pacific, while cyclones emerge from the Bay of Bengal and the Arabian Sea. In the western South Pacific and south-east Indian Ocean, these phenomena are called severe tropical cyclones, while the term "tropical cyclone" is used in the south-west Indian Ocean. Should the tropical cyclone be too harsh in terms of deaths and costs, its name is retired (WMO 2020b). To retire the name of a tropical cyclone implies that the name will no longer be used in referring to the event as it will have been replaced by a new name. The responsibility for retiring the names of tropical cyclone falls on the shoulders of annual WMO Tropical Cyclone Committees. Some of the cyclone names that were retired include Mangkhut (the Philippines, 2018), Irma and Maria (the Caribbean, 2017), Haiyan (Philippines, 2013), Sandy (the USA, 2012), Katrina (USA, 2005), Mitch (Honduras, 1998) and Tracy (Darwin, 1974). Other names retried from the typhoon basin and their replacement names include Rusa (2002 replaced with Nuri, Republic of Korea), Longwang (2005 replaced with Haikui, China), and typhoon Xangsane (2006 replaced with Leepi after making multiple landfalls in Philippines, China, Vietnam, Cambodia and Thailand) (Lei and Zhou 2012). In addition, names are predefined from a list validated by the cyclone basin concerned (WMO 2020b) and can run up to five cyclone seasons. The tropical cyclones have in severe negative effects on human activities and ecosystems (Da Rocha et al. 2019).

However, what is in the name and why reserve space to deliberate it? Down the years, huge strides have been made in refining the naming of both natural and human-made phenomena that include geographical places, human beings, diseases, pandemics, cars, corporate entities, wars and natural phenomena. Among natural phenomena are artefacts such as rivers, mountains, animals, volcanoes, hurricanes, tropical cyclones, cyclones, typhoons and many more. Rasmussen et al. (2020) maintain that disease naming is critical in the medical field as this will create the common terminology needed for generating and sharing information among healthcare providers, researchers, patients and families. The authors give the example of controversy in naming syphilis which has since the fifteenth century had up to 400 names, as well as other controversies surrounding the naming of Covid-19 (Ratner et al. 2020). Hsu et al. (2020) indicate that it took more than a month for the novel coronavirus notified to the World Health Organisation (WHO) by the Chinese government on 31 December 2019 to get its reference and globally accepted name. It was only on 11 February 2020 that the official name, Severe Acute Respiratory Syndrome Coronavirus 2 (SARS-CoV-2) and the disease (coronavirus disease 2019) abbreviated to Covid-19 were announced both by the Coronaviridae Study Group of the International Committee on Taxonomy of Viruses (CSG-ICT) and the WHO. The CSG-ICTV (2020, p. 535) revealed that:

> Based on phylogeny, taxonomy and established practice, the CSG recognizes this virus as forming a sister clade to the prototype human and bat severe acute respiratory syndrome coronaviruses (SARS-CoVs) of the species Severe acute respiratory syndrome-related coronavirus and designates it as SARS-CoV-2. In order to facilitate communication, the CSG proposes to use the following naming convention for individual isolates: SARSCoV-2/host/location/isolate/date.

Gastorn (2020) picks up issues with the naming of Covid-19. In his view, there remains a need to highlight the rules that govern the naming of viruses and their associated epidemics and pandemics. Therefore, drawing from the WHO 2015 Best Practices for the Naming of New Human Infectious Diseases, there had to be some minimum negative impact caused by a disease name "on trade, travel, tourism or animal welfare, and avoid offending any cultural, social, national, regional, professional or ethnic groups" (Gastorn 2020, p. 2). To this end, the WHO refused to use the name SARS as this could have resulted in panic and SARS-induced fear, particularly in Asia following the 2003 SARS. Therefore, Covid-19 seem to fulfil the naming guidelines from the WHO. In the end, Gastorn concludes that the final name could also have been agreed based on those with the most political power, rather than the rationale of the scientific basis for it. One may therefore not exonerate the Chinese

power and influence from all of this. Names such as the Wuhan coronavirus were also surfacing (University of Michigan 2020).

In the past, there have been some bad names for diseases according to Gastorn (2020). Such bad names include acquired immune deficiency syndrome (AIDS), swine flu (which led to some farmers slaughtering their pigs), Middle East Respiratory Syndrome (MERS), Hendra (the name of a suburb in Brisbane, Australia) and Ebola (the name of a river in Africa). Such names had negative social and economic implications. Of late, a reference to Covid-19 as the "China virus" by President Donald Trump of the USA has caused social, political and trade tensions (Sparke and Anguelov 2020. p. 501). Such provocation should be condemned in the strongest of terms as this only makes the situation worse.

Naming is also an issue in patterns, and Callaghan (2020) agrees that anybody who has embarked on a naming journey, be it to name a toy or child, can testify to how difficult the endeavour can be in getting the right name. Even though the public opinions may be sought, the results may not be the desired ones. Callaghan then concludes his story with the difficulty they faced when they finally settled for a new journal name, *Patterns*. The underlying principles for the journal name, Patterns, were that: "We wanted it to be distinctive; short and snappy for preference, without needing any other explanation. A name that sums up what the journal is all about" (Callaghan 2020, p. 1). What is discussed here is in direct opposition to the case of how the naming of non-communicable diseases (NCDs) emerged. Allen and Feigl (2017, p. e129) complain that the name NCD is a "longwinded non-definition that only tells us what this group of diseases is not". This does not help the medical and political fronts to urgently resource the fighting against NCDs which are the world's largest killer. The authors go further, highlighting that SDG 3 (health and wellbeing) has provided such urgency in the lead up to the 2030 Agenda for Sustainable Development (United Nations 2015).

In investigating lake names in the USA, Soranno et al. (2020, p. 1) start by reflecting on the quote by William Shakespeare "What's in a name? That which we call a rose, by any other name would smell as sweet". This quote, from the play *Romeo and Juliet*, Act II, Scene II, Lines 45 and 46, is also featured by Paige (2020, p. 1) in discussing Covid-19. Soranno et al. (2020) go on to analyse 479,950 official names from the USA Geographic Names Information System. It emerged that 83% of the lakes were nameless, and most of such lakes were small, mainly less than four hectares. However, of the 83,115 named lakes, it was revealed that lake names reflected local communities' everyday lives. It was also found that lakes could inspire creativity, with indigenous languages playing a role, and that names varied by region. Furthermore, and unfortunately, there were also lakes with derogatory names. The authors recommended that more than 400,000 unnamed lakes be given names and those with derogatory names be re-named. Related to the naming of lakes is the naming of rivers. Busch et al. (2020, p. 1) are of the view that "[r]ivers that cease to flow are globally prevalent. Although many epithets have been used for these rivers, a consensus on terminology has not yet been reached. Doing so would facilitate a marked increase in interdisciplinary interest as well as the critical need for clear regulations".

Naming storms assists in many respects, including forecast communication and the quick identification of such storms (Asokan and Girija 2018). Charlton-Perez et al. (2019) find additional advantages in naming storms, as was the case with Storm Doris that hit Ireland and the UK in February 2017. Naming the storm provided researchers with a useful and easily collected target to investigate the awareness of the public in understanding extreme weather events. Lin et al. (2018) highlight that naming storms permits individuals to use heuristic processing to determine and understand the severity and potential threat of the storm. Furthermore, if the storm has a more menacing or iconic name, this could motivate people differently compared to a more subdued or unemotional name.

Holmes (2016) attempts to take an historical perspective in the naming of tropical storms. Upfront, the author highlights that the storms were named after wives, girlfriends and disliked politicians. One Clement Wragge, a British meteorologist, is said to have started the tradition of describing weather systems around 1887. The naming project went into hibernation following Wragge's death in 1922. This resulted in storms being confusing and having general descriptions that included their location or things they hit. Names such as the 1911 Ship Cyclone and the 1938 New England Hurricane were among such. With too much confusion in terms of reference, the Wragge storm naming project was resuscitated during World War II, and this led to air force and navy meteorologists naming tropical cyclones after their wives and girlfriends they had left behind at home. The British National Weather Bureau then introduced some form of phonetically alphabetised list of names in 1945 that included "Able", "Baker", "Charlie" and "Dog", which did not last long as they only had 26 such names, and it ran out in 1953. In 1954, the US government embraced the naming tradition that used the first names of women. Similar approaches were also adopted by Australia and New Zealand in 1963, spreading the world over in subsequent years.

Drawing from lived experiences, names are contextual. For example the Nguni people of southern Africa have names depicting the desire to end dominant sex for their children, or end childbearing altogether. Many cultures and traditions also have names pointing to the power of God, spiritualism and other religions. The Shona people in Zimbabwe have names associated with freedom fighters that toppled the Rhodesian regime in 1979. Many other names in postcolonial times are now associated with changes from such regimes. Table 4.1 presents a sample of the names being addressed in this work.

In conducting a multidisciplinary qualitative ethnographic study to determine the most popular female anthroponyms of the Shona people of Zimbabwe's seven out of ten provinces, Makondo (2013) found that the top five most popular names were *Chipo* (Gift), *Tendai* (Be thankful), *Tsitsi* (Mercy), *Chiedza* (Light) and *Vimbai* (Trust). Taking *Chipo*, for example, the name "is a shortened version of *Chipochangu* (my gift), *Chipochedu* (our gift), *Chipochedenga* (heaven's

Table 4.1 What is in a name—the contextual setting

Nguni names for people depicting the desire to end child sex/childbearing	Shona names for people depicting the power of and walk with God	Shona names for people depicting the troubles of polygamy	Abandoned Rhodesian city names and their replacements on independence	Zimbabwe freedom fighters' names
Anele	*Munashe*	*Kurauwone*	*Gatooma (Kadoma)*	*Bvumazvipere*
Banele	*Nashe*	*Muchineripizano*	*Gwelo (Gweru)*	*Gabarinocheka*
Bandile	*Panashe*	*Mativenganisa*	*Umtali (Mutare)*	*Mabhunumuchapera*
Kwanele	*Sheunesu*	*Ndakaitei*	*Que Que (Kwekwe)*	*Maperembudzi*
Sanele	*Tadiwanashe*	*Nhamoinesu*	*Salisbury (Harare)*	*Teurairopa*
Zandile	*Tinashe*	*Tinongogara*	*Shabani (Zvishavane)*	*Tichatongachete*

Source: Authors

gift), *ChipochaMwari* (gift from God), *Chipochatapihwa* (the gift we have been given), *Chipondechedu* (the gift is ours) and *Chipochake* (his/her gift), among others (Makondo 2013, p. 115). Additional meanings could also be drawn. A name such as *Chipochangu* depicts a namer (agent) who has been censured regarding what has happened. It could have been a child born with a disability, which could result in stigma from relatives and society at large.

Against the background provided, the main objective of this chapter is to unearth and document protocols and procedures for naming tropical cyclones that both the general public and learned colleagues should understand. The chapter also sets the objective of exploring the feelings of the affected local communities in the event that tropical cyclone names have meanings understood differently by the local communities, as was the case with Tropical Cyclone Idai in parts of Mozambique and most parts of Zimbabwe.

4.2 Materials and Methods

To address the outlined main research objectives, a mixed method research design was preferred. The research design included the use of documents and critical discourse analysis, interviews (with more than 25 interviewees), focus group discussions (with five participants) and an offline QuestionPro administered household survey (*n* = 155). Geographical Information System (GIS) was also utilised in generating the broader

study area (Fig. 4.1) (Martínez-Hernández and Yubero 2019). The methods identified have also been used in studying naming in general (Makondo 2013;), names and naming of cyclones (Asokan and Girija 2018; Charlton-Perez et al. 2019), naming of rivers (Busch et al. 2020), naming in the medical field and patterns (Allen and Feigl 2017; Callaghan 2020; CSG-ICTV 2020), as well as regarding contestations of female names (Skilton 2018).

The analysis started during fieldwork as the interviews were being conducted and preliminary themes started emerging. Detailed analysis of the interviews involved having them transcribed and applying the fundamentals in the document and critical discourse analysis that drew insights and some elements from the grounded theory (Rieger 2019; Peiter et al. 2020). The findings were then supported by relevant and recent literature. Some data from QuestionPro were imported into MS Excel and graphs were plotted, while photos taken during the fieldwork were also used to buttress the emerging narratives.

4.3 Presentation and Discussion of Findings

This section is arranged in three subsections as follows: discussion on the naming of tropical cyclones; the naming of Tropical Cyclone Idai; and lastly, other local names for Tropical Cyclone Idai in Chimanimani, including the most common, *Dutumupengo,* in the Shona (local ChiManyika) language.

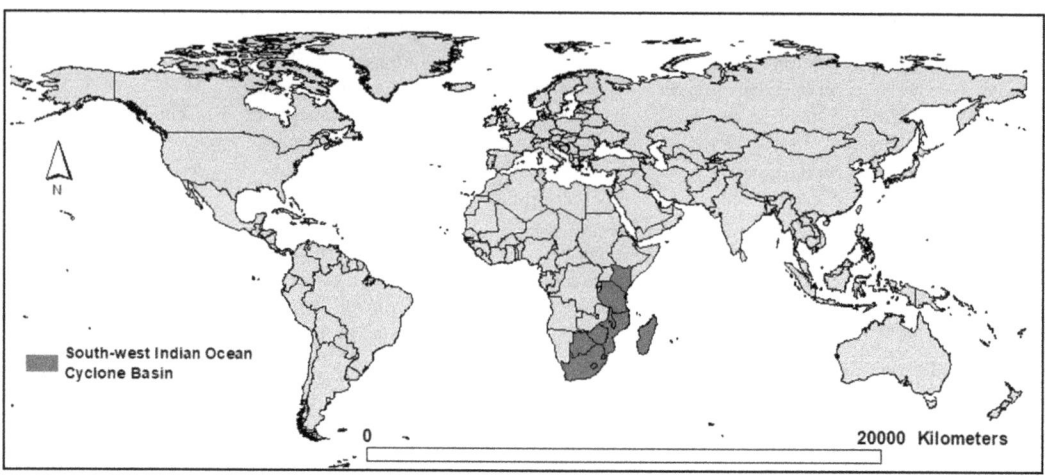

Fig. 4.1 The South-western Indian Ocean Cyclone Basin. Source: Authors

4.3.1   The Naming of Tropical Cyclones

The WMO is the custodian of rotating lists of tropical cyclone names (WMO 2020b). Such a list is appropriate to each tropical cyclone basin. The practice of naming cyclones emerged because names are far easier to remember and make reference to than numbers and technical terms. The names also facilitate the issuing of early warnings, reporting by the media and relief calls. The WMO highlights that "[i]n the beginning, storms were named arbitrarily. An Atlantic storm that ripped off the mast of a boat named Antje became known as Hurricane Antje. Then the mid-1900s saw the start of the practice of using feminine names for storms" (WMO 2020b online). This practice was later replaced by a system that would identify storms using alphabetically listed names. Male storm names only started featuring in the late 1990s, particularly in the Southern Hemisphere.

The USA National Hurricane Centre was tasked with naming the Atlantic tropical storms as early as 1953 (WMO 2020b), although the names are now in the custodianship of the International Committee of the WMO which works with ten tropical cyclone basins (Riddle 2019). Each of the basins has names that are arranged in alphabetical order. Men's names only started featuring in 1990s, and they now alternate with the women's names (WMO 2020b). The Atlantic tropical basin has six lists of the set of names that are used in rotation. Hence, the 2019 list will be used once more in 2025. The names of tropical cyclones for the South-west Indian Ocean cyclone basin for the seasons 2019/2020, 2020/2021 and 2021/2022 are presented in Table 4.2.

A new phenomenon in naming winter storms has been gaining momentum (Nuessel 2015; Rainear et al. 2017; Charlton-Perez et al. 2019). In the USA, the Weather Channel® initiated the practice of naming winter storms in November 2012. The naming is also done systematically, with only disruptive storms being given names (Nuessel 2015). From the USA National Oceanic and Atmospheric Administration, names cannot be given to winter storms as it is difficult to pinpoint where one starts and ends, unlike in the case of hurricanes. Some of the names of winter storm for the 2014/15 season are shown in Table 4.3.

What is of interest from the 26 winter names for the 2019/2020 season is that both male and female names have been included for use. This should have been informed by the long battles that led to the overturning of the female-only hurricane names of yesteryear. However, there has been criticism of the naming of winter storms, with some indicating it is nothing more than a marketing initiative (Rieger 2019).

Table 4.2 Names of tropical cyclones 2019–2022

2018/2019		2019/2020		2020/2021		2021/2022	
Contributor	Name	Contributor	Name	Contributor	Name	Contributor	Name
France	ALCIDE	Malawi	AMBALI	France	ALICIA	Mozambique	ANA
Comoros	BOUCHRA	Mauritius	BELNA	Tanzania	BONGOYO	Zimbabwe	BATSIRAI
Madagascar	CILIDA	South Africa	CALVINIA	Lesotho	CHALANE	Madagascar	CLIFF
South Africa	DESMOND	Botswana	DIANE	Mauritius	DANILO	Eswatini	DUMAKO
Lesotho	EKETSANG	Zimbabwe	ESAMI	Seychelles	ELOISE	Comoros	EMNATI
Eswatini	FUNANI	Mozambique	FRANCISCO	Kenya	FARAJI	South Africa	FEZILE
Tanzania	GELENA	Eswatini	GABEKILE	Mozambique	GUAMBE	Tanzania	GOMBE
Mauritius	HALEH	Seychelles	HEROLD	Botswana	HABANA	Malawi	HALIMA
Zimbabwe	IDAI	Madagascar	IRONDRO	Mauritius	IMAN	Kenya	ISSA
Mozambique	JOANINHA	Kenya	JERUTO	Lesotho	JOBO	Mauritius	JASMINE
Seychelles	KENNETH	Zimbabwe	KUNDAI	South Africa	KANGA	Seychelles	KARIM
Kenya	LORNA	Lesotho	LISEBO	Malawi	LUDZI	Lesotho	LETLAMA
Botswana	MAIPELO[a]	France	MICHEL	Tanzania	MELINA	Botswana	MAIPELO
Malawi	NJAZI[a]	Comoros	NOUSRA	France	NATHAN	Malawi	NJAZI
France	OSCAR[a]	Mauritius	OLIVIER	Zimbabwe	ONIAS	France	OSCAR
Tanzania	PAMELA[a]	Malawi	POKERA	Madagascar	PELAGIE	Tanzania	PAMELA
Kenya	QUENTIN[a]	Seychelles	QUINCY	Comoros	QUAMAR	Kenya	QUENTIN
Comoros	RAJAB[a]	Botswana	REBAONE	Seychelles	RITA	Comoros	RAJAB
Mozambique	SAVANA[a]	Comoros	SALAMA	Eswatini	SOLANI	Mozambique	SAVANA
Eswatini	THEMBA[a]	France	TRISTAN	Mauritius	TARIK	Eswatini	THEMBA
Botswana	UYAPO[a]	Kenya	URSULA	South Africa	URILIA	Botswana	UYAPO
Mauritius	VIVIANE[a]	South Africa	VIOLET	Lesotho	VUYANE	Mauritius	VIVIANE
South Africa	WALTER[a]	Mozambique	WILSON	Kenya	WAGNER	South Africa	WALTER
Madagascar	XANGY[a]	Madagascar	XILA	Malawi	XUSA	Madagascar	XANGY
Zimbabwe	YEMURAI[a]	Eswatini	YEKELA	Botswana	YARONA	Zimbabwe	YEMURAI
Lesotho	ZANELE[a]	Tanzania	ZAINA	Mozambique	ZACARIAS	Lesotho	ZANELE

[a]Names not used

Source: Authors, based on WMO (2020b), Becoming the Muse (2020)

Table 4.3 Selected winter storm names 2014/2015 and 2019/2020 seasons in the USA

2014/2015 season		2019/2020 season	
Name of the winter storm	Meaning of the name	Names	
Astro	In Greek, it means star	Aubrey	Nash
Bozeman	In honour of the Miss Shupe's Bozeman High School Latin class, which provided the 2013–2014 list of winter storm names	Bessie	Odell
		Caleb	Pearl
Cato	The name of a Roman statesman and his great-grandson, who were both known for integrity	Dorothy	Quincy
		Ezekiel	Ruth
Kari	A Finnish name derived from the Greek name Makarios from old-Greek meaning blessed or happy	Finley	Sadie
		Gage	Thatcher
		Henry	Upton
Quantum	From the Latin word quantus, meaning how much	Isiah	Veronica
Thor	From Scandinavian mythology, god of thunder and rain	Jacob	Wyatt
Xander	Dutch form of Latin name, Alexandrus	Kade	Xandra
Zelus	From Greek mythology, personifies dedication, envy, jealousy and zeal	Lamont	Yates
		Mabel	Zachariah

Source: Authors, based on Nuessel (2015, p. 244), The Weather Chanel (2020)

4.3.2 How Did the Name Idai Emerge?

The researchers were fortunate in being able to hunt down and to locate the originator of the name Idai. The namer of Idai is one of the officials at the Zimbabwe Meteorological Services Department (ZMSD). From that in-depth and an open-ended interview was conducted with him, details emerged that the name was aligned to the WMO way of naming tropical cyclones. The account had it that the world is divided into cyclone basins, and Idai was a name from the South-west Indian Ocean Cyclone Basin (SWIO-CB). The respondent highlighted that every 2 years, the Tropical Cyclone Committee in which the SWIO-CB countries are included meets and look at the rules and regulations of how that information can be taken care of, including the naming of tropical cyclones.

The key informants also brought up the naming of tropical cyclones using female names, which was so strong up to the 1980s. One of the reasons for giving female names according to the respondent was that the tropical cyclones tend to be temperamental and unpredictable like the way some women behave. The notion of female naming is picked up by Skilton (2018) who reviews the naming of hurricanes between the 1950s and the 1970s. In her view, the introduction by the USA Weather Bureau of the phenomenon of female-only naming led to sexualised descriptions of the hurricanes. Smith (2016), Christensen and Christensen (2014), and also Jung et al. (2014) add to the equation by indicating that female-named hurricanes were considered to be deadlier since authorities and the general public did not take them seriously. However, the conclusion, according to Smith (2016), was based on questionable statistical analysis from limited data sets. Christensen and Christensen (2014) also disputed some of the earlier findings as many male-named hurricanes caused more deaths than female-named hurricanes. Hence, the protests from feminists took a pathway lasting 25 years, leading to the modern system of alternating male and female names on the basin lists (Skilton 2018). However, even with the recognisable advancement and progress removing the female names-only mentality, the battle continues. Following Hurricane Katrina in 2005, Macomber et al. (2011, p. 525) made shocking revelations regarding the t-shirts prints they saw in New Orleans, 7 months after the disaster. Prints such as "That Bitch Katrina … and No wonder they name hurricanes after women!"

According to the respondent, the WMO recommends that each basin sit down and nominate arbitrary names, which each country is supposed to advance to the committee. Each country chooses two names that are arranged in alphabetical order. As the tropical cyclones develop throughout the summer season, a country may discover that the tropical cyclone it named such as Idai could have occurred somewhere and just dissipates. Hence, that country may not even realise its name has been used. However, coincidentally, tropical cyclone Idai, which the respondent was privileged to name following the SWIO-CB committee meeting in Seychelles in 2017, happened to make multiple landfalls in Malawi, Madagascar, Mozambique and Zimbabwe in March 2019. The other piece of information that emerged from the interview was the confirmation of the retiring of hurricane and cyclone names. In the respondent's view, Tropical Cyclone Idai was likely to be retired in the future. What came out clearly from the discussions was that:

> There is no specific scientific naming system. It's just names. But having said that, when we collect all these names, some names in one country might mean something not nice in the other country. For example, I remember Tanzania gave another name, which I can't remember and we refused because it means something to us and we had to reject it as a committee.

Another interesting but uncommon phenomenon is when cyclone names on the agreed roaster all get used in one season. This happened in 2020 when active hurricane names all got used. The predetermined list of all 21 names for the season is shown in Table 4.4. On 21 September 2020, the WMO (2020c) issued a statement indicating that the US National Hurricane Centre had raised advisories on five hurricanes on 14 September 2020. The advisories covered five hurricanes

namely Paulette, Rene, Sally, Teddy and Vicky. This left the name Wilfred as the last name to be used in the season. The advisories tied the record for the highest number of hurricanes in that basin in one season, a record last set in September 1971. With the likelihood that Wilfred will be used (which happened), the use of the Greek alphabet (Alpha, Beta, Gamma, Delta, etc.) had to kick in. As predicted, Hurricane Delta made landfall in Louisiana on 12 October 2020 (McLaughlin and Maxouris 2020). The use of the Greek alphabet only happened once in 2005, with six names from the Greek Alphabet used (WMO 2020c).

Knowing the meaning of Idai (love) in the local Shona language in Zimbabwe, a question was asked if people were not stopping the respondent in street or calling asking why the name Idai for such a bad tropical cyclone. The response was: "Unfortunately, some overzealous stations broadcasted that and people still ask me questions." To this end, there were times when the respondent had to go on air to talk about Tropical Cyclone Idai and how it was behaving. Then there would be calls from high-ranking officials indicating the respondent gave it the name Idai. Obviously, there was a gap in understanding because it was mere coincidence that the name came from Zimbabwe and that it was the next one on the list from the WMO protocol.

The challenge with the name Idai is that what the tropical cyclone left behind did not come anywhere near its local meaning. The name is translated as "love" or to "to love". Hence many would wonder what kind of love was shown when 1000 of people perished and many more went missing in Mozambique. As this chapter was being finalised in September 2020, some bodies that had washed away from Kopa into Mozambique had not been repatriated, and many families were still in emergency shelters in four camps in Chimanimani. The numbers as per the 20 April 2020 update from IOM Displacement Tracking Matrix (2020) amounted to 224 households and 859 individuals. The following paragraphs are reserved for discussing alternative names for Tropical Cyclone Idai from the local and grassroots level.

Table 4.4 Atlantic Tropical Cyclone Basin (Hurricane) names for 2020 season

Name	Name	Name
1. Arthur	8. Hanna	15. Omar
2. Bertha	9. Isaias	16. Paulette
3. Cristobel	10. Josephine	17. Rene
4. Dolly	11. Kyle	18. Sally
5. Eduard	12. Laura	19. Teddy
6. Fay	13. Marco	20. Vicky
7. Gonzalo	14. Nana	21. Wilfred

4.3.3   Cyclone Idai's Other Names from Chimanimani, Including *Dutumupengo*

The silent contestation and possible disapproval of the name Idai from the local chiefs and communities in Chimanimani came up indirectly during fieldwork. From all the three chiefs interviewed and many other key informants, including the focus group discussions, the response when one asked about Tropical Cyclone Idai would be, "okay, do you mean *Dutumupengo?*" This common response revealed that Tropical Cyclone Idai had a local name constructed from two phrases "*Dutu*" (great winds and/or storm with great winds) and "*Mupengo*" (madness or someone insane). Therefore *Dutumupengo* can be translated as a merciless and uncontrollable violent storm. This kind of naming is common in traditional set-ups in Zimbabwe, across Africa and in many other regions of the world, as events are usually referred to in generic terms. In other words, the phrase *Dutumupengo* can still be used for another similar tropical cyclone in the future.

However, confusion results when it comes to the global interpretation of such a tropical cyclone, which is why the WMO way has gained global acceptance. In this case, should another tropical cyclone of similar magnitude hit Chimanimani in the future, the references to the two tropical cyclones will be something like *Dutumupengo ra* 2019 (from 2019) and possibly *Dutumupengo ra* 2020 (from 2020). When the authors of this chapter were growing up in Zimbabwe, the elders would gather us around the fires and start recollecting several natural disasters. Because the recollection was so problematic the disasters were referred to as "the year of armyworm", "the year of measles", "the year of floods", etc. Having a local name to tropical cyclone sis not something unusual, Typhoon Goni as per the WMO, known locally as Typhon Rolly, made landfall in Philippines on 1st October 2020 (Thornton & Westcott 2020). Rolly, left at least 17 people dead and more than two millions impacted. Goni was the known world's known strongest typhoon in 2020. The typhoon hit the country with the magnitude of Idai's destructive force and came in the midst of COVID-19 too.

Investigating further, it became clear that naming was and still is a big issue in Chimanimani like many other places across the world. One of the chiefs interviewed told stories about places in Chimanimani. To start with, one of the areas that was severely affected by Tropical Cyclone Idai called Ngangu (Fig. 4.2) has a history, as does Kopa Township (Fig. 4.3). The name Ngangu is said to have come from the time when there were wars between the Ndebele tribes running away from King Shaka in the KwaZulu-Natal province of South Africa which settled in Bulawayo in Zimbabwe. The chief explained that it happened that one of the Shona warriors from the Chimanimani area had his private parts removed by the arrows from a Ndebele warrior, which were found hanging on the spear. In comedic fashion, somebody had to ask whose private parts they were, with the response in the Shona language, specifically the ChiManyika dialect, being *Ngeangu* (meaning they are mine). In everyday speech and Shona language construction, the "e" in *Ngeangu* is swallowed into *Ngangu*, which became the name of both the mountain and the area where the Ngangu Township sits to this day. Similar explanations emerged regarding the Kopa Township, a name derived from the fact that the houses there were part of the cooperative scheme. Hence, the locals used the ChiManyika/Shona English derived name Kopa. The other story about naming came from the chief was that of the Chimanimani mountain range from which the Chimanimani district gets its from. It was explained that the name came from a very tight gap [*pakamanika (kidzika)* also known as *pamukaha* in ChiNdau] between the mountains that is used as a pass between Zimbabwe and Mozambique. There were many other stories regarding names from the Chimanimani area, including those for rivers, other mountains and clan names. Hence, a name means a lot to people, especially local communities.

To tease out further insights regarding the name Idai and potential local alternatives, a

Fig. 4.2 The remains from Ngangu Township. Source: Authors

Fig. 4.3 One of the houses that miraculously remained standing in Kopa Township. Source: Authors

question was asked towards the end of many interviews on perceptions of the name Idai and to give an alternative name based on the experiences. Another question was also included in the household interview to gauge the appropriateness of the name. All these investigations were made possible only because the WMO name Idai coincided with the local Shona/ChiManyika name Idai. The findings from the household survey are presented in Fig. 4.4. The findings show that 40.64% of the household respondents thought the name Idai was not appropriate. The same percentage was of the view that the name Idai was appropriate. When asked why, those that indicated the name Idai was appropriate in from the interviews highlighted that the nature of the damage meant that both the world and the local communities had to show love in such times of need. This understanding was probably confirmed by the huge numbers of Tropical Cyclone Idai victims who were taken in by relatives and other community members when their houses

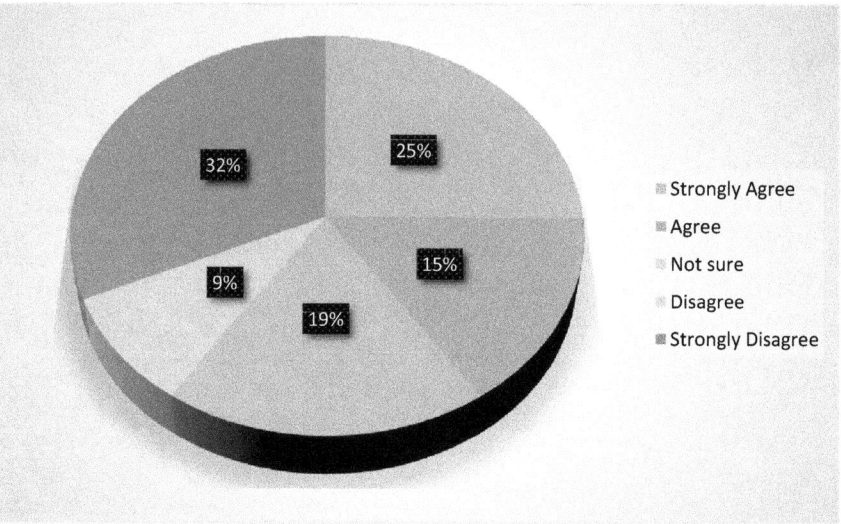

Fig. 4.4 In your view, was the name Idai appropriate in terms of what transpired? ($n = 155$). Source: Authors, Household Survey 2019

were destroyed. This act of Tropical Cyclone Idai kindness not only included providing shelter, but food, clothing and other necessities as well.

As was indicated earlier, the key informants were asked to give an alternative name to Idai as per their understanding of the situation and experience. The possible alternative names that came out are presented in Box 4.1. In the naming, the categories of names selected fell mainly into something to do with suffering and/or to cause suffering like *Tambudzai*, *Rwadzisaii* and *Dambudzo*. In another category, names refer to destruction or a destroyer or to kill, such as *Paradzai*, *Muparadzi*, *Chirakashi*, *Muroyi*, *Maondonga*, *Pondai* and *Mauraye*. The other category was for madness and war, with such names as *Mupengo*, *Mvuramupengo* and *Hondo*. The remaining phrase names were, for example, *Kudzidza Hakuperi*, which means we continue learning, and *Pachedu Tinokunda*, which means together we can conquer. Such is the diversity in terms of what other names or phrases that could have been used locally to refer to Tropical Cyclone Idai, and these names and phrases are a true reflection of how those interviewed felt regarding the damage caused by Tropical Cyclone Idai.

The use of names such as Idai and *Dineo* (a Tswana and Sotho name meaning "we are gifted") are perplexing as these two tropical cyclones caused extensive damage and pain. These lovely and beautiful names are given to such a heartless, heinous phenomenon that wiped out villages and towns. What gift or love is there to talk about when lives and livelihoods are lost in floods and violent storms? These are the challenges that communities face when they do not understand or even get involved in the processes in naming tropical cyclones.

Box 4.1 Alternative Names to Tropical Cyclone Idai

1. *Tambudzai X 3*	9. *Pondai*
2. *Paradzai*	10. *Dambudzo X 3*
3. *Muparadzi*	11. *Mauraye X 2*
4. *Hasha*	12. *Hondo*
5. *Maondonga*	13. *Kudzidza Hakuperi*
6. *Rwadzisai*	14. *Pachedu Tinokunda*
7. *Muroyi*	15. *Mvuramupengo*
8. *Mupengo X 2*	16. *Chirakashi*

Source: Author, Fieldwork 2019

4.4 Conclusion and Recommendations

What is in the name? As the chapter moves towards a close and in drawing some conclusions and recommendations for the future, we come back to the starting point. What comes out clearly is that the world is divided into ten tropical cyclone naming and management basins, of which Tropical Cyclone Idai originated in the South-western Indian Ocean Cyclone Basin. The names of hurricanes, typhoons, severe tropical cyclones and cyclones are all decided prior to the tropical cyclone happening. There is also no logic in the naming as this is arbitrary. Hence, the name can be anything really, including the names of one's children, historical events, biblical names and the like. The work further revealed that the naming started in 1953 in the USA, where all hurricane names were exclusively female. This sparked protests and activism from feminists that continues for about 25 years. Only then were male names included in hurricane and other tropical names. The reasoning behind associating tropical cyclones with female names is believed to be in how many women were perceived to be temperamental and destructive, an aspect that aligned to the nature of the tropical cyclones. However, several publications disputed this thinking as more work started emerging which questioned biased samples and also presenting the truth (of late) that male named tropical cyclones are as destructive.

Once the cyclone is adjudicated by the WMO committee as having caused untold destruction and suffering in the affected region, the name is retired and replaced with another. Examples of many such retired names were given in the paper. As for Tropical Cyclone Idai, it was a mere coincidence that the name is from Zimbabwe and the Shona/ChiManyika language. Furthermore, the name was given by a Zimbabwean meteorologist, which caused some resentment, given that Idai means love in the local language. The question posed by Shakespeare comes up again. What is in a name? For many who suffered the untold destructive nature of Tropical Cyclone Idai, there would be no love to talk about. Hence, the cyclone had another local name *Dutumupengo*, meaning a berserk, uncontrollable and destructive storm. Tropical Cyclone Idai was like a demon-possessed person in the eyes of the locals. Many other names that tell the real story of the destructive nature of Tropical Cyclone Idai emerged once the local respondents were given a hypothetical opportunity to name the tropical cyclone. However, one of the challenges with the name *Dutumupengo* is that it is generic, forcing one to refer to it as *Dutumupengo from 2019* for example. Such a reference point has other challenges in that after many years people will fail to make direct reference because there will still be many other tropical cyclones with the generic name *Dutumupengo*. Needless to say, the name Idai was presented and approved in the South-western Indian Ocean Cyclone Basin committee during a meeting that took place in Seychelles in 2017.

Moving forward, the work recommends the scaling up of educating citizens globally on how the tropical cyclone names are decided. However, knowing that names have meaning and must be contextual, there are many suitable names to describe tropical cyclones. However, from the point of view of not having names aggravate the situation and prolong psychological suffering and trauma, those in positions of power and those naming these tropical cyclones may need to consider neutral names and also regional names that still have meaning for one of the localities of the cyclone basin. Given the generic nature of the name *Dutumupengo*, there is an opportunity to work with residents and citizens in getting easy to refer to names also exists. This is an aspect that needs to be taken up by the responsible authorities so that an alternative local name can be found and agreed to for archiving and local referencing. This pathway would then address both the modern way of naming tropical cyclones and also harmonise with what local people would have agreed. Therefore, there is always something in the name, especially local names like Idai.

References

Allen, L.N. & Feigl, A.B. (2017). What's in a name? A call to reframe non-communicable diseases. The Lancet, 5: e129-e130.

Asokan, T. & Girija, P. (2018). Names of cyclone and its origin: A Review. International Journal of Computational Engineering Research (IJCER), 8(2), 42-45.

Becoming The Muse. (2020). Of cyclones, hurricanes, typhoons and their names. Retrieved from https://becomingthemuse.net/2019/03/16/of-cyclones-hurricane-typhoons-and-their-names/ (Accessed 25 August 2020).

Busch, M.H., Costigan, K.H., Fritz, K.M., Datry, T., Krabbenhoft, C.A., Hammond, J.C., Zimmer, M., Olden, J.D., Burrows, R.M., Dodds, W.K., Boersma, K.S., Shanafield, M., Kampf, S.K., Mims, M.C., Bogan, M.T., Ward, A.S., Rocha, M.P., Godsey, S., Allen, G.H., Blaszczak, J.R., Jones, C.N., Allen, D.C. (2020). What's in a Name? Patterns, Trends, and Suggestions for Defining Non-Perennial Rivers and Streams. Water, 12, 1980; doi:https://doi.org/10.3390/w12071980.

Callaghan, S. (2020). What's in a Name? How We Named Patterns. Patterns, https://doi.org/10.1016/j.patter.2020.100048.

Charlton-Perez, A.J., Greetham, D.V., Hemingway, R. (2019). Storm naming and forecast communication: A case study of Storm Doris. Meteorological Applications, 26, 682–697 doi:https://doi.org/10.1002/met.1794.

Christensen, B. & Christensen, S. (2014). Are female hurricanes really deadlier than male hurricanes? PNAS, 111(34): E3497-E3498. www.pnas.org/cgi/doi/10.1073/pnas.1410910111.

CSG-ICTV (Coronaviridae Study Group of the International Committee on Taxonomy of Viruses). (2020). The species Severe acute respiratory syndrome-related coronavirus: Classifying 2019-nCoV and naming it SARS-CoV-2. Nature Microbiology, 5: 536-544. https://doi.org/10.1038/s41564-020-0695-z.

Da Rocha, R.P., Reboita, M.S., Gozzo, L.F., Dutra, L.M.M., de Jesus, E.M. (2019). Subtropical cyclones over the oceanic basins: A review. Ann. N.Y. Acad. Sci. 1436, 138–156. doi:https://doi.org/10.1111/nyas.13927.

Gastorn, K. (2020). To name a new coronavirus and the associated pandemic: International Law and Politics. Chinese JIL. doi:https://doi.org/10.1093/chinesejil/jmaa024.

Holmes, T.T. (2016). Tropical storms were once named after wives, girlfriends, and disliked politicians. Retrieved from https://www.atlasobscura.com/articles/tropical-storms-were-once-named-after-wives-girlfriends-and-disliked-politicians (Accessed 5 September 2020).

Hsu, L.Y., Chia, P.Y., Lim, J.F.Y. (2020). The novel coronavirus (SARS-CoV-2) pandemic. Annals, 49(3), 105-107.

IOM Displacement Tracking Matrix. (2020). Tropical Cyclone Idai Response Multi-Sectoral Location Assessment (MSLA) - Round 6: Chimanimani, Manicaland. Retrieved from https://displacement.iom.int/system/tdf/reports/Zimbabwe_DTM_Multi-Sectorial%20Location%20Assessemnt_April%20 2020%20rev.pdf?file=1&type=node&id=8979 (Accessed 24 August 2020).

Jung, K., Shavitt, S., Viswanathan, M., Hilbe, J.M. (2014). Female hurricanes are deadlier than male hurricanes. PNAS, 111(24): 8782-8787. www.pnas.org/cgi/doi/10.1073/pnas.1402786111.

Lei, X. & Zhou, X. (2012). Summary of retired typhoons in the western North Pacific Ocean. Tropical Cyclone Research and Review, 1, 24-32.

Lin, X., Rainer, A.M., Spence, P.R., Lachlan, K.A. (2018). Don't sleep on it: An examination of storm naming and potential heuristic effects on Twitter. Weather, Climate and Society, 10: 769-779. https://doi.org/10.1175/WCAS-D-18-0008.1.

Macomber, K., Mallinson, C., Seale, E. (2011). "Katrina That Bitch!" Hegemonic representations of women's sexuality on Hurricane Katrina souvenir t-shirts. The Journal of Popular Culture, 44(3), 525-544.

Makondo, L. (2013). The most popular Shona female anthroponyms. Stud Tribes Tribals, 11(2), 113-120 (2013).

McLaughlin, E.C. & Maxouris, C. 2020. Hurricane Delta leaves one dead in Louisiana and brings tornado threat to Carolinas. Retrieved at https://edition.cnn.com/2020/10/11/weather/hurricane-delta-sunday/index.html (Accessed 19 October 2020).

Martínez-Hernández, C. & Yubero, C. (2019). Explaining urban sustainability to teachers in training through a geographical analysis of tourism gentrification in Europe. Sustainability, 12, 67. doi:https://doi.org/10.3390/su12010067.

Nuessel, F. (2015). A note on the names for winter storms. Names, 63(4): 242–245. doi10.1179/0027773815Z.000000000119.

Paige, J.T. (2020). What's in a name? Simulation and technology enhanced learning uses and opportunities in the era of COVID-19. BMJ Stel 2020;0:1–2. https://doi.org/10.1136/bmjstel-2020-000681.

Peiter, C.C., Santos, J.L.G., Kahl, C., Coelli, F.H.S., Cunha, K.S., Lacerda, M.R. (2020). Grounded theory: Use in scientific articles published in Brazilian nursing journals with Qualis A classification. Texto Contexto Enferm, 29:e20180177. https://doi.org/10.1590/1980-265X-TCE-2018-0177.

Rainear, A.M., Lachlan, K.A., Lin, C.A. (2017). What's in a #Name? An experimental study examining perceived credibility and impact of winter storm names. Weather, Climate and Society, 9, 815-822. https://doi.org/10.1175/WCAS-D-16-0037.1.

Rasmussen, S.A., Hamosh, A., Amberger, J., Arnold, C., Bocchini, C., O'Neill, M.J.F., Stumpf, A. (2020). What's in a name? Issues to consider when naming Mendelian disorders. Genetics in Medicine. https://doi.org/10.1038/s41436-020-0851-0.

Ratner, L., Martin-Blais, R., Warrell, C., Narla, N. (2020). Reflections on resilience during the novel coronavirus disease (COVID-19) pandemic: Six lessons from working in resource-denied settings. American Journal of Tropical Medicine and Hygiene, 00(0): 1–3. https://doi.org/10.4269/ajtmh.20-0274.

Riddle, B. (2019). Hurricane season. Science Scope, 76-79.

Rieger, K.L. (2019). Discriminating among grounded theory approaches. Nursing Inquiry, 26:e12261. https://doi.org/10.1111/nin.12261.

Soranno, P.A., Webster, K.E., Smith, N.J., Vázquez, J.D., Cheruvelil, K.S. (2020). Limnology and Oceanography Bulletin, February: 1-7.

Skilton, L. (2018). Gendering natural disaster: The battle over female hurricane names. Journal of Women's History, 30(3): 132–156.

Smith, G. (2016). Hurricane names: A bunch of hot air? Weather and Climate Extremes, 12: 80-84. https://doi.org/10.1016/j.wace.2015.11.006.

Sparke, M. & Anguelov, D. (2020). Contextualising coronavirus geographically. Transactions of the Institute of British Geographers, 45, 498–508. https://doi.org/10.1111/tran.12389.

The Weather Channel. (2020). Winter storm names for 2019-20 Revealed. Retrieved from https://weather.com/storms/winter/news/2019-10-02-winter-storm-names-2019-2020 (Retrieved 26 August 2020).

Thornton, C. & Westcott, B. (2020). At least 17 dead as typhoon Goni impacts more than 2 million people in the Philippines. Available at https://edition.cnn.com/2020/11/01/asia/typhoon-goni-philippines-aftermath-intl-hnk/index.html (Accessed 3 October 2020).

United Nations. (2015). Transforming our world: The 2030 Agenda for Sustainable Development. New York: United Nations Secretariat.

University of Michigan. (2020). Wuhan Coronavirus. Retrieved from https://www.uofmhealth.org/health-library/ack8845 (Accessed 8 September 2020).

WMO (World Meteorological Organisation). (2020a). Tropical Cyclones. Retrieved from https://public.wmo.int/en/our-mandate/focus-areas/natural-hazards-and-disaster-risk-reduction/tropical-cyclones (Accessed 25 August 2020).

WMO (World Meteorological Organisation). (2020b). Tropical Cyclone Naming. Retrieved from https://public.wmo.int/en/our-mandate/focus-areas/natural-hazards-and-disaster-risk-reduction/tropical-cyclones/Naming (Accessed 25 August 2020).

WMO (World Meteorological Organisation). (2020c). 2020 hurricane season exhausts regular list of names. Retrieved from https://public.wmo.int/en/media/news/2020-hurricane-season-exhausts-regular-list-of-names (Accessed 19 October 2020).

Part III

Impacts of Tropical Cyclone Idai on Selected Sectors

Abstract

The challenges of illegal gold mining are well known; environmental damage, violence from syndicates, poor sanitation, deaths and ill-health are among such. From natural hazards that include extreme floods, tropical cyclones and earthquakes, new and/or expanded gold deposits can be exposed causing a rush. The chapter investigates the impact of Tropical Cyclone Idai on the illegal gold mining and related activities from the Chimanimani area in Zimbabwe. Through the use of elements of citizen science, interviews, focus group discussions, on the ground observation, and geographical information systems (GIS), the findings show that there was a gold rush in the aftermath of Cyclone Idai. The gold rush led to extensive degradation of biodiversity and the natural Chimanimani (Mawenje) Mountains' aesthetics. Illegal gold miners came from far and wide, colonising the gold hotspots such as the Chimanimani Mountains, Nyabamba, Tarka Forest, Chimanimani National Park and Blocky. Water pollution, dying aquatic life, damaged cultural and traditional sites, as well as extensive deforestation also emerged. Illegal gold mining and its associated activities were a real threat to the tourism industry as many tourists would hike along trails in the Chimanimani Mountains. To this end, the chapter recommends that the government finds lasting solutions to the prob-lem, including scaling up patrols and rooting out corruption from its security officials. There is also a need to capacitate and educate the illegal miners to protect the environment, and use a carrot and stick approach to address the situation.

Keywords

Illegal gold mining · Tropical Cyclone Idai · Chimanimani · Mozambique

5.1 Introduction and Background

The problem of uncontrolled and illegal gold mining in Zimbabwe first came to light in the late 1980s—mainly due to its adverse environmental impacts. Musingwini and Sibanda (1999) highlighted that river stretches in excess of 4600 km had at that stage been affected by this menacing practice. However, widespread illegal mining in Zimbabwe was triggered by the 2000 government land reform programme when many black Zimbabweans resettled on former white farms, many of which had rich mineral deposits, gold in particular (Mkodzongi and Spiegel 2020). In a way, the fast-tracked land reform programme changed livelihoods across the country, as there were many incidences of gold and other mineral rushes. Artisanal mining, a form of legal small-scale mining, has grown so big in the country.

However, this has not prevented an even higher growth of illegal artisanal mining activities, mainly for gold exploration and mining. Zolnikov (2020) observes the challenges associated with artisanal and illegal gold mining that led to the government of Ghana placing a ban on this in 2017. Mambondiyani (2017) reports that in the Kwekwe area of Zimbabwe, illegal gold miners dug deep tunnels underneath roads, railways and buildings. However, as Tropical Cyclone Idai made landfall in Mozambique and hit Zimbabwe in March 2019, other unforeseen opportunities emerged in the form of more exposed gold deposits. This resulted in a gold rush that pulled many illegal miners into the border areas of Chimanimani between Mozambique and Zimbabwe.

The use of derogatory names to identify illegal miners and their untold destruction on the environment, and, in many instances, their own lives is common across the world and in Africa. In Zimbabwe, illegal miners are known as *Makorokoza* or *Magweja* (both plural) in the Shona language (Kachena and Spiegel 2019). In other instances other names for those coming from certain localities are used, such as *Mashurugwi*, referring to those that come from the Shurugwi area in the Midlands Province of the country (Njanike 2020). In Mozambique, illegal miners are referred to as *gariemperos* (The Birdlife 2020), while in Ghana they are known as *galamsey* (Mantey et al. 2020; Forkuor et al. 2020; Owusu-Nimo et al. 2018) and *zama-zamas* in South Africa (Mhlongo et al. 2019). However, following running battles with security forces and the outcry over the manner in which illegal miners are usually handled, governments have been trying to regularise their activities and the sanitised naming of the miners and the industry to either artisanal or small-scale mining (Mantey et al. 2020; Oladipo et al. 2020). This approach still has not stopped the massive illegal mining operation, including illegal gold mining in the Chimanimani area, which is the main focus of this chapter.

The environmental, health and other challenges of illegal gold mining are also well documented in the literature. Oladipo et al. (2020) look at lead toxicoses in free-range chickens

from the Zamfara communities of Nigeria. The findings show that hundreds of children under 5 years could have died from eating such contaminated chicken in 2010. There were also reports on geese dying from the lead poisoning. Researching several countries in South America, Douine et al. (2020) and Terças-Trettel et al. (2019) identify malaria as one of the main challenges associated with gold mining, whether formal or illegal. Cases of malaria correlated with deforestation in Brazil and Colombia; gold production in Colombia; gold prices in Guyana, and/or the location of the mining region in Peru, Colombia, Venezuela and Guyana (Douine et al. 2020).

Many more authors have raised the challenge of water pollution from the mercury used by illegal gold miners. This was so because many mining activities follow watercourses where gold is washed permitting alluvial mining (Zolnikov 2020; Machacek 2020; Mantey et al. 2020). In fact, Duncun (2020) raises the notion of what he calls a dangerous combination of illegal mining and water pollution after studying the activities in Fena River in the Ashanti Region of Ghana. There is also a challenge with general global mercury emissions as was observed by Prescott et al. (2020) from northern Myanmar. From the Rutsiro district in Rwanda, the main environmental challenges of artisanal alluvial gold mining were observed to include the following:

> Changes in landscape structure, deforestation, intensification of geomorphological processes, new relief shapes (check dams, gravel benches, anthropogenic channels) and hydrological river regime, chemical pollution of soil and watercourses. (Machacek 2020, p. 1)

Similar findings as those from Rwanda were observed by Ali (2019) in Indonesia, Obeng et al. (2019) in Ghana where illegal gold mining is also very high, and Machacek (2019) in the Great Lakes Region in East Africa.

Commissioning the two Bailey bridges that were constructed jointly by the South African National Defence Force and the Zimbabwe National Army at Kopa in Chimanimani as part of Tropical Cyclone Idai recovery, former Vice President Constantino Chiwenga highlighted that the government was concerned by the rampant illegal gold mining that was taking place in Chimanimani (Maodza 2019). However, illegal mining in Chimanimani seems to have a long history, with the Mail and Guardian (2010) reporting that gold panners were allowed in the Chimanimani National Park as the economic situation deteriorated, and they needed to earn a living. This scenario later changed when the government ordered it to stop as it intended giving parts of the Chimanimani National Park and/or Mountains, particularly the gold-rich Skeleton Pass, to the Chinese. From Maodza's (2019) account, this is when the parks officials started beating up the gold panners and confiscating their belongings, thereby driving many into Mozambique. The allegation of government officials and traditional leadership, especially the chieftaincy's involvement with the Chinese in illegal and informal mining, also comes to the fore from Ghana. Boafo et al. (2019: 1) note that Chinese involvement, "creates parallel operations of formal and informal systems that promote different levels of agency and manoeuvring among actors – breeding uncertainty, bureaucratic logjams, and illegalities in the mining industry".

Following up on illegal gold mining in the Chimanimani area of Zimbabwe and also the border areas in Mozambique, Kachena and Spiegel (2019) present a picture of great complexity. The authors highlight that government officials from the Chimanimani Rural District are aware of such complexities and what was driving many to take up the risky occupation. As indicated earlier, since the year 2000, hard economic sanctions were imposed on Zimbabwe by major economies such as the USA and the UK, making livelihoods across the country very difficult. Hence, illegal miners flocked to Chimanimani from all over Zimbabwe leading to a clean-up programme called *Operation Chikorokoza Chapera* (No More Illegal Mining) that took place between 2006 and 2008. However, in the so-called operation, there was a selective approach where the local administration took advantage of cleaning out illegal miners from outside Chimanimani District, but not all illegal miners. This forced the so-called

intruders from other areas of the country to get into Mozambique where the terrain was difficult and the police and game rangers could not follow.

Given the foregoing, this chapter asks the research question: To what extent did Tropical Cyclone Idai attract illegal gold mining in Chimanimani? Linked to the research question, an objective is spelt out as follows: to investigates the extent to which Tropical Cyclone Idai pulled out illegal gold miners and related activities from the Chimanimani area of Zimbabwe.

5.2 Materials and Methods

The main research methods used to address the outlined objective included field observations in some of the areas affected by the Tropical Cyclone Idai gold rush. Further information and data were gathered from both key informant interviews and focus group discussions (Zolnikov 2020; Mkodzongi and Spiegel 2020). Social media, particularly WhatsApp, was also used, especially to conduct follow-ups to check the status in the third quarter of 2020. This method has been used in disaster studies before by Qin et al. (2020). Some elements of citizen science (Carlson and Cohen 2019) were also applied as a lot of data, including pictures, were forwarded from the archives of key informants. Notwithstanding some of the challenges associated with ethics in citizen science, the methodology has gained popularity in environmental science and management research (Quinlivan et al. 2020; Sy et al. 2020). In this research, citizen science permitted the researchers to gain extensive and, at times, exclusive data and information far and wide across the Chimanimani area, including hotspots for illegal gold mining. To address matters of ethics, the data from the community, particularly pictures—some that revealed many identifiable faces—were not used in the write-up. In such instances, only narratives were developed to maintain privacy and anonymity.

As for the study area, Chimanimani, specifically the mountain range, these form part of the border between Zimbabwe and Mozambique in south-central Africa (Kachena and Spiegel 2019). The Chimanimani Mountains (known locally as the Mawenje Mountains, with one mountain peak also known as Mawenje) cover an area of around 1000 km^2, mostly in Mozambique. The mountain range presents "whitish jagged peaks of ancient white sandstone and quartzite interspersed with broad smooth grassy valleys containing small crystal-clear rivers" (The Birdlife 2020 online). From the Zimbabwean side, there is the Chimanimani National Park, which covers about 176 km^2 and on the Mozambique side, there is the Chimanimani National Reserve of about 660 km^2. The two establishments constitute the core of the Chimanimani Trans-Frontier Conservation Area (TFCA) that stretches an estimated 2500 km^2. The establishment of the TFCA was supported by the Southern African Development Community (SADC) in 2001 when the two countries signed a Memorandum of Understanding to amalgamate the two establishments (Kachena and Spiegel 2019).

The Birdlife (2020) confirms that alluvial gold deposits were discovered in the upland grasslands, especially along the large broad Mufomodzi Valley in Mozambique. This led to a rush that attracted up to 10 000 illegal and small-scale miners—the *gariemperos* as they are known locally. However, these illegal miners came from both sides of the Chimanimani Mountains, and they lived in caves and makeshift tents. From the Zimbabwean side, the National Parks authorities constantly harassed the miners leading to most of them operating from the Mozambique side. The Birdlife further revealed that the scale of illegal gold mining was so huge that large stretches of streambanks and river beds were dug up, becoming visible on Google Earth. Hence many endangered tree species were cut down for firewood.

An attempt was also made to identify illegal mining hotspots. Identifying illegal mining hotspots remains a useful tool for policy making, and this has been used elsewhere in Ghana's Western Region (Owusu-Nimo et al. 2018). From their mapping of these hotspots, the authors managed to cite three main illegal mining hotspot districts (out of the 11). These included the Tarkwa Nsuaem where 294 sites

and 3648 individual illegal miners were located; Amenfi East that had 223 sites and 1397 individual illegal miners; and lastly, the Prestea Huni-Valley Districts, which had 156 sites and 1130 illegal miners.

To analyse the data, protocols in dealing with qualitative interviews and focus group discussions that include the identification of codes, categories and themes were followed, including non-verbatim transcriptions. Geographical Information Systems (GIS) were also utilised to identify the illegal gold mining hotspots as observed during the 2019 fieldwork and provided from key informant interviews and focus group discussions. The next section presents and discusses the key findings.

5.3   Presentation and Discussion of Findings

The issues surrounding illegal gold panning and mining in Chimanimani are plenty. From violence, to the contamination of rivers and poisoning of livestock, to the destruction of the forests and the environment, scaring tourists, health and deaths, the list goes on... This section comes in three sub-sections that focus on: (1) describing and confirming the Tropical Cyclone Idai gold rush in Chimanimani, (2) a presentation on the nature of environmental damage and threats to tourism, and (3) matters pertaining to health and violence in illegal gold mining communities.

5.3.1   Tropical Cyclone Idai Gold Rush

Although illegal gold mining activities were occurring prior to Tropical Cyclone Idai, the activities are said to have increased tremendously as the cyclone is said to have opened up many other potential sites that had gold deposits. In fact, from several sources that included two chiefs, a focus group discussion and follow-up WhatsApp communications with key informants, there was some form of a post-Tropical Cyclone Idai gold rush in the area.

To start with, one chief highlighted that the Tropical Cyclone Idai gold rush was triggered by someone who posted on social media what seemed to be a plate with gold from the Chimanimani area. The posting and other communication channels were adjudicated as having attracted illegal miners from across the country, including notorious gangs (syndicates) code named *Mashurugwi*, some 700 km or so from Chimanimani. From one of the chief's interviews, it emerged that just before Cyclone Idai six artisanal miners were trapped and died in the Nyabamba area. The area is also marked by a confluence of the Nyahode and Nyabamba rivers. From the beliefs of illegal gold miners, when incidences of death occur, then those who are still alive are likely to get more gold, the concept of *kuchekeresa* (meaning an act that brings fortune in local Chimanyika language). The bodies of the illegal gold miners were retrieved and taken by their relatives for burial.

The account went further, indicating that soon after Tropical Cyclone Idai, huge stone boulders were moved, trees uprooted and gullies made in the Nyabamba area; all this favoured illegal gold mining there and even across the Chimanimani area. This included Chimanimani Mountains. Many of the heavy stones are usually left in their place as the illegal gold mining takes place since there is no suitable equipment to either blast or drill. Furthermore, Cyclone Idai left some pieces of gold exposed and this triggered the rush. Explaining the situation, the chief said,

> As the community and the other artisanal miners found the pieces of gold, the message spread across the country and so there was a rush from as far as Masvingo, Bulawayo and Harare. People flocked from there rushing to Chimanimani and it was exaggerated that in Nyahode River, Risitu and Chipile rivers there was gold all over. So, people came rushing with the intention that they were going to pick gold, but some few solid gold pieces were picked and when people came, it was not a true reflection of what the media presented.

To elaborate on some of the gold fortunes as a result of Tropical Cyclone Idai, the chief explained further the act of someone who had sold a significant amount of gold.

> There was a 35-year old male who picked gold and he hired a Honda Fit car. For three days he hired that car to be transported between his home and the Biriri beerhall. The first time, he bought everyone in the bar alcohol and beverages. When I was driving by, he stopped me and asked, "Chief, can I buy you a crate of beer?" I asked him to buy me soft drinks instead and he did of which I took them home with me to share with my family.

The behaviour described herein is common with illegal gold miners in Zimbabwe. The belief is that once one gets underground in search of the precious metal, there is no guarantee of coming back. Secondly, there is a belief that there is always tomorrow and more gold can be found. It is a kind of reckless living that many in such communities experience. Some buy cars without valid driver's licences and they wreck them within days. It is chaotic!

The illegal gold mining activities at Nyabamba were confirmed in a focus group discussion that referred to one old man who had stayed in the area for 19 years. Referring to *chikorokoza* (the common term given to activities of illegal gold mining), the respondent indicated that the real work was taking place close to Chimbiya Mountain. Locally, they refer to the area as Bullock, although other names such as Blocky, Roscommon and Brooke are used depending on whom you talk to. They also give it the name *Jonhi* (short for the City of Johannesburg in South Africa). In South Africa, Johannesburg is also known as the City of Gold, given that it sits on massive old and new gold mines. The respondent also distinguished the two main types of illegal gold mining that were taking place there, namely alluvial (around Nyabamba tributary) and underground mining at Bullock. The respondent also indicated that they used mercury to extract the gold. Bullock is an area where a white farmer, Mr Bullock, used to grow tea prior to the land reform programme. The area is located on the way to the Rusitu Mission School.

In follow-up fieldwork in August and September 2020, further investigations were made through WhatsApp texts with certain key informants interviewed in September and

October 2019 from the ground fieldwork in Chimanimani. The contacts included officials from government departments, security officers, community leaders, ordinary citizens and tour guides. The following text was forwarded to the respondents: "Greetings! We are working on a chapter dealing with illegal gold mining in the aftermath of Tropical Cyclone Idai in your area. From your experience, was there any change or gold rush from illegal miners? Were there any incidences of violence in the illegal gold miners areas? May you also provide a list of illegal gold mining hotspots you are aware of". Eight responses were received from the follow-ups. Some of the hotspots identified, including those from the 2019 fieldwork, are presented in Fig. 5.1. The Chimanimani Mountains, Tarka Forests, Nyabamba and Chimanimani Park were regularly mentioned by the key informants as the main hotspots. From an interview with the Chimanimani District Development Coordinator in August 2020, Guvamombe (2020 online)

quotes the administrator highlighting that there was "some serious mining activities in the Chimanimani Mountains" and the office had "since engaged ZimParks, police and other stakeholders so that illegal miners can be chased away from the mountains".

Illegal gold mining in the Tarka Forest was not happening for the first time. In 2017, Mambondiyani (2017) reported that the growing economic hardship had led to thousands of unemployed Zimbabweans invading the timber plantations owned by Allied Timbers Limited. The company had reported up to 600 hectares of prime timber to have been destroyed due to the illegal gold mining. The illegal miners were reported to be selling their gold to illegal buyers that offered very good prices from Mozambique. Buyers from Mozambique offered double the price of gold at US$60 per gram, compared to US$30 per gram on the Zimbabwean side. The economic downturn in Zimbabwe is attributed to both bad governance under the former and late

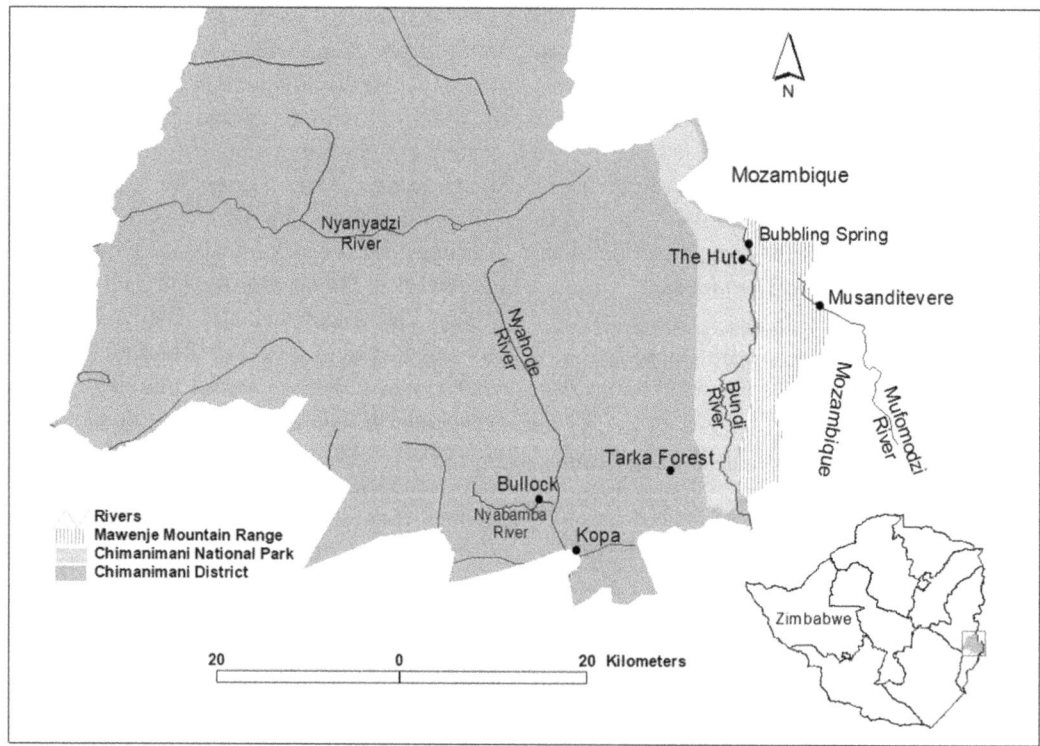

Fig. 5.1 Illegal gold mining hotspots from the Chimanimani area. Source: Authors

President Robert Mugabe and the 2000 fast-tracked land reform programme mentioned earlier. Although there were no official statistics on illegal artisanal gold miners in the country, the United Nations Industrial Development Organisation estimated the figure to be about half a million that were operating between 2007 and 2012 (Mambondiyani 2017). If the outlined economic hardship is anything to go by, this number could have more than doubled by 2020.

Concerning the gold rush post-Tropical Cyclone Idai, one respondent confirmed the rush on 31 August 2020 from a WhatsApp communication. "Yes, there has been a huge rush to come and mine gold out here. … The Cyclone kicked up a lot of minerals and the rush has been crazy, especially these last few months during the lockdown. People have been flocking to Chimanimani". In another long WhatsApp communication on 4 September 2020, one respondent provides a great summary of what transpired, and this is provided in Box 5.1.

What is of interest is the observed additional push to rush for post-Tropical Cyclone Idai mineral riches during the COVID-19 lockdown. This response is coming up 17 months after the

there are scores of locals, especially youth and the active middle aged men out in the various locations digging for gold. There appears to be less violence now than in the initial rushes, which occurred in the 2003–2007/8 period. The mining and law enforcement authorities are now tasked with ensuring that they take back control of the area and institute some order, but due to vested interests and decisions made elsewhere, this process is slow. Meanwhile, there is a lot of gold that is being bought outside of the formal structures and this represents major revenue losses to the fiscus.

cyclone. Since illegal (gold) mining results in massive environmental degradation as discussed in the background, the next section is reserved to present and discuss the findings concerning such from the study area.

5.3.2 Environmental Damage and Threats to Tourism

Chimanimani is a popular and iconic tourist destination hosting the longest mountain range in Africa, with over 90 bird species (Njanike 2020; Chiketo 2020). The Birdlife (2020) confirms a tale of two sets of "gold" in the shared Chimanimani area, namely (1) the real metallic gold, and (2) the botanical gold. However, referring to the unbelievable extent of environmental decay caused by the illegal gold mining in the aftermath of Tropical Cyclone Idai as aggravated by COVID-19, Guvamombe (2020) describes the situation in Chimanimani Mountains and National Park as human beings having lost harmony with nature. The report was filed as recently as 2 September 2020.

The evidence of continued extensive illegal gold mining in the Nyabamba area was caught on camera during the 2019 field visits, 8 months later. Figure 5.2 shows contaminated water from the illegal gold mining areas of Bullock draining

---

Box 5.1 The Pre- and Post-Tropical Cyclone Idai Summary on Waves of Gold Rush

The cyclone-induced floods dislodged a lot of alluvial gold in the belt running along the mountains, streams and rivers in the eastern parts of the Chimanimani District. There were also a lot of fresh areas opened up in and around sites where some panning activities had been going on. As the waters receded and due to the absence of previous restrictions on illegal extraction practices, the artisanal activities resumed and have escalated in scope and scale over the past 12 weeks, especially when the COVID-19 lockdown restrictions were eased to allow formal mining activities to resume. The Rusitu area, the National Park and the rivers and streams have been targeted, and

into the Nyabamba River that joins the Nyahode River. In Fig. 5.2, on the right is the dirty polluted water from illegal gold mining activities draining into Nyabamba River that joins the clear water of the Nyahode River. The broken bridge from Tropical Cyclone Idai was over the Nyabamba River. However, illegal gold mining seems to have grown following the COVID-19 pandemic with many miners anticipating earning a living from the activities as reported by Chiketo (2020) on 21 August 2020. Chiketo describes the movement of different colonising waves of illegal gold miners as a serious threat to tourists, leading to choking rivers with silt and also game poaching taking place in Chimanimani National Park.

From a study in Ghana, Zolnikov (2020) looks at the effects of the government's ban on artisanal and small-scale gold mining on women. The government's response followed a sustained media campaign against severe environmental damage, especially water pollution. The ban is said to have negatively affected all informal gold mining that perpetually used mercury to trap the precious metal. However, the ban also had unintended consequences in that more than one million women lost their income for their families. In fact, prior to engagements in illegal gold mining,

the women were portrayed as having low paying jobs compared to illegal gold mining.

Environmental damage remains one of the major worries from illegal gold mining in Chimanimani. In one of the communications with a tour guide familiar with Chimanimani resorts done on 1 September 2020, the response was desperate.

> I'm not happy about this. I don't think the president knows this. Our area is a key hot spot of the biodiversity. I got a feeling it must be highly protected. This is the Trans-frontier Conservation Area between Zimbabwe and Mozambique. I'm confused. Frogs are dying, deforestation and water pollution is now the order of the day.

The respondent went further to apportion some blame on officials from the Zimbabwe Parks and Wildlife Management Authority (ZimParks) who were thought of as not doing their work to protect the environment in their estates, especially the illegal gold panning activities in the beautiful Chimanimani Mountains. The junior rangers were said to be ineffective in policing the activities. The worries from the respondent were justified because, in one of the 11 photos forwarded to the researchers taken between 9 and 12 August 2020, there were ±100 people in an illegal gold panning pit in the

Fig. 5.2 Polluted water in the Nyabamba River draining from Bullock illegal mining sites. Source: Authors, Fieldwork 2019

Chimanimani Mountains area. Furthermore, some white tourists could also be seen passing with their backpacks in between groups of illegal gold miners. The fact that gold mining syndicates work with government officials has also been observed in Venezuela by Rosales (2020), who further brings the perspective of economic decline and poverty as sustaining such relationships. In addition, insufficient law enforcement to force illegal miners from their sites was also observed by Prescott et al. (2020) in a study in Myanmar. The authors then advised that if the government was to be successful in fighting the activities, there was a need for "either constant presence of enforcement officials at each informal mining site, or confiscating equipment every month" (Prescott et al. 2020, p. 1). Enforcement effectiveness was also noted to be heavily undermined by the corruption that came in the guise of informal payments to local authorities.

In the case of illegal gold mining in Chimanimani National Park, reports by concerned citizens and residents were met by demands of written requests from the ZimParks and the police before they could intervene in the National Park (Chiketo 2020). The report by Chiketo further indicated that some of the pristine tourist attractions within the Chimanimani Mountains such as the Redwall Cave and on the northern end of the Bundi plain had all the trees cut down for firewood. The Digby's Falls was dug up, and the famous Skeleton Pass leading into Mozambique had its highways strewn with litter. Some of the illegal gold mining activities from Chimanimani are shown in Fig. 5.3.

The Chimanimani area hosts several pristine tourism attractions. These include botanical gardens, waterfalls and pools, beautiful scenery (Fig. 5.4), rock paintings and caves, and other traditional sacred sites. Other main attractions could be listed as the Chimanimani Mountains, including Mawenje (Fig. 5.5), Tessa's Pool and the Bridal Veil Falls. The Bridal Veil Falls (Fig. 5.6) is the closest to the Chimanimani Town—just less than 5 or so kilometres. What looks like pathways on mountain and hill slopes in Fig. 5.4 are gullies

Fig. 5.3 Illegal gold mining in the Mawenje Mountain area. Source: Authors, Fieldwork 2019

Fig. 5.4 View of Chimanimani Town from the observatory. Source: Authors, Fieldwork 2019

Fig. 5.5 Parts of the Mawenje Mountain. Source: Authors, Fieldwork 2019

left after landslides from Tropical Cyclone Idai. Following media reports of illegal gold miners attempts to pan around one of the cultural, historical and tourist sites called the Bubbling Spring in the Chimanimani Mountains area, we then followed up with a series of questions to one of the tour guides who has been operating in the area for 26 years for confirmation. Hence, we wrote on 1 September 2020, "We understand gold panners tempered with the Bubbling Spring in search of gold. Is this true and to what extent?" The response was "Yes". The Bubbling Spring (*Mutsero Tseri* in local Chimanyika and Ndau languages) was used by the liberation heroes on their way to get trained and take arms from Mozambique and back into Zimbabwe. The narrative from the villagers has it that only the popping of a big fish out of the dry land being dug saved the shrine (Njanike 2020).

Several rivers and streams were indicated as being polluted from illegal gold mining activities in Chimanimani. The pollution included mercury

Fig. 5.6 Bridal Veil Falls. Source: Authors, Fieldwork 2019

and siltation. Some of the rivers mentioned included Nyahode, Nyabamba, Risitu, Musapa, Bundi, Mwatsara, Mutsangazi, Muchira, Ngaone, Chisengu and Haroni. Most of these rivers ultimately empty into Rusitu, which drains into Mozambique, changing its name to Lucite. This scenario brings other unwanted consequences of transboundary water resources management. Unfortunately, the environmental damage from extensive illegal gold mining activities extend into Mozambique as the Chimanimani Mountains are shared by the two countries. In fact, the *Musanditevera* area (where Binga Mountain is) drains into the Mufomodzi River and is also a popular gold panning area. The area remains like a no man's land, but mainly falls on the Mozambique side. Many illegal gold miners come from both countries, temporarily migrating there for their activities. The area was also heavily damaged by Tropical Cyclone Idai and was said to be in the same condition as the Zimbabwe side.

Gandiwa and Gandiwa (2012) also observe the destructive nature of illegal gold mining in the Chimanimani National Park. The authors identify over a dozen impacts of the illegal gold mining activities on biodiversity, vegetation and water resources of which the nature of damage ranged from medium to very high. Using the data from the records of arrests by the Chimanimani

National Park authorities, it emerges that more than 10,000 illegal gold miners were arrested in 2006 alone. This figure confirms the launch of the clean-up operation, code named *Operation Chikorokoza Chapera*, which was identified by Kachena and Spiegel (2019). The arrest figures for illegal gold miners then went down to 1040 in 2007, about 533 in 2008 and an estimated 324 in 2009 (Gandiwa and Gandiwa 2012).

Of all the rivers, Nyabamba was confirmed by more than three sources on the ground as one that had a lot of fish dying from gold panning. Drawing from the dirty water shown in Fig. 5.2 earlier, this remains a reality that is difficult to contest. Villagers, livestock, vegetation and animals in Chimanimani were also reported to be at risk as the rivers were getting heavily polluted by mercury from illegal gold mining (Njanike 2020; Chisiri 2020). The sustained media campaign to raise awareness of the fast-degrading Chimanimani biodiversity and aesthetics, as well as the illegal gold mining activities, was acknowledged by the Minister of Environment, Climate, Tourism and Hospitality Industry, Mangaliso Ndlovu (Guvamombe 2020) in September 2020. The minister then promised to have the Environmental Management Agency (EMA) and ZimParks work together towards addressing the matter. In addition, the minister also indicated that there will be staff reshuffling, especially the rangers who were alleged to be in cahoots with illegal gold mining syndicates, to allow investigations. However, from the EMA perspective, the escalation in illegal gold mining activities could have been as a result of the COVID-19 lockdown period.

5.3.3 Violence and Health-related Matters

Describing the situation concerning violence, one of the respondents from the August/September 2020 follow-ups revealed that violence was limited because the gold panners were syndicated with the government game rangers. In a 21 August 2020 report by Chiketo (2020), it emerged that indeed, the illegal gold miners were in a syndicate with senior ZimParks officials. This was confirmed by an internal investigation that resulted in the dismissal, and, in some instances, transfer of senior officials. Even some board members are said to have been implicated. However, another respondent confirmed that there have been some reports of violence among the gold panners themselves. The gold panners were also reported to be attacking people who were trying to stop them from mining. Another respondent confirmed the presence of violence in the Mawenje Mountain area and the gold rush post-Tropical Cyclone Idai and wrote, "Violence is rife over the control of mining areas and attempts to remove miners by officials are followed by resistance". Mawenje Mountain is part of the Chimanimani Mountains.

Another area that came up strongly as an illegal gold mining hotspot is one across in Mozambique known by those from Chimanimani as *Musanditevera* ("do not follow me") and *Mufumodzi* from the Mozambican side. The reason this name came about from several on the ground key interviews in 2019 that included a chief, focus group discussion group, tour guide and an official from Outward Bound, is that when illegal gold miners go there, their chances of survival are slim. Many die in the pits with their

bodies never recovered, and others die from malaria and syndicate (territorial) wars. The issue of malaria also came up from many countries in South America from the studies by Douine et al. (2020). Studies done by De Santi et al. (2016) in French Guiana reveal that even the French soldiers sent to control illegal mining activities in two sites ended up contracting malaria. In fact, 72 out of the 272 soldiers stationed there got sick. Malaria is very prevalent because the areas are usually not sprayed to control malaria spread and other diseases. Given that illegal gold miners are so mobile and even crossing borders in the case of Chimanimani, it becomes difficult to control malaria. It also means there should be an arrangement for transborder and regional cooperation, which is also complicated by the fact that the activities are illegal. The Mail & Guardian (2010) reported on 22 January 2020, after interviewing some illegal gold miners in the Masanditeera area, that going so far away from home in Chimanimani into Mozambique was the last decision someone wanted to take. It was indeed a dangerous and risky venture.

From Ghana, Nakua et al. (2019) observe high annual injury rates among illegal gold miners in the four mining districts in four different regions. The injury rate stood at an estimated 289 per 1000 workers. These injuries were mainly from machinery/tools (46.1%) and slip/falls (32.2%). Safety training from the past year remained low at 3%. Drawing from the research they did in Chimanimani, Kachena and Spiegel (2019) point out the use of multiple goldfields as a form of avoiding conflict. However, the authors also report incidences of robbery that result in many illegal miners preferring establishments and goldfields in mountains where it is difficult to reach. The local youth mafias called *gombiros* were said to operate in the Rusitu Valley, occasionally robbing the miners of the extracted gold. This has led to illegal miners retreating further into the Chimanimani National Park where such raids are less frequent. This could have been due to the fact that the Chimanimani National Park is a protected and patrolled area by rangers.

5.4 Conclusions and Recommendations

The chapter reveals that it is clear there is a long history of illegal gold mining in Zimbabwe, and in the Chimanimani area specifically. As early as 2000, illegal gold mining and multiple gold rushes emerged as livelihoods changed following the fast-track land reform programme. As people moved into exclusively private and former white-owned farm areas, mines and mineral deposits—especially gold—were found. Even the government did not have the capacity to manage the illegal mining and the gold rush. As for the Chimanimani area, the trends were similar, although the event of Tropical Cyclone Idai acted as a huge pull factor. Narratives from the key informants and focus groups, as well as on the ground observations, confirmed one great rush into the area in the aftermath of the cyclone. Many hotspots emerged, including new ones (although old spots were revamped). Tropical Cyclone Idai is said to have rolled over huge boulders, swept away some gold and deposited it into several low-lying areas as the flood water was flowing. Among the main hotspots identified were the Chimanimani Mountains, Chimanimani National Park (and the Mozambique Forest Reserve on the other side of the border), the Nyabamba area and river, Tarka Forest, Bullock and Bundi. Allegations were also made that the massive illegal gold panning was as a result of syndicates involving officials from the ZimParks. This was indeed true as some senior officials and board members ended up being dismissed.

However, what was beyond a reasonable doubt is the fact that there is massive damage to the environment and great risk to tourism. Apart from many major rivers becoming polluted from illegal gold panning, there were specific tourist attractions that were severely degraded. Among such attractions are the Redwall Cave, the northern end of the Bundi plain that had all the trees cut down for firewood, the Digby's Falls that was dug up, the Bubbling Spring that was partially dug up, and the famous Skeleton Pass leading into Mozambique that had its highways strewn with litter. Thousands of illegal miners, sometimes seen in their hundreds in large mining pits, remain a real threat to visitors. Aquatic life was also reported to be dying due to alleged mercury and other poisonings.

Based on the key findings and conclusions emerging herein, the chapter recommends some form of carrot and stick approach to the problem. Although not in many rule books, there is a need for the government to use non-governmental organisations to educate the miners on the dangers of environmental damage, especially the use of mercury, which is a heavy metal and an ecological accumulator. Furthermore, the government may need to allow the illegal miners to still sell their gold through formal channels to eliminate exploitation and/or the trafficking of the precious metal to Mozambique where the gold is bought with the US dollar. This route will also have the government acknowledging the hardship the people are facing due to the economic decline. The responsible authorities may also need to map out exclusive no-go areas for illegal gold mining, while trying to regulate operations already in place. Future plans addressing the gold rush in the aftermath of a natural disaster such as Tropical Cyclone Idai need to be put in place immediately, detailing how the movement of persons into the affected areas can be controlled. However, given that Chimanimani is a border area, the challenge of a failed region in terms of controlling illegal gold mining remains a reality.

References

Ali, M.I. (2019). The consequences of illegal mining in the environment: Perspectives behaviour, knowledge and attitude. International Journal of Environment, Engineering and Education, 1(1), 25-33. https://doi.org/10.5281/zenodo.2633654.

Boafo, J., Paalo, S.A., Dotsey, S. (2019). Illicit Chinese Small-Scale Mining in Ghana: Beyond Institutional Weakness? Sustainability, 11, 5943; https://doi.org/10.3390/su11215943.

Carlson, T. & Cohen, A. (2019). Linking community-based monitoring to water policy: Perceptions of citizen scientists. Journal of Environmental Management, 219, 168e177. https://doi.org/10.1016/j.jenvman.2018.04.077.

Chiketo, B. (2020). Gold panners overrun Chimanimani mountains. Retrieved from https://zimmorningpost. com/gold-panners-overrun-chimanimani-mountains/ (Accessed 31 August 2020).

Chisiri, T. (2020). Small scale miners destroy Chimanimani heritage during Covid-19 lockdown. Retrieved from https://sportwaynewsnet.word-press.com/2020/08/28/small-scale-miners-destroy-chimanimani-heritage-during-covid-19-lockdown/ (Accessed 31 August 2020).

De Santi, V.P., Girod, R., Mura, M., Dia, A., Briolant, S., Djossou, F., Dusfour, I., Mendibil, A., Simon, F., Deparis, X., Pagès, F. (2016). Epidemiological and entomological studies of a malaria outbreak among French armed forces deployed at illegal gold mining sites reveal new aspects of the disease's transmission in French Guiana. Malaria Journal, 15, 35. https://doi.org/10.1186/s12936-016-1088-x.

Douine, M., Lambert, Y., Musset, L., Hiwat, H., Blume, L.R., Marchesini, P., Moresco, G.G., Cox, H., Sanchez, J.F., Villegas, L., de Santi, V.P., Sanna, A., Vreden, S., Suarez-Mutis, M. (2020). Malaria in Gold Miners in the Guianas and the Amazon: Current Knowledge and Challenges. Current Tropical Medicine Reports, 7, 37–47. https://doi.org/10.1007/s40475-020-00202-5.

Duncan, A.E. (2020). The Dangerous Couple: Illegal Mining and Water Pollution: A Case Study in Fena River in the Ashanti Region of Ghana. Journal of Chemistry, https://doi.org/10.1155/2020/2378560.

Forkuor, G., Ullmann, T., Griesbeck, M. (2020). Mapping and Monitoring Small-Scale Mining Activities in Ghana using Sentinel-1 Time Series (2015–2019). Remote Sensing, 12, 911; https://doi.org/10.3390/rs12060911.

Gandiwa, E. & Gandiwa, P. (2012). Biodiversity Conservation versus Artisanal Gold Mining: A Case Study of Chimanimani National Park, Zimbabwe. Journal of Sustainable Development in Africa, 14(6), 29-37.

Guvamombe, T. (2020). Illegal mining, land degradation and dwindling biodiversity; the case of Chimanimani Mountains. Retrieved from https://spotlight.co.zw/illegal-mining-land-degradation-and-dwindling-biodiversity-the-case-of-chimanimani-mountains/ (Accessed 2 September 2020).

Kachena, L. & Spiegel, S.J. (2019). Borderland migration, mining and transfrontier conservation: questions of belonging along the Zimbabwe-Mozambique border. GeoJournal, 84, 1021–1034. https://doi.org/10.1007/s10708-018-9905-0.

Machacek, J. (2020). Alluvial Artisanal and Small-Scale Mining in A River Stream: Rutsiro Case Study (Rwanda). Forests, 11, 762; https://doi.org/10.3390/f11070762.

Machacek, J. (2019). Typology of Environmental Impacts of Artisanal and Small-Scale Mining in African Great Lakes Region. Sustainability, 11, 3027; https://doi.org/10.3390/su11113027.

Mail & Guardian. (2010). Panning beyond the pale. Retrieved from https://mg.co.za/article/2010-01-22-panning-beyond-the-pale/ (Accessed 31 August 2020).

Mambondiyani, A. (2017). Gold rush fever among poor Zimbabweans leaves trail of destruction. Retrieved from https://www.reuters.com/article/us-zimbabwe-mining-landrights/gold-rush-fever-among-poor-zimbabweans-leaves-trail-of-destruction-idUSKBN17J1CJ (Accessed 31 August 2020).

Mantey, J., Nyarko, K.B., Owusu-Nimo, F., Awua, K.A., Bempha, C.K., Awankwah, R.K., Akatu, W.E., Appiah-Effah, E. (2020). Mercury contamination of soil and water media from different illegal artisanal small-scale gold mining operations (galamsey). Heliyon 6, e04312, https://doi.org/10.1016/j.heliyon.2020.e04312.

Maodza, T. (2019). Stop Illegal Mining, Says VP Chiwenga. Retrieved from https://allafrica.com/stories/201912110196.html 9Accessed 1 September 2020).

Mhlongo, S.E., Amponsah-Dacosta, F., Muzerengi, C., Gitari, W.M., Momoh, A. (2019). The impact of artisanal mining on rehabilitation efforts of abandoned mine shafts in Sutherland goldfield, South Africa. Jàmbá: Journal of Disaster Risk Studies, 11(2), a688. https://doi.org/10.4102/jamba.v11i2.688.

Mkodzongi, G. & Spiegel, S.J. (2020). Mobility, temporary migration and changing livelihoods in Zimbabwe's artisanal mining sector. The Extractive Industries and Society, 7: 994-1001. https://doi.org/10.1016/j.exis.2020.05.001.

Musingwini, C. & Sibanda, P. (1999). The riverbed mining project: A step towards sustainable development in mining in Zimbabwe. Paper presented at the Sixth Symposium of the Research Council of Zimbabwe, Harare, 3–5 August 1999.

Nakua, E.K., Owusu-Dabo, E., Newton, S., Koranteng, A., Otupiri, E., Donkor, P., Mock, C. (2019). Injury rate and risk factors among small-scale gold miners in Ghana. BMC Public Health, 19:1368. https://doi.org/10.1186/s12889-019-7560-0.

Njanike, N. (2020). Villagers' Lives At Risk As Gold Panners Pollute Chimanimani Rivers. Retrieved from https://www.newzimbabwe.com/villagers-lives-at-risk-as-gold-panners-pollute-chimanimani-rivers/ (Accessed 1 September 2020).

Obeng, A.A., Oduro, K.A., Obiri, B.D., Abukari, H., Guuroh, R.T., Djagbletey, G.D., Appiah-Korang, J., Appiah, M. (2019). Impact of illegal mining activities on forest ecosystem services: local communities' attitudes and willingness to participate in restoration activities in Ghana. Heliyon, 5, e02617, https://doi.org/10.1016/j.heliyon.2019.e02617.

Oladipo, O.O., Akanbi, O.B., Ekong, P.S., Uchendu, C., Ajani, O. (2020). Lead Toxicoses in Free-Range Chickens in Artisanal Gold-Mining Communities,

Zamfara, Nigeria. Journal of Health and Pollution, 10(26), 1-7.

Owusu-Nimo, F., Mantey, J., Nyarko, K.B., Appiah-Effah, E., Aubynn, A. (2018). Spatial distribution patterns of illegal artisanal small scale gold mining (Galamsey) operations in Ghana: A focus on the Western Region. Heliyon, 4, e00534. https://doi.org/10.1016/j.heliyon.2018.e00534.

Prescott, G.W., Maung, A.C., Aung, Z., Carrasco, L.R., De Alban, J.D.T., Diment, A.N., Ko, A.K., Rao, M., Schmidt-Vogt, D., Soe, Y.M., Webb, E.L. (2020). Gold, farms, and forests: Enforcement and alternative livelihoods are unlikely to disincentivize informal gold mining. Conservation Science and Practice, 2:e142. https://doi.org/10.1111/csp2.142.

Qin, L., Sun, Q., Wang, Y., Wu, K.F., Chen, M., Shia, B.C., Wu, S.Y.(2020). Prediction of Number of Cases of 2019 Novel Coronavirus (COVID-19) Using Social Media Search Index. Int. J. Environ. Res. Public Health, 17(7), 2365; https://doi.org/10.3390/ijerph17072365.

Quinlivan, L., Chapman, D.V., Sullivan, T. (2020). Validating citizen science monitoring of ambient water quality for the United Nations sustainable development goals. Science of the Total Environment, 699, 134255. https://doi.org/10.1016/j.scitotenv.2019.134255.

Rosales, A. (2020). Statization and denationalization dynamics in Venezuela's artisanal and small scale-large-scale mining interface. Resources Policy, 63: 101422. https://doi.org/10.1016/j.resourpol.2019.101422.

Sy, B., Frischknecht, C., Dao, H., Consuegra, D., Giuliani, G. (2020). Reconstituting past flood events: the contribution of citizen science. Hydrol. Earth Syst. Sci. Discuss., https://doi.org/10.5194/hess-2019-188.

Terças-Trettel, A.C.P., de Oliveira, E.C., Fontes, C.J.F., de Melo, A.V.G., de Oliveira, R.C., Guterres, A., Fernandes, J., da Silva, R.G., Atanaka, M., Espinosa, M.M., de Lemos, E.R.S. (2019). Malaria and Hantavirus Pulmonary Syndrome in Gold Mining in the Amazon Region, Brazil. International Journal of Environmental Research and Public Health, 16, 1852; https://doi.org/10.3390/ijerph16101852.

The Birdlife. (2020). Mountain Gold – Conservation in the Chimanimani Mountains, Mozambique. Retrieved from https://www.birdlife.org/worldwide/mountain-gold-%E2%80%92-conservation-chimanimani-mountains-mozambique (Accessed 31 August 2020).

Zolnikov, T.R. (2020). Effects of the government's ban in Ghana on women in artisanal and small-scale gold mining. Resources Policy, 65: 101561. https://doi.org/10.1016/j.resourpol.2019.101561.

Energy Infrastructure and the Building Back Better Concept: Lessons from Tropical Cyclone Idai

6

Abstract

The impact of tropical cyclones and hurricanes on critical infrastructure that includes energy is at times equated to that of weapons of mass destruction. From the electricity side, the literature indicates that both wooden and steel transmission poles go down due to severe winds and weakened ground from the cyclones, raising matters pertaining to their design and at times their age. This research focuses on assessing the severity of the destruction of energy infrastructure from Tropical Cyclone Idai in Chimanimani, Zimbabwe. The cyclone hit Zimbabwe on 15 March 2019 as a category 3 cyclone. It emerges that there were severe electricity outages that plunged Chimanimani into total darkness, with power only partially restored after a month. Destroyed powerlines further resulted in negative knock-on effects that included the disruption of public utility services such as water supply, telecommunications and businesses, all of which had a negative effect on livelihoods. Three private mini hydropower stations were extensively damaged, one washed away completely. Liquid fuel distribution was also disrupted as the roads network was destroyed which rendered the Chimanimani area inaccessible for more than a week. During the record period of the restoration of power, the building back better (BBB) concept was applied, resulting in the installation of thicker transmission lines and stronger transmission poles. The work recommends that new energy infrastructure should be installed that is resilient to tropical cyclones and other extreme weather events.

Keywords

Cyclones · Southern Africa · Energy infrastructure · Cyclone Idai · Chimanimani

6.1 Introduction and Background

The 2030 Agenda for Sustainable Development (AfSD) is unequivocal about the desire to have economies driven by sustainable energy. From the 17 Sustainable Development Goals (SDGs) embedded in the 2030 AfSD, SDG 7 looks at ensuring "access to affordable, reliable, sustainable and modern energy for all" (United Nations, 2015). SDG 7 goes further to outline several targets, including ensuring universal access to affordable, reliable and modern energy services, increasing the share of renewable energy in the energy mix and doubling energy efficiency. One of the fundamental measures for the attainment of SDG 7 therefore is the proportion of the population that has access to electricity (IAEG-SDGs

2018). Hence, disasters such as tropical cyclones will certainly retard progress towards universal access to electricity. Linked to energy production is the climate action SDG—SDG 13 that also focuses on promoting resilience and adaption of energy-related infrastructure. The climate action SDG further demands the reduction of greenhouse gas emissions that lead to global warming and cause climate change. Such a pathway then promotes the transitioning of economies and societies towards greener and low carbon energy sources such as solar, hydropower and natural gas (Nhamo 2013) putting all economies on pathways moving away from fossil-based fuels.

Tropical cyclones (in other regions known as hurricanes or typhoons) are recurring and causing extensive damage to infrastructure (Mohanty et al. 2020; Frame et al. 2020). Such disturbances also delay households and community's recovery as was observed 8 months after Hurricane Irma (Mitsova et al. 2019). Disruptions in electricity and water supply were noted as major drivers of a loss of well-being during Hurricane Wilma in 2005 (Chatterjee and Mozumder 2015). The hurricanes further result in the loss of lives. Hurricane Mathew (AON 2017), reported as the most intense October category 4 hurricane on record in the Atlantic, had three landfalls in the Caribbean and the USA that resulted in the loss of lives. At least 604 people died, with the official death toll in Haiti put at 546, although the unofficial figures estimate over 1300 deaths in the country. More than 200,000 houses were destroyed in Haiti.

Hurricanes are also known to leave trails of destruction running into billions of United States dollars. Hurricane Harvey's damage, for example, was estimated at US$90 billion (Frame et al. 2020), while Hurricane Odile's damage was approximated at US$1.654 billion (Murià-Vila et al. 2018). Hurricane Matthew's (AON 2017) bills added up to US$15 billion across the impacted countries. In Haiti, the estimated economic loss was US$2.8 billion, the Bahamas (US$600 million to 1.0 billion), Canada (US$150 million) and Cuba (at least US$100 million). In 2005, Hurricane Wilma was estimated to have resulted in damages of up to US$20.6 billion

(2005 rate) in Mexico, Cuba and Florida State in the USA (Chatterjee and Mozumder 2015).

The energy infrastructure is always hit after hurricanes and cyclones. This happened during Hurricane Rita in 2005 (Reed et al. 2010). Other hurricanes whose damage to energy infrastructure is well documented in the USA history by the U.S. Department of Energy (2009) include Katrina (2005), Rita (2005), Gustav (2008) and Ike (2008). Electricity power systems usually consist of three subsectors of generation, transmission and distribution (Reed et al. 2010). The last two subsectors are usually badly damaged during hurricanes. Figure 6.1 presents the impact of selected hurricanes that made landfall in the USA between 2005 and 2012. Overall, the damage by Hurricane Katrina surpassed the others.

Hurricane Maria's major known damage was to Puerto Rico's electric grid (Kishore et al. 2018). The electricity power outage initially affected the entire island and lasted more than 10 months. From similar instances in the past, governments and other stakeholders are forced to quickly repair the damage by mobilising huge numbers of workers, specialised vehicles and emergency mobile plants to generate electricity for critical public utilities such as hospitals, emergency shelters, water treatment plants and schools (Keellings and Hernández-Ayala 2019). The impact of Hurricane Matthew on the electricity grids of the Caribbean Islands was extensive, although temporary (AON 2017). In St. Lucia, up to 70% of the population (120,000 people) were left in the dark and the Cuban government reported about 50,000 power outages. In the USA, an estimated 3.15 million customers lost their electricity supply. These were in Florida, Georgia, South Carolina, North Carolina and Virginia. The breakdown of customers that lost power in the most affected state of Florida is presented in Table 6.1. In 2005, Hurricane Wilma left 4.3 million customers in the dark (Chatterjee and Mozumder 2015). Other hurricanes that caused extensive damage to electricity distribution and transmission infrastructure in the USA and other countries are Irene (August 2011) and Sandy (October 2012) (U.S. Department of Energy 2013). Hurricane Irene left 6.69 million customers in South Carolina and Mine with no

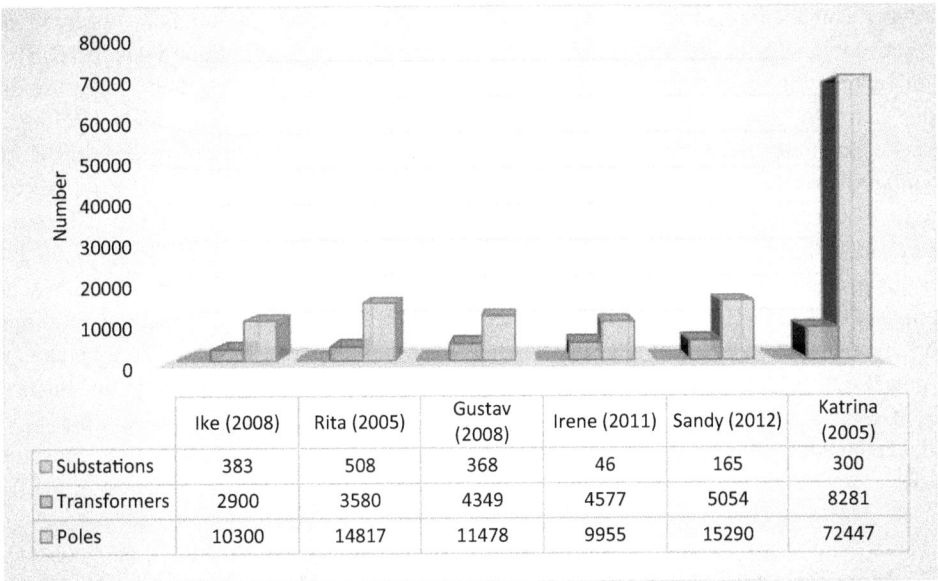

	Ike (2008)	Rita (2005)	Gustav (2008)	Irene (2011)	Sandy (2012)	Katrina (2005)
▢ Substations	383	508	368	46	165	300
▢ Transformers	2900	3580	4349	4577	5054	8281
▢ Poles	10300	14817	11478	9955	15290	72447

Fig. 6.1 Electricity infrastructure damage during selected hurricanes (2005–2012). Source: Authors, Data from U.S. Department of Energy (2009; p. 8, 2013, pp. 9–10)

Table 6.1 Distribution of power outages in Florida due to Hurricane Matthew

County	Power outages (Customers)	Power outages (% of Total)
Volusia	255,048	90
Duval	215,748	52
Brevard	159,298	52
St. Johns	78,610	90
Seminole	64,606	31
Flagler	58,536	99
Orange	50,877	9
St. Lucia	41,149	27
Indian River	41,000	47
Clay	28,214	20

Source: AON (2017, p. 26)

power, while Sandy left 8.66 million customers in the dark in North Carolina, Maine, Illinois and Wisconsin.

Tropical Cyclone Idai ravaged Malawi, Madagascar, Mozambique and Zimbabwe in March 2019. The cyclone left more than 1000 people dead and many more injured, homeless and affected (UNECA 2019; Yu et al. 2019). In Zimbabwe, Tropical Cyclone Idai arrived on 15 March 2019, badly affecting the Chimanimani and Chipinge districts, among other places, and left 344 people dead (UNECA 2019). In Chimanimani, both Ngangu and Kopa townships were the most severely affected, with parts of the townships destroyed leaving no sign that there were once houses, shops and other related infrastructure (MPDCO 2019). At the time of finalising this chapter, many more people from Kopa Township were still unaccounted for, with some bodies having been swept into Mozambique and buried by authorities there. Most families were still in temporary tents awaiting resettlement, an aspect that was compounded by the complexities surrounding the COVID-19 pandemic in terms of logistics under lockdowns and the required financial resources that were needed to address the pandemic. As a category 3 tropical cyclone, the heavy rains and storms damaged infrastructure, including energy supplies (ZMSD 2019). Generally, both renewable and non-renewable energy sources and infrastructure remain vulnerable to tropical cyclones. Of critical concern is the extensive damage that may be caused to electricity supply systems that include dams, coal plants, solar and wind farms, oil fields and refineries, fuel service stations, gas pipes, electricity transmission and distribution lines (Frame et al. 2020).

Drawing from the foregoing, this research sets the objective to assess the severity of the destruction of energy infrastructure from Tropical Cyclone Idai in Chimanimani, Zimbabwe. The focus is on electricity, solar and liquid fuel in the form of petrol and diesel. The study further considers the extent of damage to supporting infrastructure including roads and bridges. What makes this research more valuable is the fact that the damage Tropical Cyclone Idai caused reached its peak in the middle of the night amidst the large power outages, especially in Chimanimani. The next section is dedicated to presenting the material and methods used in data generation and analysis.

6.2   Materials and Methods

The main methods used for generating and analysing data included a household survey, interviews and focus group discussions, document analysis, on the ground field observations and Geographical Information Systems (GIS). These methods have been used in similar studies by Kishore et al. (2018) and Keellings and Hernández-Ayala (2019) in studying Hurricane Maria, Mitsova et al. (2019) in researching Hurricane Irma, Hurrican Wilma (Chatterjee and Mozumder 2015) and also Murià-Vila et al. (2018) in investigating hurricane Odile's damage to the infrastructure of Baja California Sur, Mexico. The location of the study area is shown in Fig. 6.2. Among the key documents with authentic information were the monthly Manicaland Provincial Development Coordinator's progress reports on Tropical Cyclone Idai response and recovery. On the ground observations captured on camera showed destroyed infrastructure that included mini hydropower plants, transmission poles, cables, transformers and other artefacts of interest. Key informant interviews were granted by traditional leaders that included chiefs, officials from the

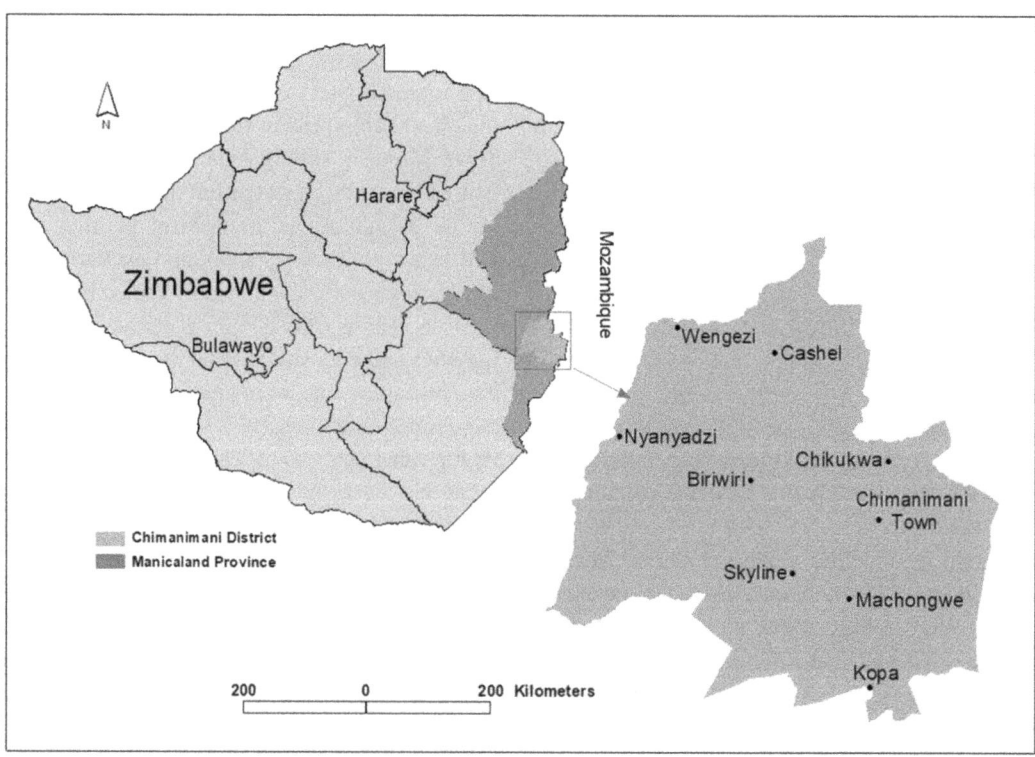

Fig. 6.2 Location of the study area. Source: Authors

Zimbabwe Electricity Transmission and Distribution Company (ZETDC) and telecommunications companies.

The survey also had questions asking household respondents to evaluate the nature of damage to roads, bridges and energy infrastructure (power lines, poles, transformers, etc.). Five possible response options were given which included (1) complete damage (no longer usable), (2) partial damage, (3) no damage, (4) I do not know and (5) not applicable. The analysis of such data was done in both QuestionPro and MS Excel with descriptive statistics used and graphs plotted. As for interview and focus group discussion data, these were transcribed and analysed using elements of grounded theory, document and critical discourse analysis as advised by Rieger (2018) and Peiter et al. (2020).

6.3 Presentation and Discussion of Findings

This section consists of two main subsections. These include a focus on national grid electricity infrastructure, as well as impacts on liquid fossil fuels (including petrol and diesel) and solar energy. The findings will be buttressed by concurrent discussions based on the literature.

6.3.1 Impacts on, and Building Back Better Grid Electricity Infrastructure

Although the main focus of the work was on Chimanimani, data on the overall damage to the electricity grid and related infrastructure was available for the whole of the affected districts. This information is summarised in Table 6.2 based on the Manicaland Provincial Development Coordinator's Office—MPDCO report of 6 December 2019 (MPDCO 2019). Overall, there was approximately 50 km of powerlines to replace, 106 poles to replace and 26.5 km of powerlines to repair. Similar damage occurred in Cuba during Hurricane Matthew in 2017 when the power authorities reported 2500 poles, 300 transformers and 884 km of wires damaged (AON 2017). In the USA, Florida Power and Light Company, the biggest energy company had to repair damages to 400 km of wire, more than 900 transformers and more than 400 poles as a result of Hurricane Matthew. In addition, the Duke Energy Company in Carolina had to repair the damage to more than 115 substations, 800 transformers, 58 transmission towers, about 2000 poles and several kilometres of power lines. Lastly, Georgia Power had to repair 1000 poles and 130 km of wire. Drawing from recurring cyclones in the Indian State of Odisha, Mohanty et al. (2020) identify the energy sector as one that needs to continue strengthening its resilience to cyclones. The power sector needs to be immediately restored after any disaster as all other critical services including search and rescue, water and sanitation, telecommunications, healthcare and businesses depend on the availability of power. In addition, many activities in the disaster recovery stage also require power. Following Hurricanes Irene (2011) and Sandy (2012), power outages which affected 95% of the millions of customers as highlighted earlier were restored in 5 and 10 days, respectively (U.S. Department of Energy 2013).

Power was cut off after transmission poles fell, cut lines and were sometimes washed away. The greater part of Chimanimani and Chipinge districts were without power for about 3 weeks. Asked if the Tropical Cyclone Idai caused a blackout in Chimanimani, 95.68% of the household surveyed ($n = 185$) indicated it did. Because Chimanimani is a forest and commercial plantation area, battles with farmers that resulted in mature and tall plantation trees grown very close to powerlines meant the heavy storm and winds from Cyclone Idai uprooted and broke many trees, which then fell over powerlines breaking them. This account came from one of the interviewees, who is a traditional chief in the area and was confirmed by the officials from the ZETDC. The wires also broke as these were thin and not strong enough to resist the forces of falling trees. Heavy rains that flooded parts of Ngangu and Kopa Townships in Chimanimani swept away several transformers and substations. The

Table 6.2 Summary of Cyclone Idai damage to the electricity grid

District/Region	Section	Nature of the damage and required construction work
Chimanimani	Charter to Chisengu	Construction of 2 km of 33 kV overhead line
	Chisengu turn-off to Machongwe	Construction 3 km of 33 kV overhead line
	Machongwe turn-off to S177	Construction 5 km of 33 kV overhead line
	Charter to Chimanimani village	Construction 5 km of 11 Kv overhead line
	From switch U2S8 to Wanganella S1	Construction 3.5 km of 33 kV overhead line
	Chipinge to Makondo	Maintenance on 29.5 km line section
	Ngangu Residential	Replacement of 20 poles on the Medium Voltage lines. Reconstruction of 2 × 100 kVA 11/0.4 kV substations, 1 × 25 kVA 11/0.4 kV substation, and 1 × 200 KVA 11/0.4 kV substation (total of 4 substations)
Chipinge	South feeder	Construction of 1 × 50 KVA 11/0.4 kV and the replacement of 8 poles on the 11 kV overhead line
	North feeder	Replacement of 7 poles on the 11 kV overhead line
	Bamboo Creek	Reconstruction of 2 km of 33 kV overhead line
	Dimire to Mt Selinda	Line section maintenance and supplies restoration
	New Year's Gift	Replacement of 2 poles on the 11 kV overhead line
	Jersey feeder	Replacement of 3 poles on the 11 kV overhead line
	Davora 11 kV feeder	Replacement of 2 poles on the 11 kV overhead line
	Devon to Mt Selinda	20 km of line construction
	Chipinge 33 kv line	Replacement of 5 poles on the 11 kV overhead line

(continued)

Table 6.2 (continued)

District/Region	Section	Nature of the damage and required construction work
Middle Sabi	Masapi feeder	Replace 11 poles on the 11 kV overhead line
	Buhera	Replace 19 poles on 33 kV overhead line
	Nyanyadzi—Mbuya Nehanda	Reconstruct 3.3 km of 33 kV overhead line
	Humani	Replace 9 poles on 33 kV overhead line
Rusape	Nyazura	Replace H pole in Nyazura 11 kV Replace 2 LT poles in Nyazura
	Vengere G section	Replace 10 poles on Medium Voltage overhead lines (1 km)
	Dorowa 1	Replace 2 poles from D1 turn-off to Mezoitne
	Chimbi 11 kV feeder (Moyomakaza)	Repair broken conductors and restrung 0.5 km of 11 kV overhead line
	Rukweza 11 kV feeder	Replace 2 poles on overhead line
	Masvosva clinic, Munemo, chief Chiduku	Replace 5 poles on overhead line
Nyanga	Nyamaropa 33 kV line	Repair 1 km of 33 kV overhead line
	Rodel 11 kV line	Repair 0.5 km of 11 kV overhead line
	Troutbeck 11 kV line	Repair 1.2 km of 11 kV overhead line
	Bonda 11 kV line	Repair 0.5 km of 11 kV overhead line
	Honde 33 kV line	Repair 0.5 km of 33 kV overhead line

Source: Authors, based on MPDCO (2019)

contribution of topography to cyclones damage is also confirmed from Mexico where Hurricane Odile destroyed electricity infrastructure in Baja California Sur (Murià-Vila et al. 2018). The damage to both transmission and distribution lines was extensive. The wooden and steel transmission poles were destroyed, reflecting their limited resistance to the strong winds of the hurricane. The authors conclude that the lack of resistance could have been due to underestimated wind speed values at design and fatigue from repeated past strong wind events and age. Falling trees also contributed to the damage to electricity infrastructure, especially wires and poles during Hurricane Matthew in Georgia. An estimated 1800 trees fell over (AON 2017).

Although the Zimbabwe Meteorological Services Department—ZMSD (2019) predicted rainfall of below 100 mm to be received in 24 h on the eve of Tropical Cyclone Idai, information from one of the citrus exporting farms in Chimanimani revealed that the rainfall received in 48 h was more than 1950 mm. Such abnormal amounts of rainfall are not unusual as Hurricane Maria deposited record-breaking rainfall over Puerto Rico resulting in unprecedented floods and landslides when it made landfall on 20 September 2017 (Keellings and Hernández-Ayala 2019), a category 4 hurricane. During field observations, it became clear that the damage from Cyclone Idai on the electricity grid was massive relative to the total infrastructure and the small economic muscle of the country. Some of the artefacts from the field are shown in Figs. 6.3 and 6.4. From one of the key informants (who had lived in Chimanimani for his entire life which

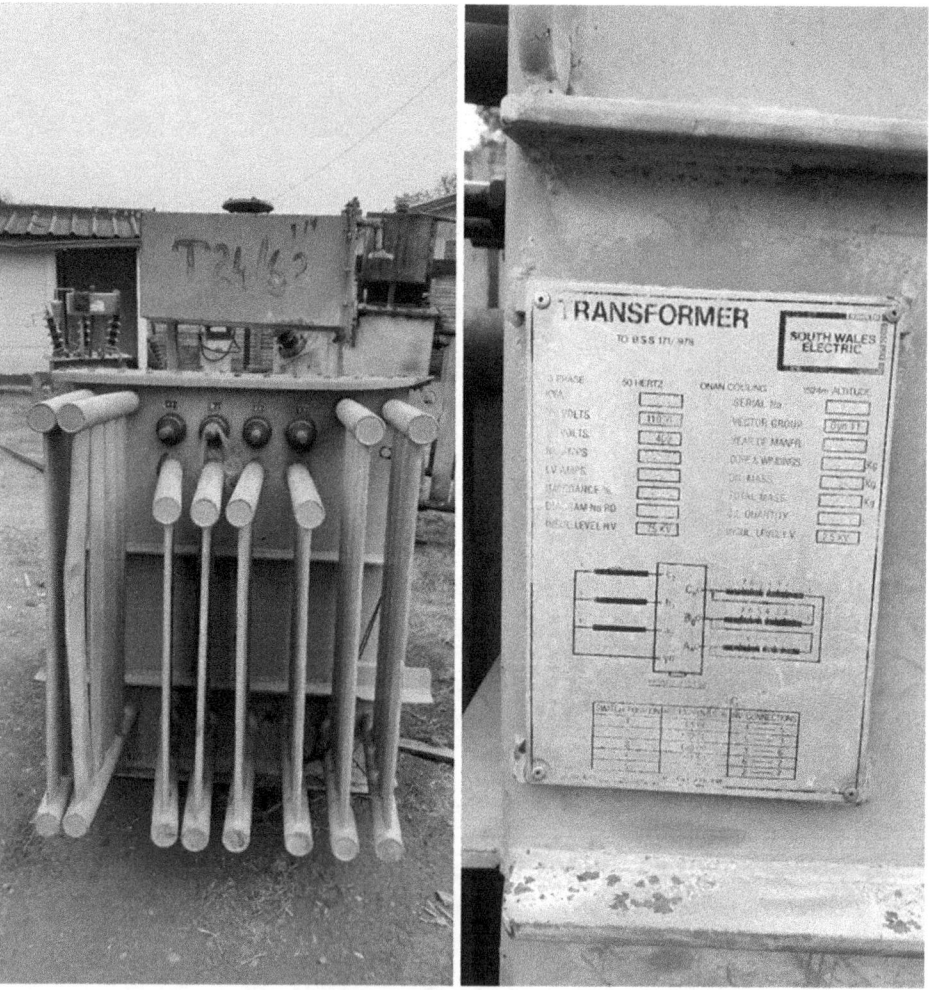

Fig. 6.3 Some damaged transformers. Source: Authors, Fieldwork 2019

Fig. 6.4 Some damaged electricity wires. Source: Authors, Fieldwork 2019

spanned more than 60 years), the electricity went off on Thursday, 14 March 2019, just before the downpours. From their household rain gauge, they got about 25 mm an hour on the night, and at some point, on Friday, 15 March 2019, they recorded 70 mm in 40 min.

From one of the interviews with an official from the ZETDC in Chimanimani office, the first fault happened on 14 March 2019 in an area called Rose Common. The technicians indicated that they spent the entire day trying to fix the fault, but they were eventually overwhelmed as many trees started falling over transmission lines and breaking them. The reports on the faults on the transmission lines kept coming in one after the other. The following day, the staff tried to find a chainsaw contractor, but the conditions would not allow it as roads were impassable and some were completely blocked. The heavy storm affected almost 75% of the power lines in Ngangu Township. As indicated earlier, both Ngangu and Kopa were the locations in Chimanimani that were severely affected. Figures 6.5 and 6.6 show some of the damage from the two areas. Many poles were swept away, with many more damaged. The same happened to the wires for both the 11 kv and 3 kv. The respondent also compared Cyclone Idai's damage to Cyclone Eline that took place in 2000

and concluded that it was not comparable. Since his stay in Chimanimani for 15 years, he had never seen such a magnitude of damage to the electricity infrastructure.

From the survey, a question requested households to indicate sources of their electricity. The responses ($n = 159$) revealed that 45.91% of the households used power from the national electricity grid, 44.65% were not connected to the grid, while 9.43% were connected to another source of electricity. From the household survey (Fig. 6.7), it emerged that there was extensive damage to energy and related infrastructure in Chimanimani. Up to 97.45% of responding households ($n = 157$) concurred that Tropical Cyclone Idai either completely or partially damaged roads. The trend was the same for bridges and energy infrastructure that included power lines, poles and transformers. An estimated 98.04% of households ($n = 153$) indicated that there was either complete or partial destruction of bridges, while 94.16% of the households ($n = 154$) were of the view that Tropical Cyclone Idai either totally or partially destroyed energy infrastructure. The findings confirm in the field, official reports and interview narratives. The damage to the roads meant that electricity restoration equipment and poles could not be transported immediately.

Fig. 6.5 Extent of damage, with heaps of stones left where there were houses in Ngangu. Source: Authors, Fieldwork 2019

Fig. 6.6 Extent of damage in Kopa where almost all houses were swept away. Source: Authors, Fieldwork 2019

From one of the interviews with a key informant (confirmed by the ZETDC), initial surveying of damage to the electricity grid was done by helicopters that were dropping relief packages (food, blankets, etc.). In addition, helicopters in turn dropped electricity poles and other equipment and relief. The key informant had this to say:

> Sometime during that time we had a meeting with the ZETDC and we decided that we would ask the guys bringing food to carry ZETDC poles because they couldn't move the poles, there were no roads

but we needed power. The ZETDC guy was flown from Chimanimani along power lines to establish how many poles were down and how many [there] needed to be so they came to us and told us that they needed to get poles there.

Once the plan of transporting the poles for one of the damaged key section was designed, the key informant highlighted that there were about 30 poles needed. Then the black hawk helicopter had to transport them.

> So we would have a long rope and we would suspend the poles like a sling and we would carry

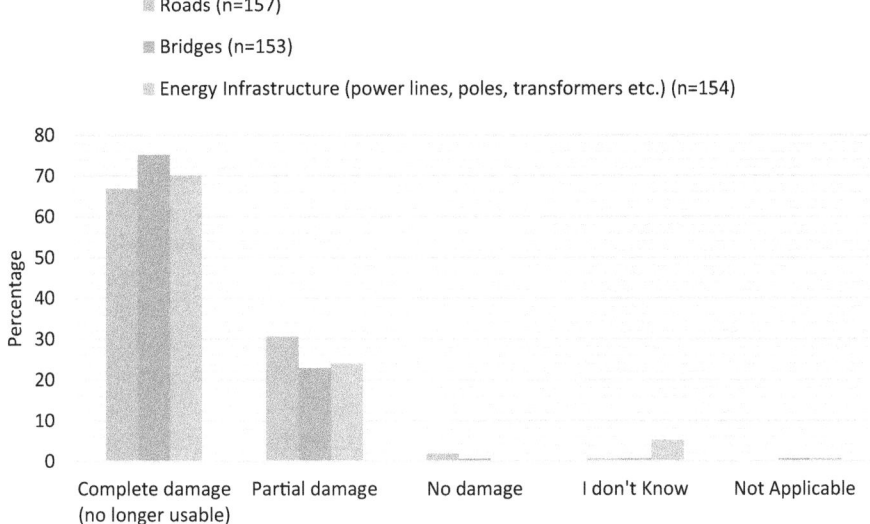

Fig. 6.7 Extent of damage to key energy and related infrastructure. Source: Authors, Household Survey 2019

them, hanging below the helicopter and then drop them where they needed to be dropped. We didn't want to jeopardize the food distribution so we had to do that in the background. So I had one helicopter and we were flying next to them, showing them where to drop by landing where we wanted the poles to be, we had the ZETDC guy as well. We would get the poles at Kopa where the helicopters would refuel (indicated the key informant).

The speed at which the ZETDC mobilised additional resources and manpower for Tropical Cyclone Idai reconstruction cannot go unnoticed. An interview granted by the regional manager indicated that there was streamlining of procurement services and mobilisation of technical and other staff from all over Zimbabwe and beyond. Such similar quick and massive mobilisation of resources to repair damaged energy infrastructure took place in Mexico after the landfall of Hurricane Odile (Murià-Vila et al. 2018). To swiftly repair the damages in the electricity sector,

The Federal Electricity Commission reported that it had mobilised about 6,200 workers, 2,221 specialized vehicles, and 451 emergency mobile plants with generation capacities between 4 and 500 kW. With these emergency plants, they primarily attended hospitals, shelters, water-treatment plants and public lighting, while reparation of the affected infrastructure was carried out. The rapid restoration work executed by the Federal Electricity Commission, allowed the provisional reestablishment of power service, as well as water supply (Murià-Vila et al. 2018, p. 975).

The need to deploy as much repair staff and equipment as quickly as possible is common during hurricane times. When Hurricane Sandy hit the USA in 2012, the U.S. Department of Energy (2013) led an Integrated Energy Restoration Task Force (IERTF). The purpose of the IERTF was to ensure efficiency in the coordination and deployment of needed resources. Bureaucratic bottlenecks had to be unplugged to facilitate the swift movement of utility workers and issuing of appropriate permits for trucks crossing state borders. The IERTF was also needed for deploying new communication systems. As a result, 235 staff were seconded to attend to Sandy damages, and 200 pieces of equipment were also deployed.

Tropical Cyclone Idai also resulted in extensive damages to mini hydroelectric plants. During the fieldwork, visits were made to two such plants, the Rusitu Power Plant and the Kupinga Power Station. The sheer volume of water from Cyclone Idai swept away the entire Rusitu Power Plant. All that remained at the site was a small metal strip and overhead powerlines presenting what once was a small hydropower plant. Even the operator's house was swept away. A sign pointing to where Rusitu Power was off the main road (Fig. 6.8) and the now historic site where the power plant used to be (Fig. 6.9).

The authors also visited another small hydropower plant called Kupinga Power Station. This

Fig. 6.8 Direction to Rusitu Power Plant. Source: Authors, Fieldwork 2019

plant is a joint venture between Kupinga Renewable Energy (Pty) Ltd and Old Mutual. The power plant (Fig. 6.10) was commissioned on 24 August 2017. From the description based on the interview granted by one of the staff members at the power plant, Tropical Cyclone Idai storm surges flooded the property completely. Water was flowing from both the river and the mountains. As a result, in the recovery phase post-Cyclone Idai, the company had to construct barriers (Fig. 6.11). All the original fencing was swept away during the flood. What prevented

massive damage, as per the interview, were two containers that acted as a barrier against the huge volume of water. In fact, one of the containers was moved and blocked by the tree (Fig. 6.12).

In trying to BBB a tropical cyclone resilient grid electricity infrastructure in the Chimanimani and surrounding areas of Manicaland Province, the ZETDC brought in modern accessories and stronger wires. They changed the use of thin wires to those that are thicker and much stronger. The concept BBB is usually elevated in disaster recovery and draws our attention to have energy infrastructure (in this case electricity infrastructure) become more resilient (World Bank 2018). It also embraces restoring services quickly and efficiently. Based on the turnaround done by the ZETDC to have electricity back in about a month (despite the extensive damage), there were strong indicators of BBB. The concept of BBB is part and parcel of the United Nations Sendai Framework on Disaster Risk Reduction that runs from 2015 to 2030 (UNDRR 2015). The Kupinga Power Station repairs also incorporate the BBB concept, trying to create a buffer for possible future floods.

6.3.2   Impacts on Liquid Fossil Fuels and Solar Energy Infrastructure

The key challenge concerning liquid fuel (petrol and diesel) was the extensive damage to all major roads in Chimanimani district. In fact, for about 2 weeks, normal road traffic to Chimanimani was completely cut off. The first teams that went for relief only managed to do so 3 days later and they were transported by helicopters. Later, it was only the army vehicles that managed to navigate the damaged roads through detours after collective efforts. The army and community members had to cut and remove trees and massive mounds of mud that closed the roads. From one of the interviews granted by the owners of Glo Petroleum (Pty) Ltd in Chimanimani town, for 3 months there were no fuel deliveries at all. This was because detour roads did not permit fuel tankers to manoeuvre. On average, the service

Fig. 6.9 The site where Rusitu Power Plant used to be. Source: Authors, Fieldwork 2019

Fig. 6.10 Kupinga Power Plant. Source: Authors, Fieldwork 2019

station received about 40,000 L of fuel per month (half for diesel and the other half for petrol). Glo Petroleum further indicated that they had only received three deliveries after Tropical Cyclone Idai. This had a serious impact on their business and all other businesses that require fuel to function.

The reported shortage of fuel was similar to other instances elsewhere across the world. During Hurricane Matthew in 2017, the USA

State of Florida had 775 fuel service stations reporting severe shortages (AON 2017). This was as a result of eight main petroleum-importing ports on the East coast being closed. These ports distributed on average 381,000 barrels daily. Hurricane Matthew had brought high waves and storm surges that made it impossible to continue with business. Additional details provided indicated that of the 7107 fuel stations in Florida, 775 had no fuel at all as of 10 October, while in

Fig. 6.11 Building back better flood barriers after Cyclone Idai. Source: Authors, Fieldwork 2019

Fig. 6.12 One of the containers that blocked Cyclone Idai floodwaters. Source: Fieldwork 2019

Georgia, 19 out of 6024 were in a similar situation. In South Carolina, 25 out of 2957 fuel stations had nothing, while out of the 5276 fuel stations in North Carolina, 21 did not have fuel (AON 2017). During Hurricane Sandy in 2012, which had a huge impact on petroleum infrastructure, up to 61% of the services and infrastructure was back online 9 days later (U.S. Department of Energy 2013). The petroleum refining capacity of up to 238,000 barrels per day was shut during Irene, while the capacity

of 308,000 barrels per day was shut during Sandy. In addition, 57 petroleum product terminals were shut during Irene, compared to 25 for Sandy.

Damage was also reported to road infrastructure across all affected states in the USA. Hurricane Matthew resulted in 446 roads and 35 bridges being closed as of 10 October 2017 (AON 2017). More than a month later, 21 roads and four bridges remained closed as they had collapsed. Other hurricanes to hit the East Coast with similar, but less pronounced impacts on road and electricity infrastructure were Hurricanes David (1979) and Floyd (1999). While there were similar, but longer delays in repairing the roads and bridges in Chimanimani, the literature from the USA show that this remains a global challenge. In fact, two of the major bridges that needed replacement in Kopa were donated by the South African government in a joint operation between the South African and Zimbabwean defence forces resulting in their successful replacement. As indicated earlier, Hurricane Maria's damage to Puerto Rico's electric grid (Kishore et al. 2018) resulted in a power outage that lasted more than 10 months.

As the detour roads became available, there was a great deal of logistics to bring relief by road, with helicopters that eventually increased to more than 14 doing rounds. These helicopters and other road transport vehicles needed fuel too. Tankers carrying up to 30,000 L of fuel started driving in as big helicopters such as the black hawks needed 650 L an hour, said one of the key informants who was operating one helicopter. The helicopters were also used to take diesel to various mobile network providers' base stations as there was no power and many had run out of diesel. Two base stations that benefitted from this were one at Biriri on top of Nyamasandu and another in Rusitu.

Although some solar installations were damaged along with the houses they were mounted on in both Ngangu and Kopa, some pockets were left operating in the Chimanimani Village suburb and also at the Chimanimani Rural District Council Hospital. Some solar installations from the Chimanimani Rural District Council Hospital that remained functional are shown in Fig. 6.13. Up to 362 injured people were treated at the hospital. From a focus group discussion with staff at the hospital, it was clear that the solar provided many services to the communities including charging mobile phones and other devices. One of the respondents indicated that even during the night of the disaster, the lights stayed on. The solar system at the hospital was installed in 2017 sponsored by a donor and took about a month to commission. Solar allows the hospital and staff to use fridges, computers, heaters and provides power to the female ward, pharmacy and main building.

In their publication which considers the potential for using solar PV to supply critical loads during hurricanes in the USA, Cole et al. (2020) conclude that solar remains an alternative to feed local sources while the damaged grid transmission systems are being repaired. This is because hurricanes are the main reason for outages in the USA electricity sector. When the authors studied

Fig. 6.13   Solar installation at Chimanimani Rural District Council Hospital. Source: Authors, Fieldwork 2019

18 hurricanes that made landfall in the USA between 2004 and 2017, they discovered that even if there is usually cloud cover during the hurricane period, solar PV generation still ranges from 18 to 60% of clear sky potential. This range increases drastically with solar PV generating 46–100% of clear sky potential. Solar power storage makes the option even more feasible.

6.4   Conclusions and Policy Recommendations

The documentation of the impacts of Tropical Cyclone Idai on energy and related infrastructure remains a huge contribution to the body of knowledge in this field, especially literature from southern Africa. This is because very little could be retrieved about the impacts of past tropical cyclones on energy resources. Among the past tropical cyclones that reached Zimbabwe are Cyclone Favio in 2007 and Cyclone Eline of 2000. It is clear that Tropical Cyclone Idai caused massive damage to energy infrastructure, especially the electricity grid, as well as some small hydropower plants, including the Rusitu Power Plant that was completely washed away. Although there were no major generation plants damaged as these are outside the Manicaland Province and the hard-hit Chimanimani and Chipinge districts, there was significant damage to the transmission and distribution subsystems. As witnessed in other countries that were hit by cyclones, including the USA, transmission poles, transformers and substations remain vulnerable to tropical cyclones. The ZETDC tasked with the mandate to provide electricity in Zimbabwe did very well to restore power within a reasonably short period of about 1 month. This conclusion was arrived at given that both the country and the company had never experienced damage to the power sector of such a huge magnitude and comparing to delays of up to 10 months experienced in other countries as documented herein. Hence, it was a learning by doing for everyone involved. Staff and equipment were mobilised both from within and outside Zimbabwe. It is also important to note that in line with the United Nations Sendai Framework on DRR that encourages the BBB concept, stakeholders did their best. The ZETDC went further to replace thinner wires with thicker ones that will likely be able to withstand similar stress from falling trees and poles and remain intact. Connections of the wires to the poles have also been improved, with new substations situated in relatively safer areas.

Based on the findings and conclusions of this study, several recommendations are made that will be useful in building a resilient and adaptive electricity sector in the impacted areas and in Zimbabwe overall. Disaster procurement protocols need to continue to be refined and streamlined. The knowledge gained should be documented by the entities concerned, particularly the ZETDC. A registry of the staff and expertise that worked on the impressive repaired and rehabilitated electricity system in Tropical Cyclone Idai-affected areas should be kept for organisational memory and future disasters. Regarding solar, the resilient installation at Chimanimani Rural District Council Hospital provides a good case study worth replicating in the impacted areas. The fact that the panels were not cracked and kept generating power that assisted in saving the lives of many remains a marvel. It was a pocket of joy in the midst of a dark Chimanimani.

References

AON (2017). Hurricane Matthew: Event Recap Report April 2017. New York: AON.

Chatterjee, C. & Mozumder, P. (2015). Hurricane Wilma, utility disruption, and household wellbeing. International Journal of Disaster Risk Reduction, 14, 395–402. https://doi.org/10.1016/j.ijdrr.2015.09.005.

Cole, W., Greer, D., Lamb, K. (2020). The potential for using local PV to meet critical loads during hurricanes. Solar Energy, 205, 37–43. https://doi.org/10.1016/j.solener.2020.04.094.

Frame, D.J., Wehner, M.F., Noy, I., Rosier, S.M. (2020). The economic costs of Hurricane Harvey attributable to climate change. Climate Change, 160, 271–281. https://doi.org/10.1007/s10584-020-02692-8.

IAEG-SDGs (United Nations Inter-agency and Expert Group on SDG Indicators). (2018). Tier Classification for Global SDG Indicators 15 October 2018. New York: IAEG-SDGs.

Keellings, D. & Hernández-Ayala, J. J. (2019). Extreme rainfall associated with Hurricane Maria over Puerto Rico and its connections to climate variability and change. Geophysical Research Letters, 46, 2964–2973. https://doi.org/10.1029/2019GL082077.

Kishore, N., Marqués, D., Mahmud, A., Kiang, M.V., Rodriguez, I., Fuller, A., Ebner, P., Sorensen, C., Racy, F., Lemery, J., Maas, L., Leaning, J. Irizarry, R.A., Balsari, S., Buckee, C.O. (2018). Mortality in Puerto Rico after Hurricane Maria. The New England Journal of Medicine, 379, 62–70. https://doi.org/10.1056/NEJMsa1803972.

Mitsova, D., Escaleras, M., Sapat, A., Esnard, A.M., Lamadrid, A.J. (2019). The Effects of Infrastructure Service Disruptions and Socio-Economic Vulnerability on Hurricane Recovery. Sustainability, 11, 516; https://doi.org/10.3390/su11020516.

Mohanty, S.K., Chatterjee, R., Shaw, R. (2020). Building Resilience of Critical Infrastructure: A Case of Impacts of Cyclones on the Power Sector in Odisha. Climate, 8, 73; https://doi.org/10.3390/cli8060073.

MPDCO (Manicaland Provincial Development Coordinator's Office). (2019). Manicaland Province Cyclone Idai Disaster consolidated report as at 20 May 2019. Mutare: MPDCO.

Murià-Vila, D., Jaimes, M.Á., Pozos-Estrada, A., López, A., Reinoso, E., Chávez, M.M., Peña, F., Sánchez-Sesma, J., López, O. (2018). Effects of hurricane Odile on the infrastructure of Baja California Sur, Mexico. Natural Hazards, 91, 963–981. 10.1007/s11069-017-3165-z.

Nhamo, G. (2013). Green economy readiness in South Africa: A focus on the national sphere of government. International Journal of African Renaissance. 8(1), 115–142.

Peiter, C.C., Santos, J.L.G., Kahl, C., Coelli, F.H.S., Cunha, K.S., Lacerda, M.R. (2020). Grounded Theory: Use in scientific articles published in Brazilian nursing journals with Qualis A classification. Texto Contexto Enferm, 29:e20180177. https://doi.org/10.1590/1980-265X-TCE-2018-0177.

Reed, D.A., Powell, M.D., Westerman, J.M. (2010). Energy Infrastructure Damage Analysis for Hurricane Rita. Natural Hazards Review, 11: 102-109. DOI: https://doi.org/10.1061/(ASCE)NH.1527-6996.0000012.

Rieger, K.L. (2018). Discriminating among grounded theory approaches. Nursing Inquiry, 26:e12261. https://doi.org/10.1111/nin.12261.

UNECA (United Nations Economic Commission for Africa). (2019). Building back better: planning workshop for climate resilient investment in reconstruction and development in cyclone affected regions of Malawi, Mozambique and Zimbabwe. Retrieved from https://www.uneca.org/building-back-better (Accessed 19 August 2020).

United Nations. (2015). Transforming our World: The 2030 Agenda for Sustainable Development. New York: United Nations Secretariat.

UNDRR (United Nations Office for Disaster Risk Reduction). (2015). Sendai Framework on Disaster Risk Reduction (2015-2030). New York: UNDRR Secretariat.

U.S. Department of Energy. (2009). Infrastructure Security and Energy Restoration. Washington DC.: U.S. Department of Energy.

U.S. Department of Energy. (2013). Comparing the Impacts of Northeast Hurricanes on Energy Infrastructure. Washington DC.: U.S. Department of Energy.

World Bank. (2018). Building Back Better: Achieving resilience through stronger, faster and more inclusive post-disaster reconstruction. Washington DC: World Bank.

Yu, P., Johannessen, J.A., Yan, X.H., Geng, X., Zhong, X., Zhu, L. (2019). A Study of the Intensity of Tropical Cyclone Idai Using Dual-Polarization Sentinel-1 Data. Remote Sensing, 11, 2837; https://doi.org/10.3390/rs11232837.

ZMSD (Zimbabwe Meteorological Services Department). (2019). Review of the Meteorological Services Department's capacity to observe, forecast and respond to future extreme weather events: An assessment of Tropical Cyclone Idai. Harare: ZMSD.

Floods in the Midst of Drought: Impact of Tropical Cyclone Idai on Water Security in South-Eastern Zimbabwe

7

Abstract

The simultaneous occurrence of floods and drought has not been well researched. The aim of this chapter is to assess the impact of tropical cyclone Idai on water security in the south-eastern parts of Zimbabwe. Secondary data and document analysis were the prime sources of data for the research. In-depth interviews with key informants as well as earth observation techniques were also employed. Results show that the south-eastern parts of Zimbabwe had received below-normal rainfall when tropical cyclone Idai occurred. The area also witnessed higher than normal temperatures during the 2018/19 rainfall season which, together with the below-normal rainfall, led to widespread crop failure. Due to the cyclone, the Save valley and south-eastern Lowveld of Zimbabwe, which are mostly drought stricken, benefited from improved water security. This was evident by significant improvements in available surface water and groundwater resources. On the eastern side of the catchment, there was significant damage to water infrastructure and water security. This saw the complete damage to urban and peri-urban water reticulation systems, damage to and silting of dams as well as irrigation infrastructure. It is strongly recommended that the build back better concept be adopted in the recovery, rehabilitation and reconstruction of the water infrastructure destroyed by the cyclone.

Keywords

Drought · Floods · Cyclone Idai · Water security · South-eastern Zimbabwe

7.1 Introduction and Background

Floods are the most frequent of the natural hazards globally (40%), followed by tropical cyclones (20%), earthquakes (15%) and drought (15%) (Nelson 2008; UNISDR and CRED 2018). According to Chingombe et al. (2015), floods occur when the volume of discharge in a river surpasses its capacity, resulting in overflow of the banks. Flooding and its high flows represent an extreme occurrence within river systems. The degree by which the high flow exceeds stream capacity quantifies the intensity of the flood event. The common practice of humans to construct settlements close to rivers and other water bodies has contributed to the devastating consequences of floods (Nelson 2008). The direct impact of floods includes loss of life, damage to infrastructure, disturbances to the natural environment and loss of clean and safe water for domestic purposes. Floods also destroy farmland, making it unusable and preventing the planting or harvesting of crops, leading to food insecurity (Sidek et al. 2016). The indirect impact associated with flooding includes the emerging and spreading of water-related diseases, such as malaria, or waterborne diseases, such as typhoid and cholera. Furthermore, floods may lead to emotional damage to the victims, especially where deaths, serious injuries and damage to assets are caused (Chatiza 2019). Economic hardships may also result from food shortages, rebuilding of damaged infrastructure and resettlement of the affected population (Mahmood 2017). Floods may, on the other hand, bring some positive results, such as significant groundwater recharge and alluvial deposits that improve soil nutrient fertility. As observed by Berz (2000), flood waters can bring water security to water-stressed regions where rainfall is not evenly distributed throughout the wet season and can also reduce pest infestations on the farmlands. Freshwater floods, in particular, help to maintain riverine ecosystems and are an important factor in supporting floodplain biodiversity.

There is no universally accepted definition of a drought, and each definition must be related to the area of concern and be explicit to a particular application. In this study drought is defined as a period of protracted moisture deficit which deviates from the mean climatic conditions of any given area, resulting in widespread loss of crop yields and rangelands (National Drought Mitigation Center (NDMC) 2020). Droughts can be categorised as meteorological, agricultural, hydrological and socio-economic. They are usually ranked according to their intensity, duration and spatial extent, with intensity being defined by the degree of precipitation deviation from the mean or the severity of impact linked to the deviation. Southern Africa is vulnerable to droughts that occur at different spatial scales for periods varying from weeks to decades (Bayarjargal et al. 2006) and is also susceptible to the El Niño effect. Compared with other natural disasters such as floods and tropical cyclones, droughts are noted to be the most costly, affecting more people and larger areas. Droughts are also observed to be a creeping phenomenon (Gillette 1950) with a slow onset and its effects accumulating slowly over a period of time. The onset and end of a drought episode are therefore not easy to determine, but whether it becomes a disaster or not is hinged on its effect on local people and the environment, hence the need to understand it in both its natural and social dimensions (Gillette 1950). Glantz (2003) maintains that droughts are a regular part of climate, rather than a rare and random event. There is therefore a need for concerted efforts towards preparedness plans targeted at those populations, economic sectors, regions and ecosystems most vulnerable. In Southern Africa there is a noted need for effective policies to cope with drought that emphasise preparedness rather than being reactive to the disaster. This is because most of the rural communities in the region rely on dryland subsistence agriculture for their income and livelihoods (Chikodzi 2018).

The World Water Council (WWC) (2018) defines water security as the obtainability of adequate quantities and quality of water to sustain socio-economic development, livelihoods, health and ecosystems. Water obtainability is a multifaceted subject and a precondition for livelihood security, with its scarcity impacting on all levels of society, deepening poverty levels and holding back socio-economic development (Shrestha et al. 2018). Together with drought, water scarcity is the most serious stumbling block to agricultural productivity and poses significant threats to dryland ecosystems (Moumen et al. 2019). In Southern Africa, water scarcity due to drought has been observed to be a leading factor influencing social, ecological and economic hardships because water supply plays a vital role in influencing social welfare. Drought management plans should therefore be a top national priority. Water security is interrelated with the success of other sectors such as energy, food, health and education, either directly or indirectly, and when we can guarantee water security, we can guarantee security in all these other sectors (WWC 2017).

The United Nations 2018 annual report on the water-related Sustainable Development Goals (SDGs), such as access to drinking water and sanitation (Goal 2), reducing famine (Goal 6) and the resilience of cities (Goal 11), highlights that, while notable progress has been witnessed in the past years, the pace of progress is not enough to meet the 2030 deadline (United Nations 2018). Achievement of the SDGs therefore risks being compromised by the lack of water security. For example, over 70% of exploitable global water resources are used in agriculture, and water security is key in tackling famine and achieving zero hunger by 2030. Due to the interlinkages and the integrated nature of SDGs, the WWC (2018) has defined water security as the glue that holds Agenda 2030 together.

In 2019 the south-eastern parts of Zimbabwe suffered from contrasting crises in one season: deadly flooding and bitter drought. The flooding was a result of tropical cyclone Idai, and the drought was induced by the El Nino phenomenon. In this chapter the impact of tropical cyclone Idai on water security in the south-eastern parts of Zimbabwe is assessed. Specifically, the severity of drought before tropical cyclone Idai and the livelihoods put at risk by the drought are unravelled. The water infrastructure damaged by the cyclone and changes in water security observed post-cyclone Idai are determined. The results help to devise innovative strategies in improving water security for communities in the context of both drought and flooding which may impact on the progress and achievement of SDGs in the area.

7.2   Literature Survey

Water is a key asset essential for the existence of human societies and also for supporting ecosystem integrity. The obtainability of sufficient of freshwater resources is a key factor determining the quality of social and ecological systems on which life depends (Gain et al. 2016). As part of the United Nations 2030 Agenda, SDG 6 aims at safeguarding the accessibility and effective management of water and sanitation for all (United Nations 2015). The major challenges in achieving SDG 6 include issues of water scarcity, flood hazards, access to clean drinking water, water quality, water quantity, sanitation and management of transboundary watercourses. To achieve SDG 6 and monitor its targets, Kreibich et al.

(2019) advocate for an all-inclusive quantification of global water resources, taking into account multiple challenges in an integrated manner. They note that very few studies have attempted to develop integrated assessment frameworks for water resources.

UNISDR and CRED (2018) highlight droughts and floods as having a widespread impact across the globe. They note that for the period 1998–2017, floods affected over 2 billion people and droughts affected over 1.5 billion people. In Europe, 1998–2009 saw floods and droughts causing economic losses of €52 billion and €5 billion, respectively (European Environment Agency 2013). Zhou and Liu (2018) recognise the widespread appreciation of the negative impact of extreme climatic events worldwide by academics, policy-makers and the general public media. However, they argue that the joint occurrence of extreme events such as flooding and drought, and heatwaves and drought, has not been well researched, with most studies focusing on each phenomenon independently. Several studies have shown the importance of investigating joint climatic events instead of isolated ones (Zscheischler and Seneviratne 2017). Years of successive droughts have been shown to lead to catastrophic floods when water management practices adapt to the trend of reduced water resources without considering that high flows can still reach the highest levels (Watanabe et al. 2018). For example, the 2011 floods in Brisbane, Australia, in 2011 occurred after periods of prolonged severe drought (Watanabe et al. 2018). This underlines the need to identify and study the likelihood of various compound extreme events occurring and to formulate multi-hazard management plans.

Kreibich et al. (2019) posit the need for developing effective and adaptive risk mitigation policies for coping with droughts and floods. This could be done through experts in both fields working together in joint studies aimed at impact and vulnerability assessment. To make scientific progress in this regard, they suggest that accessible databases, together with data gathering processes, need to be reinforced and further enhanced by developing universal standards for information gathering, gathering even more comprehensive data, using big data and broadening the collected data to address the issue of potential drivers. A complementary approach to the collection of many single or multiple paired-event case studies from diverse hydro-climatic and socioeconomic settings from the world over is also necessary. There is also a noted need for conducting in-depth studies on shifts in the effects and drivers of both floods and droughts and undertaking comparative studies in order to unravel generalisable and transferable conclusions (Kreibich et al. 2019).

Moumen et al. (2019) find that countries or regions situated in places of high climatic contrast such as flush flooding and widespread drought usually suffer from water insecurity. According to Grey and Sadoff (2007), there is a strong connection between hydrologic legacy and poverty levels. Their study highlights that regions that have an easy hydrologic legacy have high water security and attain relatively strong economic progress; on the other hand, those with a difficult hydrologic legacy are water-stressed and usually have strained economic growth. Although the concept of water security is not easy to define, UNESCO's International Hydrological Program's Strategic Plan (2012, online) defines it as "the capacity of a population to maintain sustainable access to acceptable quantities and quality water for supporting livelihoods, human well-being, and socioeconomic development". The concept also ensures the protection of water users from water contamination and hazards linked to water and the conservation of ecological systems in an environment of harmony and political tranquillity (Shrestha et al. 2018). This therefore means that water security is a system with implications for numerous interrelated and co-dependent sectors. The pillars that support the concept of water security are hydrological, geographical, economic, environmental, social, political, legal and monetary which can operate at local, national, regional and global scales (Moumen et al. 2019). Van Beek and Arriens (2016) highlight the significance of connecting water security to food

and ecological security, and view this a stepping stone to achieving sustainable development.

7.3   Materials and Methods

The study area and the methods that were used in gathering and examining data to answer the research objective are explained in this section. The research objective was to assess the impact of tropical cyclone Idai on water security in the south-eastern parts of Zimbabwe.

The south-eastern parts of Zimbabwe are the areas that were mostly affected by the 2019 tropical cyclone Idai. These areas include the Chimanimani, Chipinge, Mutare, Buhera, Mutare, Bikita, Chiredzi and Gutu districts. The areas are predominantly within the Save and parts of the Runde Catchments. The main rivers in the study area include the Save, Macheke, Pungwe, Runde, Devure, Nyazvidzi, Nyanyadzi, Budzi, Devure, Umvumvumvu, Rusitu, Nyahode, Tanganda, Haroni and Odzi.

The area contains all of the five natural agroecological regions of Zimbabwe, with region 1 in the Eastern Highlands being the most productive, having the most rainfall (over 1000 mm per season) and deep fertile soils. Region 5, on the other hand, has the lowest agricultural potential, receiving the least reliable and lowest rainfall (less than 450 mm per season) and also containing poor soils (Chikodzi et al. 2020). South-eastern Zimbabwe is of vital agricultural importance for the whole country with huge areas of economically important large-scale irrigation in Chisumbanje, the middle Save and Tanganda. Important smallholder irrigation schemes in the area include Devure, Nyanyadzi, Chakohwa, Mutema and Mutambara (Chikodzi 2018). All these key irrigation schemes are in arid/semi-arid agro-ecological regions 4 and 5.

Mambo and Archer (2007) observe that southeastern Zimbabwe exhibits serious contrasts, for example the lush, cooler and elevated areas mainly on the eastern side contrasts sharply with the unproductive, water-stressed and hotter lower parts mainly in the west. The high precipitation areas on the windward Eastern Highlands contrast with the dry, semi-arid and leeward sections that receive less than 500 mm/year of precipitation on average, with high interseasonal fluctuations. The exotic pine and wattle tree plantations in the Chipinge and Chimanimani mountains contrast sharply with the extensive sugarcane plantations found in the lowveld around Middle Save which are grown under irrigation. The large-scale commercial farms, which are well managed, look ecologically healthier, with less degradation when compared to the heavily degraded and mostly overpopulated communal areas in the Save valley (Van der Zaag et al. 2001). The harmonious co-existence of communities from these different land uses and contrasting climatic conditions is determined largely by water security in the area. Figure 7.1 depicts the south-eastern districts of Zimbabwe.

Secondary data and document analysis were used as the prime sources of data during the research. Kumar et al. (2017) define document analysis as a procedure of systematically collecting, categorising and evaluating content

and reaching conclusions from documents. Secondary data included already available records of rainfall and run-off before, during and after the tropical cyclone, official government and institutional reports and records of the impact of tropical cyclone Idai, as well as scientific reports. The main benefit of using these methods is that they are cheap because the information is already available and permits analysis of changes over time. Using records offers important cross-validation of some of the collected data, either supporting or contradicting it.

In-depth interviews with key informants were held in the affected areas. This qualitative data collection instrument gave the opportunity to capture rich, descriptive information about the extent of drought and how the cyclone impacted water security. The prime benefit of interviews is that they give much more detailed information than what is attainable through questionnaires. The interviews provided some history on the drought, floods and water security in the area and changes that occurred in water security due to tropical cyclone Idai since they were conducted

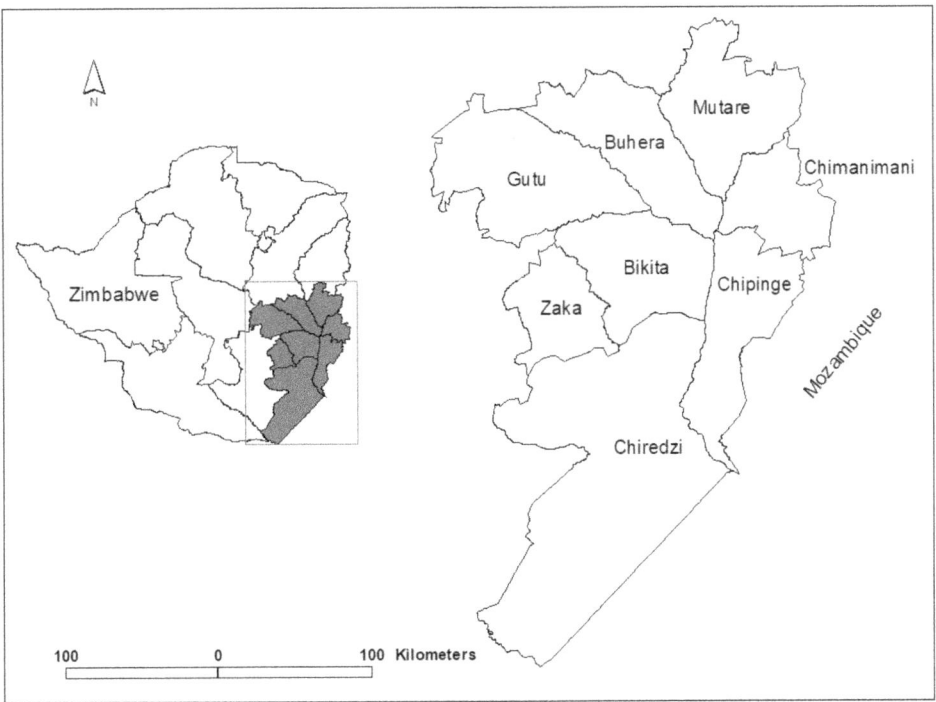

Fig. 7.1 Map of the study area. Source: Authors

with participants who had been resident in the community for more than 20 years. The participants included local traditional leaders, individual water users, officials from the Zimbabwe National Water Authority (ZINWA), local government officials, community-based organisations and NGOs operating in the study area. These participants were chosen because they were individuals or part of institutions that directly or indirectly have an impact on water resources use and management.

Earth observation was used to observe the levels and extent of drought and surface water bodies before and after the tropical cyclone in order to note any observable changes. Data from interviews was examined using fine grain analysis. This was done by making several readings and evaluative note-taking of what was said and what was not said. This was followed by grouping the data into wide-ranging themes and examining the similarities and differences within and between interviews. The themes were then itemised, and content analysis was conducted to make sense of the transcriptions. In-depth analysis of finer themes which might have been previously overlooked was then done. Eventually, a written narrative of the results was made using qualitative descriptors.

7.4   Findings and Discussion

Using precipitation anomalies as a signature mark for drought. Figure 7.2 shows that Zimbabwe and most of Southern Africa received below-normal rainfall, leading to drought in the 2018–2019 rainfall season. The south-eastern parts of Zimbabwe, in particular, had by February 2019 received between −100 mm and −200 mm of their average rainfall, leading to the occurrence of drought. The term "precipitation anomaly" implies a deviation of precipitation from the long-term average. Any negative deviation therefore implies the presence of a meteorological drought, and the extent of deviation shows the severity of drought. Depending on the crop, a deviation of -200 mm of precipitation can lead to a significant decline in the final harvest.

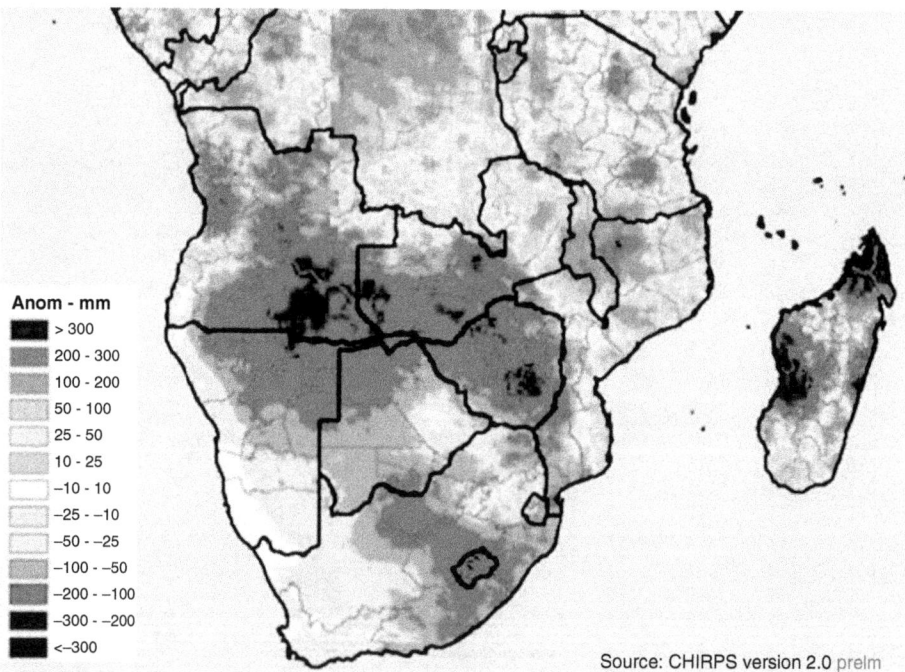

Fig. 7.2 Precipitation anomalies over Southern Africa, October 2018–February 2019. Source: Authors, data from CHIRPS (2019)

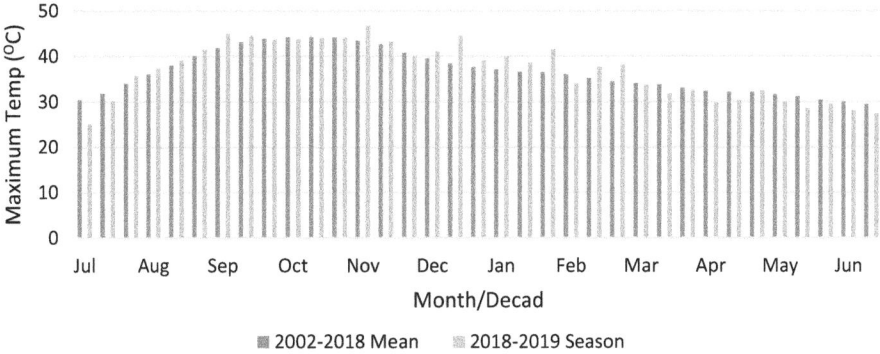

Fig. 7.3 Decadal maximum temperatures for Bikita district compared to the mean. Source: Authors

Figure 7.3 shows the decadal average temperatures for Bikita district in the western parts of the Save Catchment. The average temperatures for the decadal 2018/2019 season are compared to the long-term mean for the period 2002–2018. Each decad shows the average temperature at 10-day intervals with the first decad of the season starting in the first 10 days of January and decad 36 on the last days of December. Figure 7.3 shows that from the first decad of November to the first decad of March, maximum temperatures were mostly above the normal long-term average. In the context of below-normal rainfall, these higher-than-average temperatures can increase the levels of evapo-transpiration and the wilting of crops due to soil moisture deficiency. Where supplementary irrigation is not available, as is the case in most parts of communal south-eastern Zimbabwe, crop failures occur.

Figure 7.4a, b shows the decadal NDVI values for Bikita district in the west and Chipinge district to the east of the Save Catchment, respectively. NDVI was used to represent productivity in plant life in the areas. The two areas represent the climatic differences that occur in the study area, with the east being humid and the west sub-humid. The median NDVI was chosen because as a measure it is more resistant to distortions due to outliers. In Bikita for October during the start of the rainfall season all the decads had vegetation well above the long-term median as computed from 2003 to 2017 from Modis satellite imagery. The vegetation used to calculate the NDVI was for both crop fields and rangelands.

From November all the way to March, which is the normal end to the rainfall season, vegetation productivity was way below the normal expected for that time of the year. This shows the occurrence of drought during that period, which also implies reduced yields in the fields and also less productive rangelands including the pasturelands.

Figure 7.4b shows the decadal NDVI values over Chipinge district. The same pattern of vegetation productivity as Bikita can be observed, with values falling below the normal expected for all the decads from November through to March. In all the sample districts, during the month of March, which coincided with the occurrence of tropical cyclone Idai, vegetation became more productive and healthier as NDVI values were now way above the median values expected for that time of the year. Inasmuch as most crops had been written off or yields reduced because of the observed drought, it can be argued that the tropical cyclone improved the net primary production of other plants, pastures and rangelands, which are also equally important in the livelihoods of the communities settled in the area.

Key informants indicated that the semi-arid and lower lying areas in the study area, such as those on the Save valley and south-eastern Lowveld of Zimbabwe, benefited a great deal in terms of improved water security as a result of tropical cyclone Idai. They highlighted that there was a significant improvement in the availability of both surface water in the rivers and dams and also in groundwater after the tropical cyclone.

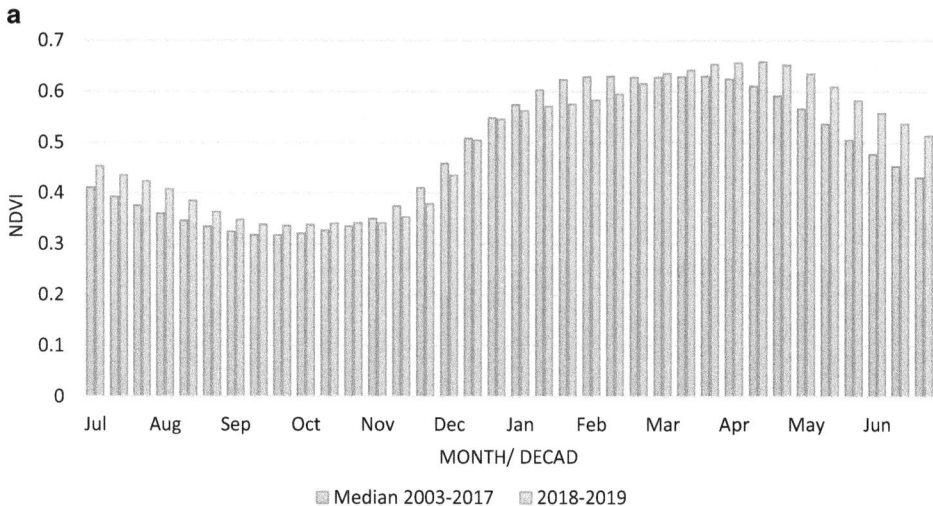

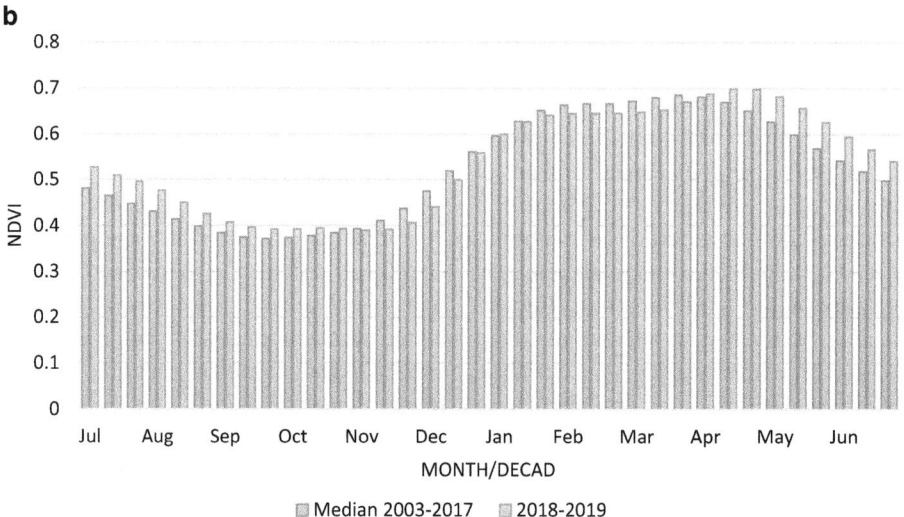

Fig. 7.4 (**a**) Decadal NDVI values for Bikita. (**b**) Decadal NDVI values for Chipinge. Source: Authors, data from FEWS NET (2020)

Key informants indicated that over the years, water scarcity and insecurity had been on the increase for both productive and domestic uses, whereas demand had been on the increase. This was attributed to the frequent droughts in the area, population growth and increase in livelihoods constructed through water abstraction. The state of preparedness of both the communities and water resource managers to deal with water insecurity was observed not to match the increasing demand, and declining river flows was failing to sustain the demand.

Small-scale irrigation schemes in the area, for example Nyanyadzi, Chakohwa, Devure and Musirwizi, are located in semi-arid areas vulnerable to both water and food insecurity. Water abstraction here is done mainly by diverting water using gravity through furrows and polythene plastic. This water can be extracted from up to 10 km from the point of abstraction to the irrigation fields. The Nyanyadzi irrigation scheme, which is about 450 ha, draws most of its water from the Nyanyadzi River via gravity. The river usually dries up and stops flowing around September and serious water shortages start affecting the scheme. Water conflicts between the scheme and upstream users and also among the members of the irrigation scheme start to occur, causing disharmony within the catchment. The alternative source of water for the scheme during this deficit period will now be the Odzi River, but the water will be more expensive since electricity is needed to run motorised pumps.

In the 2019 season after cyclone Idai hit the catchment, as late as November, water was still flowing in the Nyanyadzi River and available for use by the irrigation scheme via gravity. No water conflicts had been registered by October 2019. The same situation applies to the Umvumvumvu River which supplies water to Chakohwa irrigation scheme, which is about 100 ha. This therefore means that the supply side of water was improved as a result of tropical cyclone Idai in the western parts of the Save Catchment, which are mostly drought stricken, and in the rain shadow area of the Chimanimani mountains. Further to this, the problem of streambank cultivation along the major rivers in Chimanimani was completely eliminated due to the cyclone. The problem had reached critical levels and was threatening the health of most rivers in the area, and conservation authorities had almost given up on enforcement against this practice because of the magnitude of the problem. All the gardens on the river banks were completely swept away during the cyclone, and several kilometres of the rivers were cleared. The chances of the problem recurring are very marginal since all the fertile soils were washed away and replaced with rock boulders, where cultivation is impossible. Most key informants agreed that as much as streambank cultivation was a source of livelihood for some of the locals, it was destructive to the environment due to siltation of the water bodies.

The Nyanyadzi irrigation scheme is also a good example of "building back better" when it comes to climate-smart irrigation schemes. In 2000, the scheme became dysfunctional for up to 5 years after tropical cyclone Eline caused damage. The heavy downpours caused structural damage to the irrigation infrastructure and filled the canals with silt, which blocked the flow of water to the community and the farming plots. However, in 2019, the scheme suffered minimal damage due to tropical cyclone Idai. Although the cyclone brought significant flooding to the scheme, silt traps (gabions) built in vulnerable areas throughout the scheme helped capture and reduce the amount of silt which could have been washed into the canals.

It was also noted that the areas hit by tropical cyclone Idai relied mostly on groundwater resources for both domestic and productive uses. The areas especially to the west of the Save Catchment had been experiencing years of progressive groundwater decline due to low rainfall and also overuse. This situation had led to an increase in the incidences of borehole failure, especially during the dry season, and drying up of wells. Most people would travel very long distances to have access to clean water, a situation compromising their health and hygiene. The cyclone, however, recharged the groundwater resources to levels that the communities observed would take up to three seasons to deplete to levels where they were before the cyclone. This then means that water security was improved considerably after the tropical cyclone.

On the other hand, areas mainly to the eastern side of the Save Catchment were the hardest hit by damage to water infrastructure and water security. Most of these areas are generally water secure in terms of both surface water and groundwater. Most of the rivers which flow into Mozambique and to the drier western part of the catchment originate in this area. They are therefore still at their youthful stages and fast flowing and have plenty of erosive power. At Dombera farm, for example, cyclone Idai destroyed three small earth dams which were critical to the operations of the farm. At Dombera, a key informant remarked that "water is at the heart of our operation. Without water we have no crops, no crops means no income and ultimately the 1000 people supported by Dombera wages will have to find an alternate source of income". Figure 7.5 illustrates Magarasadza Dam before and after tropical cyclone Idai damage. The destruction of the reservoir and the widening of the river downstream are evident.

The threat to water security at Dombera farm can also be linked with a threat to the livelihoods of the workers at the farm and a loss of foreign exchange for the country since the farm exports its produce to the international markets, mainly the European Union. Further, repairing the water infrastructure back to normalcy would cost an estimated US$125,000–$160,000.

Fig. 7.5 Magarasadza Dam before (21/11/2018) and after (25/03/2019) Idai damage. Source: Authors, data from Google Earth

Water reticulation systems for urban settlements such as Ngangu in Chimanimani were completely destroyed by tropical cyclone Idai, leaving the survivors with no clean source of water for both domestic and productive needs. The local authority had to quickly step in to restore water supply and for close to 6 months had to stop billing the residence and users of the water. This constituted a significant loss of revenue for the local council since water billing is one of their important sources of income. This impacted the capacity of the local authority to deliver some other important services such as the permanent relocation of the flood victims to safer areas. Key informants highlighted that a total of 18 urban and peri-urban water supply systems were completely destroyed as a result of cyclone Idai. This has placed a huge negative burden on public health in the affected communities in terms of the provision of clean water and sanitation facilities to the communities.

Tropical cyclones have also had a huge negative impact on the management of water resources in the Save Catchment. Hydrological gauging stations infrastructure that had been destroyed by tropical cyclone Eline in 2000 were not resuscitated, and Idai destroyed the few remaining operational stations. This came at a time when hydro-metrological data was very key for decision-making and modelling in order to improve the management of water resources. The measurement of run-off within river systems is also important in the allocation and permitting of water storage and abstraction, for determining levels of in-stream or ecological flow requirements of rivers, yield analysis of proposed reservoirs, water quality management, estimating the potential renewable water resources and designing engineering works, evaluation, planning and management of the water resources (Mazvimavi 2004).

The tropical cyclone also changed the morphology of rivers, especially in the eastern side of the Save Catchment. This was done mainly through the widening of previously narrow and deep river courses. The rivers were also left shallow and littered with boulders, accompanied by total destruction of the riverine vegetation, which is important in returning moisture on riverbeds. The widening of previously narrow streams can increase the rate of water loss from the river through an increase in both discharge and levels of evaporation. This may change streams from being perennial as was the case on the eastern side of the Save Catchment, to flowing only during the rainfall season and a few months afterwards. This compromises water security for riparian communities and threatens their livelihoods, for example cash crops such as bananas could not be irrigated. Figure 7.6 depicts the Haroni-Zunguni river confluence close to

Mawenje Lodge before cyclone Idai on 21 November 2018 and the Idai impact on 29 March 2019. The river, which was hardly 10 m wide, was widened to almost 114 m. This situation typifies most of the rivers in the area.

Cyclone Idai also compromised the affected community's ability to have clean and secured water supplies by directly damaging up to 250 boreholes and community water points. This left the areas temporarily without clean sources of water and vulnerable to waterborne diseases. Some small-scale irrigation schemes such as Musirwizi and Shinja had their source of secured water damaged by the cyclone. Some of their irri-

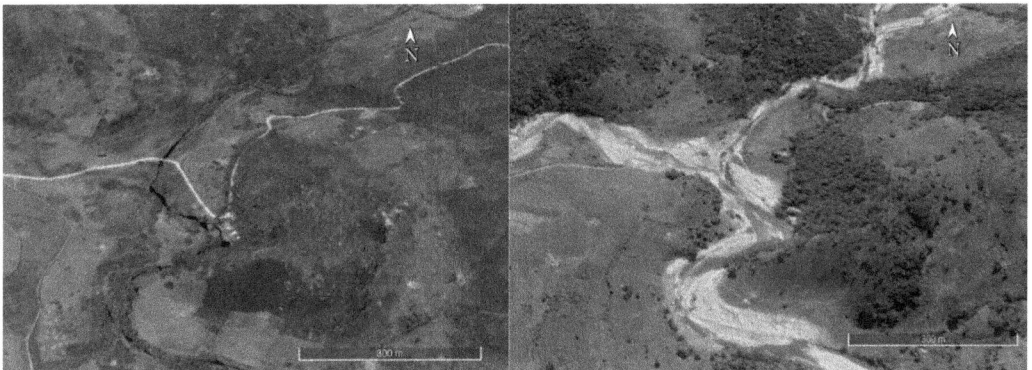

Fig. 7.6 Haroni-Zunguni river confluence before (21/11/2018) and after (25/03/2019) cyclone Idai. Source: Authors, data from Google Earth

Table 7.1 Contingent multi-hazard management plan for drought and floods in south-eastern Zimbabwe

Action	Activities	Responsible institutions
Flood and drought hazard risk maps	Develop agro-ecological rainfall deficit maps Map flood risk areas and levels of risk Conduct comprehensive assessment of water deficit and availability for both domestic and productive use in drought-prone areas every year Develop local capacity to prepare and use vulnerability and risk maps	Meteorological Services Department (MSD), ZINWA, Agritex, interdisciplinary academia, local community
Monitoring, early warning and disaster declaration	Improve flood and drought forecasting Adopt impact-based forecasting Prepare detailed advisories on water conservation and crop management measures based on drought conditions Monitor key drought indices at all levels of administration and develop a composite index of various drought indicators relevant to each area Develop thresholds for declaring drought and/or flood disaster from the indices Central authorities should issue a formal declaration of drought and/or floods specifying affected areas Communities impacted must be assisted in initiating drought/flood response measures Disseminate warnings to all affected and interested stakeholders Regularly update people in areas at risk Adopt mechanisms to activate the space and major disasters charter	MSD, ZINWA, Agritex, district authority, Department of Civil Protection

(continued)

Table 7.1 (continued)

Action	Activities	Responsible institutions
Vulnerability assessment of risk	Provide guidelines on vulnerability assessment covering social, economic, ecological, gender and fairness Model changes in vulnerability and risk under different scenarios of climate variability and change	Agritex, ZINWA, NGOs, local authorities, interdisciplinary academia, Department of Civil Protection
Research	Promote research on water conservation, water harvesting, coping with drought and soil management under drought/flood conditions and crop selection in drought-/flood-prone areas Conduct pilot studies in drought-/flood-prone areas for suggesting long-term mitigation measures Encourage the use of indigenous knowledge systems	Interdisciplinary academia, Agritex, local communities
Social protection	Ensure rainwater harvesting and storage in drought-prone areas especially for vulnerable groups Improve rural water supply through drilling more boreholes Provide food aid for the most affected Provide multi-use evacuation shelter for the flood victims	Ministries of Local Government & Social Welfare, NGOs and development partners, Department of Civil Protection
Water conservation, integrated water resources management	Promote water-saving irrigation systems Promote micro-irrigation systems Promote local-level forum for natural resource management Promote rehabilitation of degraded landscapes	Ministries of Agriculture, Environment, NGOs, Environmental Management Agency, private sector
Awareness creation	Carry out mass media campaigns, participatory learning, informal and formal education initiatives Promote culture of disaster risk reduction within communities Use local traditional leadership Use community radio stations	NGOs, media, schools, local political/religious/traditional leadership, private sector
Empowering women, marginalised communities and disadvantaged persons	Incorporate gender-sensitive and equitable approaches in capacity development at all levels of society	Ministry of Youth and Gender, NGOs, local authority
Drought and flood management plans	Prepare the national, provincial, district, ward and village drought management plans based on climatic outlook and projections of water deficit in the affected area Encourage building back better of destroyed infrastructure	Department of Civil Protection, ZINWA, Ministry of Agriculture, affected communities
Integration of drought and flood management in development plans	Government ministries, agencies and departments to integrate disaster risk reduction as part of their mandate in development plans	All ministries of government, NGOs and local communities, private sector

gation canals were damaged, and essential irrigation valves were swept away and wide gullies were formed. In addition, most weir dams were silted, causing water insecurity through a reduction in the supply of water to fields.

In the light of drought and flooding taking place in the study area, Table 7.1 illustrates the proposed multi-hazard contingent plan for the study area to deal with these two identified disasters.

7.5   Conclusions

It has been observed from the study that climate change and variability can deliver both drought and flooding in a single season. Tropical cyclone Idai was a mixed blessing in terms of its impact on water security in south-eastern Zimbabwe. The semi-arid areas mainly on the western side of the Save Catchment benefited immensely from the rainfall which resulted from the cyclone. The rainfall led to improved supply of both groundwater and surface water for both productive and domestic use. It also brought improved productivity of rangelands, including pasturelands, and reduced water-related disputes. However, on the eastern side of the Save Catchment the cyclone negatively affected water security by destroying/silting water supply dams and damaging hydrological gauging stations, boreholes and urban/peri-urban water reticulation systems.

It is strongly recommended that the "build back better" (BBB) concept be adopted in recovery, rehabilitation and reconstruction of the water infrastructure destroyed by tropical cyclone Idai. BBB is one of the four priority areas of action under the Sendai Framework for disaster risk reduction. Cyclone Idai was not the first to hit the area and neither will it be the last due to the location of the area close to the Mozambique Channel, which is a cyclone-active area. This therefore leads to the need to ensure that both rural and urban water supply projects are climate smart in order to minimise disruptions to people's livelihoods.

In the light of destroyed run-off, gauging stations and hydrometric management capacity need to be rebuilt within the Save Catchment. This needs to be done through the adoption of 4IR and emerging technologies in the area of hydrometrics. The new hydrometric gauging technology needs to support near real-time data acquisition and dissemination and at the same time be multi-purpose, for example measuring other relevant water quality parameters rather than just discharge. Since all catchments in the area are mostly transboundary and shared between Zimbabwe and Mozambique, the implementation of the Buzi, Pungwe and Save (BUPUSA) Tri-basin project would be the best vehicle for this. BUPUSA is meant to strengthen cooperation and institution building in the three river basins in order to achieve integrated water resources management. To achieve its aims, the amount of water resources available as well as their quality will have to be measured; hence the need for an operational network of gauging stations across the catchments.

In the context of Southern Africa, disasters could continue to take a huge toll on the region due to lack of readiness to manage droughts and floods on a basin-wide basis. The absence of integrated management of water resources, including data sharing and suitable institutional and regulatory frameworks, has resulted in inefficient water usage between nations and their key economic sectors. The structure for instituting and developing joint management arrangements has been in place through provisions of the SADC Protocol on Shared Watercourses. The protocol aims to develop and implement bilateral and multilateral river basin agreements, ensure sufficient sustainable water for domestic and productive needs, minimise water-related conflicts and ensure benefits to all countries involved. However, the implementation of the protocol has not been as effective as it should be.

References

Bayarjargal, Y., Karnieli, A., Bayasgalan, M., Khudulmur, S., Gandush, C., Tucker, C.J. (2006). A comparative study of NOAA–AVHRR derived drought indices using change vector analysis. Remote Sensing of Environment 105: 9–22.

Berz, G. (2000). Flood disasters: lessons from the past - worries for the future. Proceedings of the Institution of Civil Engineers. Water Maritime and Energy, 142(1) pp 3-8.

Chatiza, K. (2019). Cyclone Idai in Zimbabwe: an analysis of policy implications for post-disaster institutional development to strengthen disaster risk management. Oxfarm. www.oxfam.org

Chikodzi, D. (2018). Unusual waterscapes and precarious rural livelihoods: occurrence, utilisation and conservation of springs in the Save Catchment, Zimbabwe. Thesis submitted to the University of the Western Cape, Cape Town.

CHIRPS. (2019). Southern Africa Food Security Alert February 1, 2019. Available at https://fews.net/sites/default/files/documents/reports/Southern%20Africa%20Alert_Febraury%202019_FINAL.pdf. (Accessed March 2020).

Chikodzi, D., Tevera D., Mazvimavi, D. (2020). SDG 15 and Socioecological Sustainability: Spring Waterscapes and Rural Livelihoods in the Save Catchment of Zimbabwe. In G. Nhamo et al. (eds.), Scaling up SDGs Implementation, Sustainable Development Goals Series, https://doi.org/10.1007/978-3-030-33216-7_4.

Chingombe W., Pedzisai, E., Manatsa, D., Mukwada, G., Taru, P. (2015). A participatory approach in GIS data collection for flood risk management, Muzarabani district, Zimbabwe. Arab J Geosci, 8:1029–1040. https://doi.org/10.1007/s12517-014-1265-6.

European Environment Agency. (2013). Towards a potential European flood impact database. EEA – JRC – ETC/CCA Joint Technical Paper. Version 5.0.

FEWS NET. (2020). EWX Lite Next Generation Viewer. Retrieved from: https://earlywarning.usgs.gov/fews/ewx_lite/index.html?region=af. (Accessed 20 December 2019).

Gain, A.K., Carlo Giupponi, C., Wada, Y. (2016). Measuring global water security towards sustainable development goals Environmental Research Letters. 11, 124015 https://doi.org/10.1088/1748-9326/11/12/124015.

Gillette, H.P. (1950). A creeping drought under way. Water and Sewage Works March:104–105.

Glantz, M.H. (2003). Climate Affairs: A Primer. Washington, D.C.: Island Press.

Grey, D., Sadoff, C.W. (2007). Sink or Swim? Water Security for Growth and Development. Water Policy, 9(6), 545–71.

Kreibich, H., Blauhut, V., Jeroen, C.J.H. Aerts, J.C.J.H., Bouwer, L.M., Van Lanen, H.A.J., Mejia, A., Mens, M., Van Loon, A.F. (2019) How to improve attribution of changes in drought and flood impacts, Hydrological Sciences Journal, 64:1, 1-18, https://doi.org/10.1080/02626667.2018.1558367.

Kumar, A., Singh, N., Neelu J.A. (2017). Learning styles based adaptive intelligent tutoring. International Journal of Cognitive Research in Science, Engineering and Education, Vol 5 (2)pp 83-98.

Mahmood, A. (2017). Flood hazard mapping in integrated flood risk management: importance and problems associated to Pakistan. https://ssrn.com/abstract=3057194. (Accessed 15 May 2020).

Mambo, J., Archer, E. (2007). An assessment of land degradation in the save catchment of Zimbabwe. Area, 39(3), 380–391.

Mazvimavi, D. (2004). The Estimation of Flow Characteristics of Ungauged Catchments. Unpublished PHD Thesis to the International Institute for Aerospace Survey and Earth Sciences, Enschede, The Netherlands ITC.

Moumen, Z., Idrissi, N.A., Tvaronavičienė, M., Lahrach, A. (2019). Water security and sustainable development. Insights into Regional Development, 1 (4), pp. 301-317.

National Drought Mitigation Center (NDMC). (2020). Types of drought. Retrieved from: https://drought.unl.edu/Education/DroughtIn-depth/TypesofDrought.aspx. (Accessed 11 November 2020).

Nelson, S.A. (2008). Flood Hazards, Prediction and Human Interventions. Tulane University.

Shrestha, S., Aihara, Y., Bhattarai, A.P. (2018). Development of an Objective Water Security Index

and Assessment of Its Association with Quality of Life in Urban Areas of Developing Countries. SSM Popul Health, 6, 276–285.

Sidek, L.M., Rostam, N.E., Hidayah, B., Roseli, Z.A., Majid, W.H.A.W.A., Zahari, N.Z., Salleh, S.H.M., Ahmad, R.D.R., Ahmad, M.N. (2016). Hydrology Analysis and Modelling for Klang River Basin Flood Hazard Map. IOP Conf. Series: Earth and Environmental Science 32, 012069 https://doi.org/10.1088/1755-1315/32/1/012069.

UNESCO. (2012). International Hydrological Program. Eighth Phase "Water Security: Responses to Local, Regional, and Global Challenges" Strategic Plan IHP-VIII (2014-2021). Retrieved from: https://unesdoc.unesco.org/ark:/48223/pf000021806. (Accessed 11 November 2020).

UNISDR and CRED. (2018). Economic losses, poverty and disasters: 1998–2017. Available from: http://www.preventionweb.net/files/61119_credeconomiclosses.pdf (Accessed 10 May 2020).

United Nations. (2015). Sendai Framework for Disaster Risk Reduction 2015–2030. UNISDR: Geneva.

United Nations. (2018). The sustainable development goals report 2018. New York: United Nations.

Van Beek, E., Arriens, W.L. (2016). Water Security: Putting the Concept into Practice. (TEC background papers; No. 20). Stockholm: Global Water Partnership (GWP). https://www.gwp.org/globalassets/global/toolbox/publications/background papers/gwp_tec20_web.pdf (Accessed 15 April 2020).

Van der Zaag, P., Bolding, A., Manzungu, E., (2001). Water-networks and the actor: the case of the Save River catchment, Zimbabwe. In: P. Hebinck and G. Verschoor (eds.), Resonances and dissonances in development; actors, networks and cultural repertoires. Van Gorcum, Assen; pp. 257-279.

Watanabe, T. Cullmann, J., Pathak, C.S., Turunen, M., Emami, K., GhinassI, G., Siddiqi, Y. (2018). Management of climatic extremes with focus on floods and droughts in agriculture. Irrigation and Drainage. 67: 29–42. https://doi.org/10.1002/ird.2204.

World Water Council (WWC). (2017). Striving for Water Security. WWC Triennial report. The World Water Council, Marseille.

World Water Council (WWC). (2018). Water Security, Sustainability and Resilience. WWC Strategy 2019-2021. The World Water Council, Marseille.

Zhou, P. and Liu, Z. (2018). Likelihood of concurrent climate extremes and variations over China. Environ. Res. Lett. 13, 094023 https://doi.org/10.1088/1748-9326/aade9e.

Zscheischler, J., Seneviratne, S.I. (2017). Dependence of drivers affects risks associated with compound events. Sci. Adv. 3e1700263.

Building Back Better Domestic and Irrigation Water Supply Systems in the Aftermath of Tropical Cyclone Idai

8

Abstract

The frequency and magnitude of tropical cyclones reaching Zimbabwe, from the Mozambican channel, have been on the rise. This chapter unpacks the impacts of Tropical Cyclone Idai on domestic and irrigation water supply infrastructure, as well as exploring the key elements required, and some undertaken, to Build Back Better (BBB) water supply systems from the disaster in the Chimanimani district of Zimbabwe. The research used a mixed methods approach in data collection and analysis. Methods used include a questionnaire survey, in-depth interviews, document analysis and field observations. Results show that the water supply infrastructure for both domestic and irrigation use were extensively damaged by the tropical cyclone. Although the replacements done for these damaged facilities were meant to restore minimal service to the affected communities, there are elements of BBB emerging that are worth noting for the future. The damaged weir dams as well as school and health facility piped water systems were repaired in partnership with development partners. The business case for adopting the BBB principle in Chimanimani was primed on the projected intensification in the frequency of cyclones hitting the area in the future. The research recommends the strengthening of building standards for water supply infrastructure in order to make such facilities resilient to future shocks.

Keywords

Cyclone Idai · Build Back Better · Domestic and irrigation water supply · Resilience

8.1 Introduction

Tropical cyclones cause destruction leading to widespread economic losses, loss of livelihoods, loss of lives and damage to key structures (Aquino et al. 2019). Susceptibility to tropical cyclones is becoming more noticeable due to climate change, a rapid increase in population and poorly constructed structures (Perfecto et al. 2019). Among the structures typically damaged during cyclones are domestic and irrigation water supply systems. Irrigation agriculture is a key component of food production and supports rural livelihoods around the world (Mohan 2017). The United Nations' 2030 Agenda for Sustainable Development (AfSD) has eradicating hunger and malnutrition, and access to sustainable water and sanitation supply as part of its 17 Sustainable Development Goals (SDGs) (United Nations 2015).

Due to climate change and variability, the world has, in recent years, witnessed a growing trend in the occurrence and destructive intensity of cyclones (IPCC 2007; Perfecto et al. 2019). Almost all the tropical cyclone basins have witnessed an increase in cyclones, with the Centre for Climate and Energy Solutions (2020) partly attributing this to rising sea surface temperatures and related to global warming. However, the severity and destruction of property, infrastructure and the environment due to cyclones is not evenly distributed all over the world. The biggest losses, damages and death tolls are concentrated mainly in the global south (UNECA 2019). This is partly because of poor disaster risk reduction (DRR) capacity and also the inadequate structural capacity of the built environment. Mosley et al. (2004) observes that in most cases, the financial and technical capacity of global south communities, to deal with the impact of natural disasters, such as cyclones, is limited, resulting in the dependency upon international assistance. This, then, brings the need for societies to build and take lessons from preceding disasters for the sake of building resilience to future shocks. In the context of rehabilitation and recovery efforts, after a tropical cyclone ravages a place, it is essential to consider the concept of BBB in order to make infrastructure more resilient to similar future shocks. The principle is one of the primary focus of the Sendai Framework for Disaster Risk Reduction (SFDRR) (UNDRR 2015). BBB entails reconstruction of structures, systems and communities to standards that are "stronger, safer and more resilient than what existed before the disaster struck" (UNDRR 2015).

Domestic water and agricultural irrigation are indispensable in safeguarding the well-being of communities, as well as sustaining rural livelihoods. These, therefore, fulfil an important role in the sustainable development process by giving life support services. In disaster situations, the state of water supply infrastructure is imperative for the rapid return to normality. Tropical cyclones can lead to the pollution of water sources, disturbances in the water distribution network, damage to water facilities that may result in the total or partial failure of the whole system as well as inducing water shortages. Rehabilitation of the water supply system after tropical cyclone damage may take any timeframe

ranging from a few days to several months depending on the state of preparedness of the responsible organisation (WHO 1998).

Zimbabwe is a landlocked country that is vulnerable to tropical cyclones. In recent years, the frequency and intensity of cyclones reaching the country have increased. This is directly linked with the increase in the total number of cyclones forming in the South-west Indian Ocean Basin. This saw the 2018–19 South-west Indian Ocean Tropical Cyclones Season set a new record of 10 Intense Tropical Cyclones since records began, surpassing the 2006–07 season (UNECA 2019). This brings about the need to assess Zimbabwe's capacity to BBB water supply systems and learn from such disasters.

Given the foregoing, and using lessons drawn from Tropical Cyclone Idai, which ravaged the eastern parts of Zimbabwe in March 2019, this chapter unpacks the impacts of the cyclone on domestic and irrigation water supply infrastructure in the Chimanimani district. To this end, the chapter is set to document the extent to which the key elements in BBB domestic and irrigation water supply system played out. The next section explains the BBB concept as it applies to water supply infrastructure and the nexus which exists between resilient infrastructure and the SDGs, especially the agriculture (SDG 2) and water and sanitation supply (SDG 6).

8.2   BBB in the Context of Agriculture, Water and Related SDGs

Mannakkara (2014) show BBB as a model of rebuilding and recovery of disaster impacted societies in ways that produce resilience and disaster smart solutions. The main reason for BBB is making communities stronger and more resilient following disaster events, given the increasing frequency of such. This has made necessary the need for enhanced post-disaster rebuilding and recovery standards. The BBB concept is also reflected upon by Aquino et al. (2018) who view it as aiming to improve the "physical, psycho-social and economic aspects of communities during reconstruction and recovery, to induce greater resilience". Mannakkara (2014) sees the successful implementation of BBB as involving the improvement of building designs and applying them by way of new construction regulations and land use master plans which are hazard and risk-based, empowering the at-risk population by affording them community-specific solutions, takes into consideration socio-economic aspects and stimulates and supports mental recovery. The SFDRR promotes the construction of hazard proof structures by strengthening design flaws that lead to vulnerable constructions, as well as increasing public education and awareness of disaster risk (Aquino et al. 2018; UNDRR 2015). The provision of disaster resilient water infrastructure is highlighted in the SFDRR in a statement that aims to promote the "resilience of new and existing critical infrastructure, including water, to ensure that they remain safe, effective and operational during and after disasters, in order to provide lifesaving and essential services" (UNDRR 2015, p. 21).

The concept of BBB as applicable to the water infrastructure has synergies with several SDGs. For example SDG 11 aims at making "human settlements inclusive, safe, resilient and sustainable". Target 11.5 of the SDG highlights the need to expressively lower the amount deaths and of people impacted by disasters as well as significantly lowering the economic losses that they induce (United Nations 2015). The SDG aims at protecting interests of vulnerable groups who bear the biggest burden of disasters. The aim of Goal 6 is to guarantee the obtainability and effective management of water as well as sanitation for every human being. Target 6.1 of the Goal 6 endeavours for the world to attain equal access to clean and reasonably priced drinking water for citizens by the year 2030. On the other hand, SDG 2 advocates for ending hunger, achieving food security, improving nutrition and promoting sustainable agriculture. Target 2.3 specifically focuses on doubling output and revenue from small holder farmers with special emphasis on vulnerable groups like native peoples, peasant farmers and woman by the end of 2030 (United Nations 2015). Attaining these goals will be

greatly compromised, if not impossible, when both rural and urban water supply systems, as well as irrigation agriculture continue to be damaged during disasters like tropical cyclones.

Destruction of water supplies also leads to challenges in the attainment of other targets spelt out in the 2030 AfSD (United Nations 2015). For example destroyed domestic water supplies will impact primary health delivery systems (SDG 3) as well as the delivery of quality education (SDG 4). Reduced domestic water supplies may also result in women, girls and children suffering (SDG 5) in terms of travelling further to fetch water, and also wasting otherwise productive time, thereby negatively impacting on SDG 1 (ending poverty) and even challenges with peace and security (SDG 16). From the agriculture sector, lack of irrigation water may lead to loss of employment (SDG 8) and hindrance in industry (SDG 9). Although not the focus of this chapter, destroyed dams and other water reservoirs may result in disturbances in hydroelectric plants, particularly mini-hydro power plants.

In other countries, such as Australia, the BBB has been adopted as a way of reducing vulnerability of societies disaster risk as well as accelerating the pace of getting back to normalcy in the post-disaster phase. Carroll (2015) observes that BBB is now categorised as an important concept guiding the reconstruction of structures in the aftermath of disasters. This has seen the Queensland territorial government creating the Betterment Fund as a vehicle for delivering universal best-practice in reconstruction after disasters. The BBB concept has resulted in the construction of structures that are not only climate proof but also delivering a significant decrease in future spending on infrastructure rehabilitation. This is being achieved through the restoration or replacement of previously damaged critical civic assets to standards that can withstand the force of extreme events. The Queensland Betterment Fund was born out of back-to-back years of extreme weather events having repeated destruction on the same critical public assets. Previously, the "Natural Disaster Relief and Recovery Arrangements" would, largely, pay for like-for-like remediation or reconstruction infrastructure to conditions and standards before the disaster (Commonwealth of Australia 2014). This scenario had the effect of discouraging responsible authorities from build resilience to omnipresent threats. The Betterment Fund, therefore, provided the critical connection "between recovery now and the mitigation of future disasters", hence, infrastructure can better withstand recurrent effects from extreme events. The rationale behind this approach is in lowering the monetary and social impacts of disaster recovery as well as reducing the downtime of critical infrastructure in the post-impact period (Carroll 2015).

The implementation of BBB can be best done by learning from past reconstruction experiences after disasters the world over. Since domestic water and sewer facilities are vital in safeguarding the well-being of communities, the availability of these basic services is key in determining the speed with which affected communities return to normalcy (WHO 1998). The most critical period in implementing BBB is to build systems and structures that are resilient to potential hazards. This implies that destruction is reduced through the implementation of measures to toughen assets and having robust intervention plans in the face of disaster. emergency. Box 8.1 shows the common impacts of tropical cyclones on water supply systems.

- Flood water levels which exceed the elevation of well-head walls, thus, contaminating the water,
- Run-off water flowing directly into wells and other intakes,
- Increase in water levels in sewer systems which leads to the backflow of waste into houses or public drains,
- Soil erosion which results in water pipes being exposed, dislodged or swept away,
- Damage to foundations resulting in cracks or caving-in of constructed tanks,
- Damage to pumping equipment and electrical installations,
- Destruction of intakes, dams and other surface water reservoirs,
- Sedimentation, resulting in silting up of components.

Source: Authors, Carroll (2015) and WHO (1998)

Box 8.1 Impacts of Tropical Cyclones on Water Supply Infrastructure

- Breakage of pipelines and appurtenances in exposed areas such as at rivers due to of high volumes of discharge,
- Damage and disconnection of water supply pipes in mountainous regions due to excessive overland flow and mass movement.
- Contamination of water in tanks and pipes,
- Rise in the turbidity of water and effluence from dislodged pollutants,

---

Chikodzi (2018) observes smallholder irrigation schemes as being key in combating the twin evils of poverty and food insecurity, which affect most of the rural communities in Zimbabwe. This is in the light of erratic rainfall that makes dryland agriculture a risky enterprise. It is, therefore, no surprise that irrigation development has been viewed as a vehicle for rural development and transformation and given priority by the government of Zimbabwe since independence in 1980 (Dube and Sigauke 2015). Damages to smallholder irrigation schemes during disasters, such as tropical cyclones, have been on the increase in Zimbabwe since the year 2000. This not only results in livelihood destruction for the beneficiaries but also leaves them exposed to future disasters of similar and even smaller magnitude. The next part presents the materials and methods were employed for the gathering and examination of data in the study.

8.3   Materials and Methods

The chapter aims to investigate the impacts of Tropical Cyclone Idai on domestic and irrigation water infrastructure in the Chimanimani district of Zimbabwe, and also to highlight the key elements towards BBB the water supply system in anticipation of future disasters. Studies in the field of DRR are in most cases multi-disciplinary and involve the application of a variety of factors that are in the social, economic, psychological as well as engineering realms. Given the nature of the subject of focus, it was crucial to select a combination of research methods that are grounded in the several disciplines that address the subject of study. The mixed methods approach is used in this chapter. This approach utilised, mainly, the quantitative research method backed up by the qualitative methods. This approach provided the capacity to fulfil the objectives of the study with deeper understandings that would be difficult to reach if only one method had been used (Easterby-Smith et al. 2012), and similar studies, like Aquino et al. (2019) and Mannakkara (2014), have used such an approach. The main research methods included a questionnaire survey, key informant interviews, field observation and document analysis. The household questionnaire survey was carried out in the hotspots of Cyclone Idai and included areas like Kopa and Ngangu. The questionnaire mainly had closed questions and a total of 219 were administered. Key informants were drawn from political and traditional leaders, selected victims, the local authority representatives and representatives of community-based organisations. Purposive and snowball sampling techniques were used in the recruitment of the research participants. Presentation of data is done in the form of descriptive statistics and frequencies, and also narratives. Figure 8.1 shows the studied area, some of the irrigation schemes and river systems affected by the Cyclone Idai in Chimanimani district.

The next section presents and discusses the key findings. The findings will be presented in two sections. The first section will deal with impacts on the domestic water supply system, and the second section looks at impacts on irrigation water supply.

8.4   Presentation and Discussions of Key Findings

8.4.1   Impact on Domestic Water Supply Systems

Chimanimani district was the hardest hit area by Tropical Cyclone Idai in Zimbabwe. Table 8.1 shows the baseline water points and number of users in the district before Cyclone Idai hit the area. In Chimanimani, boreholes, springs and rivers are the most utilised water points. These provide water for, mainly, household domestic needs, sanitation and different levels of productive use. Significant damage to these sources of water, as a direct result of Tropical Cyclone Idai, resulted in a reduction capacity by local communities to access water for the different use categories.

Figure 8.2 highlights the source of water for the participants of the household questionnaire survey that was carried out in the district. Over 60% of the participants drew their water, for domestic use and sanitation, from municipal piped water systems. A further 28% use water from rivers and boreholes, with another 10% using springs and wells. The results suggest that if municipal water supply systems, in the districts, are not resilient to hazards such as tropical cyclones, the majority of the population will be at risk of water shortages and exposed to the danger of water-borne infections. Chimanimani, however, managed to avoid a serious health crisis after the cyclone because of the quick response by the Ministry of Health and Child Care, and its partners, in distributing water purification tablets and enforcing strict hygiene standards at the holding centres for the internally displaced people. They also moved to acquire a million doses of cholera oral vaccine and launched widespread vaccination campaigns. This quick response did pay off because weekly disease surveillance reports, from the affected areas, did not show any

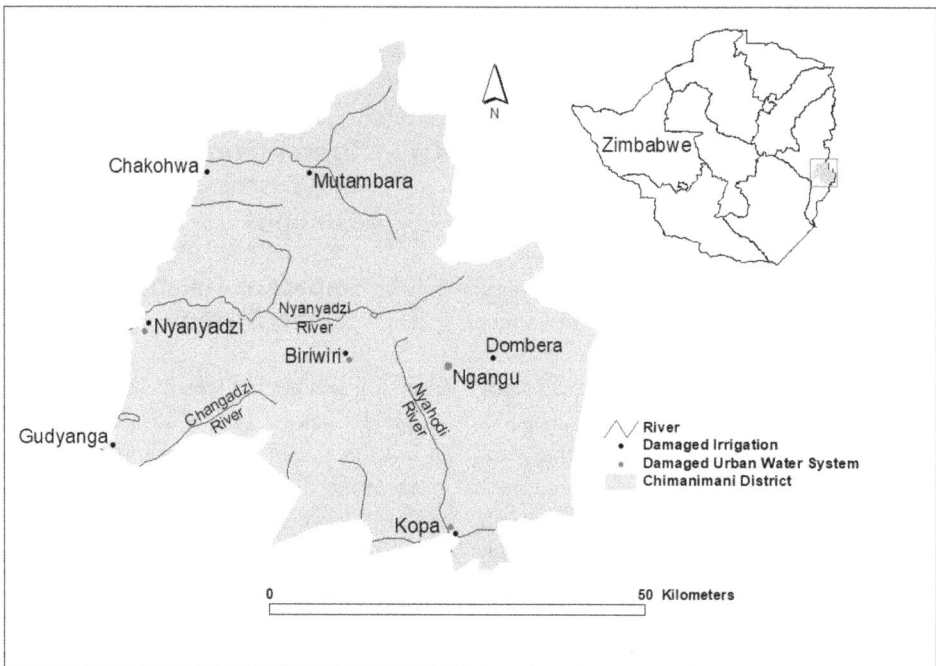

Fig. 8.1 Irrigation schemes and river systems affected by Cyclone Idai. Source: Authors

increase in diarrheal diseases that were above the normal compared to the previous years. Outbreak of diseases after tropical cyclones have occurred numerous times before. When Cyclone Sidr made landfall in south-western Bangladesh in November 2007, about 52 persons suffered from water-borne illnesses. The illnesses were significantly associated with household income, and gender and age of the survivors (Paul et al. 2010).

Figure 8.3 shows the number of water points damaged in Chimanimani district during Tropical Cyclone Idai. Damage mostly occurred on springs (642), piped water systems (438) and boreholes (61). These damaged water points also constitute the most widely used water sources, hence, potential for long-term water availability problems if these sources are not rehabilitated to become resilient to the effects of future shocks.

Figure 8.4 shows that the water supply and distribution system in Chimanimani was either completely or partially damaged during Tropical Cyclone Idai. Close to 73% of the participants saw the water supply system as being completely damaged and no longer usable. Key informants observed this to be the case for the Chimanimani

town and the Kopa area in Rusitu Valley. Damage to the water supply systems in these areas was so extensive that it required the complete laying out of new pipes, to supply not only to the old residential areas but also the new camps for the internally displaced persons. The process to reconnect the entire urban settlements took close to a month to complete, by a Rural District Council (RDC), that had lost all its inflow of revenue and could not bill the water users during the extended period of disaster recovery. Moreover, the reconnection of the water system was done hurriedly with the intention of avoiding the emergence and spread of water-borne diseases, but not necessarily to building a robust water distribution network that can withstand the impact of similar shocks in future. Close to 24% of the participants viewed the damage to water distribution systems as being partial. This means that the systems required some form of repairs before they could be safely utilised again. Less than 2% of the participants observed no damage to the water supply and distribution system in their area. These results show that the water supply system is, largely, not resilient to the impact of tropical cyclones, hence, the

Table 8.1 Water points and number of users in Chimanimani as of 28 February 2019

	Artesian Well	Borehole	Dam	Deep well	Rainwater Harvester	River	Sand Abstraction	Shallow well	Springs	Other
Number of water points	1	416	2	104	1	77	5	36	359	36
Number of households	3	17,703	3	2620	0	5,446	96	1083	7285	592

Source: Authors, data from RINA (2019)

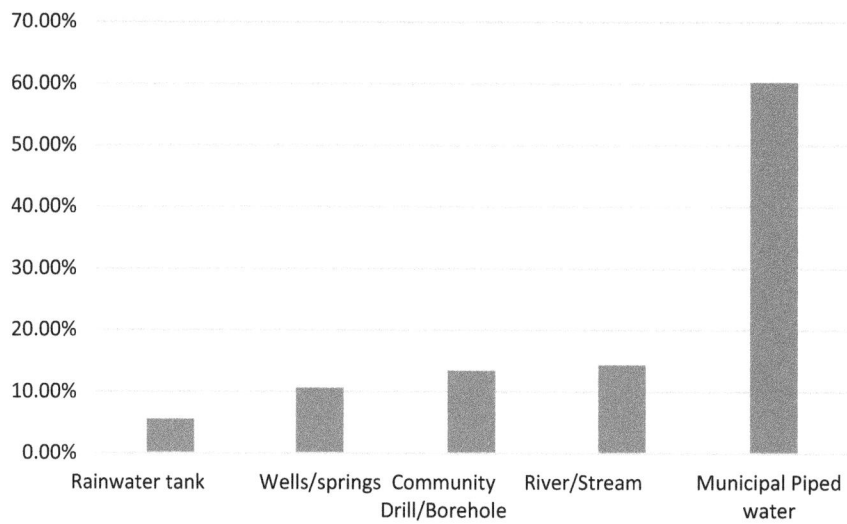

Fig. 8.2 Source of water for participants ($N = 216$). Source: Authors

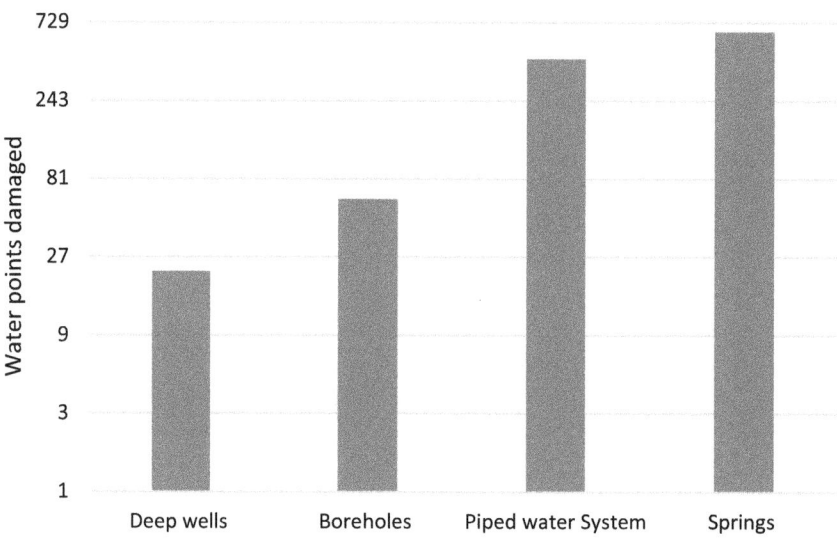

Fig. 8.3 Water points damaged during Cyclone Idai. Source: Authors, data from the Manicaland Provincial Water Sanitation Sub-Committee

need to put in place standards and building codes that promote the use of resilient materials and structures when designing and laying out the water supply and distribution systems as advocated for in the Sendai Framework.

Most household were of the view that the state of water quality in the aftermath of Cyclone Idai in Chimanimani district was poor. Close to 71.5% observed the water quality to be poor, in particular, the colour and taste, while the remaining

29.5% indicated it was still good. Key informants also observed that levels of turbidity of the water were too high. The sewer ponds at Ngangu also overflowed due to heavy rains, as well as some damages to the sewer lines that leaked raw sewage into the nearby river systems—a situation that went on for several months.

The authors had a chance to tour some of the damaged weirs that were supplying domestic water to schools (Fig. 8.5), clinics and residential

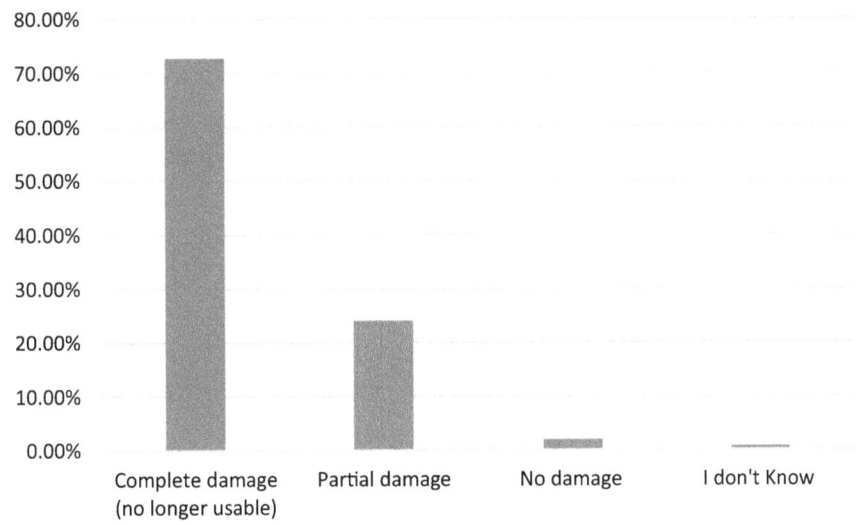

Fig. 8.4 State of water supply and distribution system ($N = 152$). Source: Authors

Fig. 8.5 Damaged Domestic Water Weir in Chimanimani Town. Source: Authors, Fieldwork 2019

areas. One organisation that was involved in BBB domestic water supply (Fig. 8.6) is the Welt Hunger Hife (WHHI), a German humanitarian organisation that has its focus on WASH (Water Sanitation and Hygiene). We toured several piped water schemes done after Tropical Cyclone Idai hit, including the Chimanimani Junior School (Fig. 8.7), Charleswood Primary School and Clinic, Ntamatanda Camp (Fig. 8.8), and Chimanimani Hospital. Other organisations that were involved in water supply restoration proj-

ects included Plan International, ADRA, GOAL and Carte International. What was also critical in BBB programming was the fact that development partners had to work with the District Development Fund (DDF), a government agency. The DDF would allocate development partner areas to work in, so as to leave no areas behind in terms of certain services, in this case, water supply. The involvement of willing and volunteering locals was also highlighted as a huge positive. In one instance, there were 495 community mem-

Fig. 8.6 Repaired and BBB Weir Supplying Chimanimani Junior School. Source: Authors, Fieldwork 2019

Fig. 8.7 Water from the BBB weir stored in overhead tanks at Chimanimani Junior School. Source: Authors, Fieldwork 2019

bers that turned up volunteering to dig the trenches for piping. This signalled how much water was needed in that community following the cyclone. One young village head in the area was singled out as outstanding, in terms of how he would always be there to support all the projects on water supply and sanitation.

From an interview with one of the field workers from WHHI, it was clear that most of the communities in Chimanimani draw their water from springs and many would tap the water individually or collectively. Hence, all the piping and other infrastructure that were in the mountains were swept away during the cyclone. From the respondent, "it was by God's grace that there were not many cases of diarrhoea, typhoid or other diseases". From the WHHI, it was strategic to start BBB water supplies for institutions like schools, hospitals and clinics. The work involved rehabilitation, and, in other instances, it was

Fig. 8.8 Nyamatanda Camp piped water scheme. Source: Authors, Fieldwork 2019

starting a completely new piped water scheme. All the schemes use gravity to take the water to the tanks. At the time of interview, WHHI had completed 20 piped domestic water schemes in Chimanimani East constituency. In addition, they also rehabilitated 20 boreholes in Chimanimani West constituency. Furthermore, in partnership with UKAID and UNICEF, the WHHI distributed 1000 water carrying buckets of up to 25 L capacity. There was also the training of borehole pump minders that was gender responsive for long-term sustainability. A total of 20 pump minders (16 men and 6 women) were trained by the time of the interview.

Studies done in other countries have shown that contamination of water in tanks, pipes and boreholes occur when the elevation of floodwaters raise above elevation of the well-head or intake valves. This sees water flowing directly into wells and other water supply intakes or sometimes raw or partially treated sewage overflows from treatment plants (Carroll 2015; WHO 1998). Other studies have also concluded that tropical cyclones have a negative effect on water quality. Mosley et al. (2004) observed at a remote island of Vanua Levu, Fiji, that after Cyclone Ami in January 2003, water had on average 50% more turbidity, 60% more coliform bacteria and more than 40% less chlorine than samples taken

before the cyclone. This means that good quality water for domestic and primary use becomes a scarce commodity in most of the cases in the aftermath of tropical cyclones. A plan for providing safe water for drinking should always be part of the plan during the pre-disaster period. Erickson et al. (2019) argue that efficient health management must take into account the probability of harmful noxious agents being present in water polluted by biological contaminants, solid waste, sewage, water-borne pathogens and even chemical spills during cyclones. The authors further observed a general absence of application of toxicity tests during cyclone-induced floods, yet, such information is key in understanding risks and preparing vulnerable communities and first responders to deal with them as part of BBB.

Figure 8.9 shows the time it took to restore domestic water supply after Cyclone Idai induced damage in the Chimanimani district. Only 15% of the participants highlighted that it took less than a week to restore the damaged water infrastructure. The rest had to rely on emergency water supply or, when not available, had to use the contaminated or unprotected sources. Many factors determined the speed at which water supply systems were replaced. This included the fact that the place could be reached only by air in the first 2 weeks of the disaster as the entire road

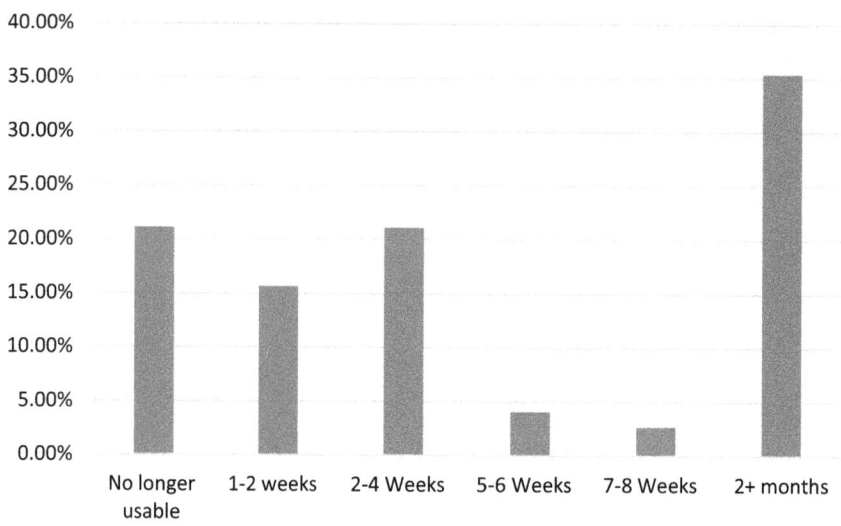

Fig. 8.9 Time taken to restore domestic water supply ($N = 147$). Source: Authors

network was destroyed. Other factors included non-availability of replacement components due to limited accessibility, the rugged terrain in the area and an overwhelmed DRR system. Most of the participants (35%) highlighted that it took authorities over 2 months to restore water supply for them. This is very worrying, given the importance of water in daily household activities and for sanitation.

The situation also exposed the survivors to deadly diseases such as cholera, typhoid and diarrhoea. In BBB of such infrastructure, there should be minimum standards set in terms of the maximum time it should take to restore the water distribution system of an affected area, in addition to the use of materials that can withstand the force of the disaster. Mosley et al. (2004) highlight that when the disaster-affected communities have limited financial and technical capacity to help themselves, assistance from outside may take months to materialise, hence, damaged water infrastructure may take long to repair.

8.4.2 Impacts of Tropical Cyclone Idai on Irrigation Water System

Irrigation systems were also heavily impacted by Tropical Cyclone Idai. In all affected districts of Zimbabwe, about 1500 micro to small-scale irrigation schemes were either completely swept away or damaged. The damage to the irrigation structures and systems was estimated at United States Dollars 4890 million, as weirs and pipelines were washed away and wells were flooded or silted (RINA 2019). The World Bank (2020) estimates that reconstruction and recovery could cost more if building back smarter costs are also factored in. In terms of creating disaster smart structures, irrigation equipment and structures are an important component of Zimbabwe's economy, as well as supporting sustainable livelihoods for the rural societies impacted by tropical cyclone Idai. This, therefore, emphasises the need for a BBB strategy to guide the speedy recovery of this priority infrastructure, in order to safeguard livelihoods of the communities at risk. Figure 8.10 shows the Zimbabwe National Water Authority (ZINWA) bulk water infrastructure affected by Cyclone Idai. Nearly 18 urban and peri-urban water supply systems were affected by compromised water security in the affected places. These included areas like Murambinda, Matendeudze, Chibuwe, Nyanyadzi. Tanganda, Rimbi, Birchenough Bridge, Zimunya, Romsely, Odzi, Checheche and Chipangayi. Some of the remedial works done by ZINWA included the replacement of swept away water distribution mains, replacement of reticulation pipelines swept away, replacement of flooded pump sets and repairing of roofs gusted away from water-

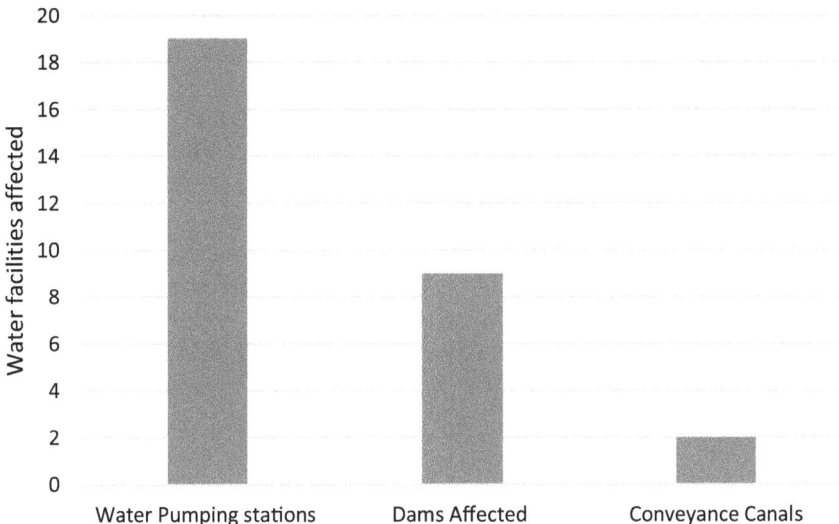

Fig. 8.10 ZINWA infrastructure affected by Cyclone Idai. Source: Authors, data from RINA (2019)

storing facilities. Some pumping units were replaced at the water intakes or were relocated to places deemed safer.

At community irrigation schemes such as Nyanyadzi, Chakohwa and Gudyanga, authorities repaired the eroded or silted conveyance canals, canal embankment, repaired access roads to canals, removed silt from the overnight storage dams and replaced washed away canal panels. However, the works that had been done by October 2019 were mainly to bring back basic water supplies, and a lot of work still to be done to ensure the refurbishment of water supply systems to their pre-disaster condition and to also make them resilient to future shocks.

Table 8.2 summarises the damages suffered at selected smallholder irrigation schemes in Chimanimani. Most of the damage was related to the erosion, silting and flooding of irrigation infrastructures such as canals, pumps and pipelines. Structures that prevent or minimise erosion, silting and flooding are, therefore, key aspects that need to be focused on when trying to build the resilience of these irrigation schemes against future tropical cyclones. According to FEWSNET (2000), similar damages occurred in Manicaland, Masvingo, Midlands, and Matabeleland South provinces in Zimbabwe when Tropical Cyclone Eline swept through the area. The subsequent flooding damaged over 33 irrigation schemes that covered a surface area of

Table 8.2 Irrigation schemes and damages suffered during cyclone Idai

Irrigation Scheme	Damage
Mhandarume	740 m pipeline swept away
Chakohwa	Total silting of canals
Nenhowe	Flooding of pumps and pumphouses
Nyanyadzi	Silting of weir dam, distribution pipes swept away, canal from Nyanyadzi River silted, 525 mm AC delivery pipeline from Odzi river swept away
Gudyanga	Fence and gabion baskets stripped, grading shade and offices silted, borehole collapsed
Tonhorai	Flooding of boreholes
Cashel Valley and Mutambara	A number of weirs and main canals washed away
Zimunda	Weir dam partially damaged and conveyance pipes washed away
Bvumbura	Weir dam partially damaged and conveyance pipes washed away
Nyabande	Distribution pipes damaged and a 300 m perimeter fence swept away

Source: Authors, data from RINA (2019)

over 4230 ha, with the magnitude of destruction varying from scheme to scheme. However, most of the destruction involved key irrigation infrastructures such as dams, weirs, intake structures, pumping stations, canals, distribution structures, pipes and irrigation plots. In Manicaland Province, 13 small dams were damaged, while 15 were damaged in the Midlands Province. The extent of damages to small dams varied from being totally washed off to breaching of the wall and complete siltation.

Irrigation as a safety net from frequent droughts is very key to rural livelihoods in Chimanimani district, especially in the drier, drought-prone and semi-arid places under natural agroecological regions 4 and 5 (Chikodzi 2018). At these places, dryland farming is unreliable and communities undertake on irrigation for guaranteed returns in terms of both nutritional security and as a source of income (Chikodzi et al. 2020). Any damages that result in the failure of small dams and irrigation infrastructure has a direct impact on food security and livelihoods of the affected households as well as the nearby societies. Attempts to build silt traps on irrigation schemes, to proof them from the dangers of tropical cyclones, have been tried before at the Nyanyadzi Irrigation Scheme. The scheme became dysfunctional in the aftermath of Cyclone Eline in 2000, when torrential downpours led to infrastructural damage at the scheme, silted the conveyance the canals which prevented the flow of water to the community and the irrigation plots (UNDP 2019). Development partners such as UNDP, Oxfam and the local community rehabilitated the scheme in BBB style in order to make it resilient to the impacts of torrential rainfall and siltation. When Cyclone Idai hit the scheme again in 2019, the scheme suffered minimal damage which was attributed to the gabions that were constructed after Cyclone Eline.

To promote the concept of BBB in the post-Cyclone Idai era, component two of the World Banks emergency fund to Zimbabwe was aimed at empowering the affected communities to build medium-term recovery from the impacts of the cyclone in the process build resilience to future shocks. The BBB component funded to the tune of US$35 m and meant to support the restore important community assets, such as water and sanitation as well as irrigation systems. This also included community-level structural mitigation efforts for DRR, such as slope stabilisation and ecological rehabilitation (World Bank 2020).

8.5   Conclusions and Recommendations

This chapter sort to document the impacts of Tropical Cyclone Idai on domestic and irrigation water infrastructure in the Chimanimani district of Zimbabwe, and to advocate for the employment of the BBB concept in the repair or reconstruction of such infrastructure. Results showed that the water supply infrastructure for both domestic and irrigation was extensively damaged by the tropical cyclone. The replacements done for these damaged infrastructures were meant to mainly restore service to the affected communities within a minimal time, which was commendable. Although the principles of BBB are visible in the cases where some projects were completed, there remains some challenges. This includes many households remaining in camps with temporary water supply systems. The business case for BBB in this area is the anticipated upsurge in the frequency and intensity of cyclones hitting the area in future, given its location, and also the importance of water availability for both domestic and productive use (irrigation).

In light of the results, the research recommends the need to strengthen and update the building codes and minimum standards for water supply infrastructure in the Chimanimani area. These standards need to be developed through an all-inclusive consultation process that takes into account realities on the ground. The standards need to balance between affordability and risk. To prevent future catastrophic damage on the water distribution system, the priority improvements need to be identified and implemented in the light of the new standards, well before the anticipated disaster occurs. Further, there is a need to educate and train the community-based builders, contractors and technocrats on the new standards, through capacity-building training. Deployment of easy-to-use tools in the process will further assist the designers, contractors and the inspecting authority to check the structural designs. There is a need to fund investment into resilience-building measures to protect women, children and other vulnerable groups. These measures can be in the form of early warning systems, early action, resilience infrastructure, social safety nets and other recovery measures that can assist the most vulnerable groups to bounce back from shocks and stresses arising from tropical cyclones.

Integrated water resources management needs to be implemented in almost all the river catchments in Chimanimani district. The need remains urgent, given the serious land degradation due to streambank cultivation, poor farming methods, deforestation and massive gold panning have led to the siltation of river systems in the area. This silt will eventually damage irrigation and water supply infrastructure during tropical cyclone-induced floods. Deforestation leaves areas vulnerable to massive erosion and mass movement. It was observed in Chimanimani that landslides affected most areas, but land that was deforested and that suffered from frequent wildfires, like the Ngangu Mountain, suffered more damage compared to similar pristine environments. Through integrated water resources management, the BBB component of environmental resilience is buttressed.

References

Aquino, D.H, Wilkinson, S., Raftery, G. M., Potangaroa R. (2019). Building back towards storm-resilient housing: Lessons from Fiji's Cyclone Winston experience. International Journal of Disaster Risk Reduction, 33, 355–364. https://doi.org/10.1016/j.ijdrr.2018.10.020.

Aquino, D.H., Wilkinson, S.J., Raftery, G., Potangaroa, R., Chang-Richards A. (2018). Challenges to building housing resilience: the case of Fiji post-cyclone Winston, Procedia Engineering. 212, 475–480, https://doi.org/10.1016/j.proeng.2018.01.061.

Carroll, F. (2015). Building it back better to reduce risks after multiple disaster events. Paper presented at the 2015 Floodplain Management Association National Conference Convention Centre, Brisbane, 19 – 22 May 2015.

Centre for Climate and Energy Solutions. (2020). Hurricanes and Climate Change. Retrieved from: https://www.c2es.org/content/hurricanes-and-climate-change/#:~:text=Frequency%20and%20intensity%20vary%20from,year%2C%20including%20about%20seven%20hurricanes. (Accessed 9 September 2020).

Chikodzi, D. (2018). Unusual waterscapes and precarious rural livelihoods: Occurrence, utilisation and conservation of springs in the Save Catchment, Zimbabwe. Thesis Submitted to the University of the Western Cape, Capetown.

Chikodzi, D., Tevera D., Mazvimavi, D. (2020). SDG 15 and Socioecological Sustainability: Spring Waterscapes and Rural Livelihoods in the Save Catchment of Zimbabwe. In G. Nhamo et al. (eds.), Scaling up SDGs Implementation, Sustainable Development Goals Series, https://doi.org/10.1007/978-3-030-33216-7_4.

Commonwealth of Australia. (2014). Australian Government Reconstruction Inspectorate Submission to the Productivity Commission Inquiry into National Natural Disaster Funding Arrangements. Retrieved from available from: http://www.pc.gov.au/inquiries/completed/disaster-funding/submissions/submissions-test/submission-counter/sub039-disaster-funding.pdf. (Accessed 5 September 2020).

Dube, K. & Sigauke, E. (2015). Irrigation Technology for smallholder farmers: A strategy for achieving household food security in Lower Gweru Zimbabwe. South African Journal of Agricultural Extension, 43(1), 1-11.

Easterby-Smith, M., Thorpe, R., Jackson, P. (2012). Management Research, London, Sage.

Erickson, T.B., Brooks, J., Nilles, E.J., Pham, P.N., Vinck, P. (2019). Environmental health effects attributed to toxic and infectious agents following hurricanes, cyclones, flash floods and major hydrometeorological events, Journal of Toxicology and Environmental Health, Part B, 22(5-6), 157-171, https://doi.org/10.1080/10937404.2019.1654422.

FEWSNET. (2000). Assessment of the Impact of Cyclone Eline (February 2000) on the Food, Agriculture and Natural Resource Sector in Zimbabwe. Retrieved from: https://fews.net/sites/default/files/documents/reports/1000050.pdf. (Accessed 9 September 2020).

Intergovernmental Panel on Climate Change (IPCC). (2007). Climate change 2007: impacts, adaptation, and vulnerability. Cambridge University Press, Cambridge.

Mohan, P. (2017). Impact of Hurricanes on Agriculture: Evidence from the Caribbean. Nat. Haz. Rev. 18(3), 04016012.

Mannakkara, S. (2014). A Framework for Building Back Better During Post-Disaster Reconstruction and Recovery, Department of Civil and Environmental Engineering, University of Auckland, Auckland.

Mosley, L.M., Sharp, D.S., Singh, S. (2004). Effects of a Tropical Cyclone on the Drinking water quality of a Remote Pacific Island. Disasters, 28(4), 393–405.

Paul, B.K., Rahman, M.K., Bankim Chandra Rakshit, B.C. (2010). Post-Cyclone Sidr illness patterns in coastal Bangladesh: an empirical study. Nat Hazards, 56:841–852. https://doi.org/10.1007/s11069-010-9595-5.

Perfecto, I., Hajian-Forooshani, Z., Iverson, A., Irizarry, A.D., Lugo-Perez, J., Medina, N., Vaidya, C., White A., Vandermeer, J. (2019). Response of

Coffee Farms to Hurricane Maria: Resistance and Resilience from an Extreme Climatic Event. Nature Scientific Reports. 9:15668. https://doi.org/10.1038/s41598-019-51416-1.

UNDP. (2019) Let it flow: Adapting the Nyanyadzi Irrigation Scheme to Climate Change. Retrieved from: https://www.zw.undp.org/content/zimbabwe/en/home/stories/let-if-flow%2D%2Dimproving-water-access-in-nyanyadzi-.html. (Accessed 9 September 2020).

UNDRR (United Nations Office for Disaster Risk Reduction). (2015). Sendai Framework on Disaster Risk Reduction (2015-2030). New York: UNDRR Secretariat.

United Nations. (2015). Transforming our World: The 2030 Agenda for Sustainable Development. New York: United Nations Secretariat.

United Nations Economic Commission for Africa-UNECA. (2019). Building back better: planning workshop for climate resilient investment in reconstruction and development in cyclone affected regions of Malawi, Mozambique and Zimbabwe. Retrieved from: https://www.uneca.org/building-back-better. (Accessed 9 September 2020).

World Bank. (2020). Zimbabwe IDAI Recovery Project (P171114). Project Information Document (PID). World Bank Group.

World Health Organisation (WHO). (1998). Guidance on Water Supply and Sanitation in Extreme Weather Events. WHO Regional Office for Europe, Copenhagen, Denmark.

Zimbabwe Rapid Impact and Needs Assessment (RINA). (2019). Zimbabwe Cyclone Idai Rapid Impact and needs assessment. World Bank, Government of Zimbabwe and GFDRR.

Part IV

Tropical Cyclone Idai Acts of Kindness

Ethical Philanthropy and Social Responsibility During Natural Disasters: The Higherlife Foundation and Tropical Cyclone Idai Interventions

9

Abstract

With increasing pressure from natural disasters that result from extreme weather events such as tropical cyclones, floods, droughts, heatwaves and extreme snowfall, the call for ethical philanthropy and social responsibility through set-up foundations has been growing. As more and more such devastating natural disasters manifest across the world; global citizens have continued fixing their eyes and ears on seeing and hearing what businesses, some repeatedly earning super profits, are doing to alleviate the suffering of affected communities. Through their foundations, corporates are expected to engage in disaster risk reduction (DRR) activities, from early warning, search and rescue to relief and recovery. In March 2019, Tropical Cyclone Idai swept across southern Africa, hitting four countries—Malawi, Mozambique, Madagascar and Zimbabwe. Lives were lost, infrastructure extensively damaged, and many residents left homeless and without livelihoods. It is in this regard that the Higherlife Foundation moved without hesitation into the affected areas of Chimanimani and Chipinge in Zimbabwe to provide support. Through a survey, interviews and field observations, it emerged that the Higherlife Foundation had many interventions. These interventions covered impact assessment, response, recovery and reconstruction. They manifested in the form of volunteerism, child protection, water, sanitation and hygiene (WASH) and education services delivery. The key activities included the distribution of hygiene packs, stakeholder engagements, procurement, assessments, food distribution, mobilisation and logistical support. As this chapter was being concluded, it was evident that the Higherlife Foundation was in the affected areas for the long haul and not to secure once-off TV camera mileage. The chapter recommends that ethical philanthropy shown by the Higherlife Foundation be developed into a global model.

Keywords

Cyclone Idai · Higherlife Foundation · Philanthropy · Social responsibility · Ethical

9.1 Introduction and Background

Traditionally, national governments, nongovernmental organisations (NGOs) and international organisations are at the coalface of disaster risk reduction (DRR) activities that include financial commitments. However, the corporate world, through their foundations, has found itself fully embedded in the DRR space, given the pressure from societies (Ballesteros and Useem 2017) and from their own realisation to be responsible citizens. Corporate philanthropic giving drastically increases during mega-events such as natural disasters that include hurricanes, cyclones, floods and droughts (Tilcsik and Marquis 2013). From a study of 206 natural disasters that took place between 2005 and 2014, which included hurricanes, floods, tornadoes and wildfires, McKnight and Linneluecke (2019) established a link between company responses to disaster type. Although good work is taking place, history does not judge some corporate philanthropic giving kindly. There are allegations that the giving is conditional, mainly done to gain the licence to operate in communities and maintain good reputation (Chen et al. 2020). The giving is also associated with unethical behaviour (Hotho and Girschik 2019) and the desire to influence positive consumer attitudes (Hwang et al. 2019). The love of media attention is further highlighted as a huge pull factor in disaster philanthropy in China and elsewhere in the world (Ouyang and Wei 2017). The mood of senior executives at the headquarters of companies also plays a major role in how many resources are released for certain philanthropy activities (Zolotoy et al. 2020).

Hartz (2017) highlights that disaster philanthropy is at a critical point as global disasters keep increasing due to climate change and other anthropogenic triggers (Chandra et al. 2016). The private sector is therefore viewed as better positioned to apply agility, its expertise and resources to mitigate the effects of these disasters as it reimagines the future of disaster philanthropy and corporate social responsibility (CSR) (Hartz 2017; Owusu-Kwateng et al. 2017). India became the first country in the world to integrate voluntary and mandatory aspects of CSR in the Companies Act 2013. Companies beyond a certain size are required to reserve and spend a certain percentage of their profits on CSR activities (Jain et al. 2020).

Ariyabandu and Hulangamuwa (2002) were concerned about CSR and natural DRR in Sri Lanka. The authors identified floods, landslides, cyclones and droughts as the most pressing. With the eastern coast of the country prone to cyclones,

November and December were observed to account for 83% of occurrences. The cyclones usually bring devastation, for instance the 1964 cyclone that ravaged through the districts of Polonnaruwa, Anuradhapura, Mannar, Batticaloa and Amparai, affecting 75,000 people, leaving 280 dead. Another cyclone in 1978 affected about one million people and left 100,000 homeless. From a study involving 15 corporates, the authors found that there were four main CSR activities, including philanthropic and charitable activities, environmental conservation, public awareness, and corporate sponsorship. Furthermore, four other key areas of CSR for DRR included unilateral initiatives for relief, rehabilitation and reconstruction; collective initiatives for relief, rehabilitation and reconstruction; long-term solutions and public awareness. The Ceylinco Group, with 200 subsidiaries, has a trust fund called Sarana (which means "help"). Apart from contributing towards food, clothing and shelter for the poor, the fund gives towards relief in times of natural disasters such as cyclones.

This chapter presents a unique case study regarding the Higherlife Foundation's ethical philanthropic engagement during and after Tropical Cyclone Idai that hit southern Africa in March 2019. The cyclone killed more than 1300 people and left many more homeless and injured in Malawi, Mozambique and Zimbabwe. Since most corporate foundations are linked to certain corporations, the matter of CSR emerges and how this can be utilised for DRR (Behl and Dutta 2019; Fernando 2006). Typically, the private sector or any other organisation could be involved in any disaster cycle that includes preparedness or preplanning, response and recovery (Chandra et al. 2016). In other instances, the response stage could be further split into search and rescue as well as relief (Jordan and McSwinggan 2012). The private sector generally contributes to disaster recovery financing through collaboration with the public sector in public–private partnerships, driving innovation and facilitating technology use, and assisting smaller communities in managing the influx of aid (Chandra et al. 2016).

With its head office in Harare, Zimbabwe, the Higherlife Foundation has other offices in

Bujumbura (Burundi), Maseru (Lesotho) and Johannesburg (South Africa). Although the Foundation has a number of partners, Delta Philanthropies, Econet Wireless and LIQUID TELECOM remain the anchors (Higherlife Foundation 2020). Other partners are the End Fund, Yale Young African Scholars Program, African Philanthropy Forum and World Vision International Zimbabwe. These partners work together with the Higherlife Foundation to further work in different communities as appropriate. Higherlife Foundation was co-founded in 1996 by African philanthropist Tsitsi Masiyiwa and partner Strive Masiyiwa, the chairman and founder of Econet Wireless, a diversified telecommunications group. The Higherlife Foundation mandate resonates in five areas, namely education, health, leadership and lifelong development, job creation and sustainable livelihoods, and girls' empowerment. The profiling of ownership in organisations and philanthropic foundations is important in that it has been found to influence how resources are deployed during disasters (Gao and Hasi 2017). As to who the Higherlife Foundation is, its vison, mission and values are presented in Box 9.1.

Since its inception in 1996 and as of March 2020, the Higherlife Foundation has recorded several achievements worth mentioning. More than 250,000 students have been supported through scholarship and leadership training, with a further 68,000 youth reached annually through mentorship under the leadership and lifelong development mandate. In terms of its health mandate, a total of US$60 million has been invested in healthcare and crisis response, especially in the programme aimed at ending chorea by 2030 in affected countries. In terms of the job creation and sustainable livelihoods mandate, a total of US$100 million has been committed to rural investment. Lastly, the Higherlife Foundation is working towards gender parity across all its programmes through the girls' empowerment mandate (Higherlife Foundation 2020).

This chapter therefore documents and analyses how the Higherlife Foundation became involved during Tropical Cyclone Idai, mainly during the relief and recovery stages. The objective is to demonstrate how the philanthropic engagements exceeded the usual window dressing and mileage-seeking endeavours commonly associated with such efforts. The magnitude of the disaster demanded that organisations such as Higherlife Foundation do something, and do it urgently, since there are usually fewer internal procedures in releasing aid (Ballesteros and Useem 2017).

Box 9.1 The Higherlife Foundation's Vison, Mission and Values

- The Higherlife Foundation is a social impact organisation that invests in human capital to build thriving individuals, communities and sustainable livelihoods.
- Vison: To provide a platform for people to fulfil their God-given purpose.
- Mission: To invest in Africa's human capital in order to build thriving individuals, communities, and sustainable livelihoods.
- Values: Faith in God, love, integrity, knowledge, accountability, collaboration, responsive leadership.

Source: Higherlife Foundation (2020 online)

9.2 Literature: The Contested Space of Disaster Philanthropy and CSR

The positive impact of corporate philanthropic activities is known (Gong and Ho 2018; Famiyeh 2017). Barnett (2019) acknowledges that the vast literature establishing a "business case" for CSR appears to find positive correlation, especially if it responds to the needs of its primary stakeholders. For example, a study on Chinese A-share listed companies revealed that corporate charity donations had a positive impact on financial performance (Liao 2020). Similar findings were made in the same country by Wang et al. (2019). However, Gao et al. (2019), and Bhardwaj et al. (2018) maintain that the relationship between corporate giving and corporate financial performance has remained inconclusive. This is so because different stakeholders may react differently to a firm's philanthropy activities. Kanji and Agrawal (2019) argue that the concept of CSR has always provided discussion points since the 1950s. To this end, the work by Howard R. Bowen published in 1953, "Social Responsibilities of the Businessman", marked the start of modern-day literature in this space.

In India, "all public and private sector companies with net worth of approximately US$69.5 million or more, or a turnover of approximately US$139 million or more, or net profits of approximately US$695,000 or more" must be engaged in CSR activities (Jain et al. 2020, p. 2). This is done by apportioning 2% of average net profit from the past three financial years (Kanji and Agrawal 2019). Furthermore, the companies should set up a CSR Committee with a minimum of three directors, including at least one independent director. Non-compliance attracts penalties ranging from US$695 and US$35,000, and defaulting individuals are liable for imprisonment of up to 3 years (Jain et al. 2020). What is of interest is that under Schedule VII of the Companies Act 2013, activities to which CSR applies are prescribed, and these include DRR (Box 9.2). Sadly, from an events study approach, Manchiraju and Rajgopal (2017) found that the Companies Act 2013 led to a 4.1% drop in the stock price of companies forced to spend the 2% on CSR when it was introduced.

As is already clear, DRR philanthropy and CSR are littered with controversy. While there is space for engagement in private sector DRR philanthropy, three common barriers to effective delivery exist, namely: (1) heightened risk of fraud, (2) complex regulatory issues and (3) public distrust of traditional institutions (Hartz 2017). These barriers demand that companies have disaster philanthropy strategies ahead of such events taking place. The sense of urgency post-disaster may result in higher risk of fraud as so-called pop-up (aid predator) organisations

> **Box 9.2 Activities that Qualify for CSR in India**
> i. Eradicating extreme hunger and poverty;
> ii. Promotion of education;
> iii. Promoting gender equality and empowering women;
> iv. Reducing child mortality and improving maternal health;
> v. Combating human immunodeficiency virus, acquired immune deficiency syndrome, malaria and other diseases;
> vi. Ensuring environmental sustainability;
> vii. Employment-enhancing vocational skills;
> viii. Social business projects;
> ix. Contribution to the Prime Minister's National Relief Fund or any other fund set up by the Central Government or the State Governments for socio-economic development and relief and funds for the welfare of the Scheduled Castes, the Scheduled Tribes, other backward classes, minorities and women; and
> x. Such other matters as may be prescribed.
> Source: http://www.mca.gov.in/Searchable Acts/Schedule7.htm (accessed 11 March 2020)

quickly set up post-disaster, emerge. Although some pop-up organisations may be genuine and attracting donations, they often lack knowledge at the local level where disasters strike to effectively deliver aid. Many pop-up organisations have been observed to exploit the disasters for personal gain, and usually philanthropic foundations with experience can easily determine whom to work with or avoid.

After Hurricane Katrina in 2005, the USA Federal Bureau of Investigation (FBI) instituted 15 investigations for illegitimate websites designed to look like charities in Florida. Such activities at times result in corporates and their foundations delaying responses or forgoing it altogether. Concerning regulatory matters, the Nepal complex regulatory system was cited as prohibiting aid to reach the affected areas on time after the 7.8-magnitude earthquake in 2015. However, regardless of the crucial role played by the private sector and their foundations in disaster financing, there is no active discussions on how best this can be avoided and how the private sector should contribute (Chandra et al. 2016).

Earlier, it was hinted that the private sector gets involved in the disaster cycle that includes preparedness or preplanning, response and recovery (Chandra et al. 2016). In the preparedness phase, the private sector could assist communities and other stakeholders in raising awareness, building resilience and building adaptive capacity to disasters. IBM, for example, became involved in long-term resilience by establishing Smarter Cities Challenge Grants in Japan after the 2011 earthquake and tsunami. The company further established an open-source database used to track people and resources for ongoing community monitoring and social and economic recovery. Airbnb partnered with Portland and San Francisco to predetermine hosts for displaced people during disasters. Alerts were also provided via web and mobile technologies. The SeeClickFix uses its citizen database requests for on-the-spot debris removal, fostering quick recovery after disasters. During the Great East Japan Earthquake, an estimated 32.75% of the sampled companies (Chandra et al. 2016) provided support for recovery. However, Hunt and Eburn (2018) discovered that even though Australian business is greatly involved in disaster relief and recovery, there still remains a gap in exploring opportunities to enhance firm involvement in building disaster resilience.

What is also of interest are the pathways of decision-making involvement in DRR, and in this case, intervening in a drought situation in Sri Lanka. In ten companies sampled, the managing director or chairperson or chief executive officer (CEO) or the board of directors made the main decisions (Ariyabandu and Hulangamuwa 2002). In two other companies, the employees made the decisions. In unilateral interventions, collected food items included bottled water, medicine,

clothing and other essential supplies, which were self-transported to affected areas in own vehicles, with own drivers and distributed. Other companies collected the items and donated them to relief workers and charities for distribution. At times, services of company doctors and other specialists were offered to join medical camps, with most of the unilateral relief efforts being one-day programmes. Collection of goods would last about 2 weeks. In rare occasions did distribution operations exceed a day. However, during a severe drought, Indra Traders embarked on relief work for about 2 months, only leaving when there was no further need for them to stay. Much of the money would come from companies, although employees could contribute, and amounts varied from about US$3500 to US$17,000. Contributions were usually aimed at the most impacted sections of the community, with beneficiary views taken into account. Some challenges were however encountered—from the difficulty in identifying beneficiaries with many companies involved and not coordinated to the government not being structured and with no clear plans. After the involvement, companies highlighted some of the benefits gained, including making employees work as a team and be more productive, building relationships with communities, obtaining tax relief, image enhancement, and potential to attract new customers.

After the Asian tsunami of December 2004, there was a surge in CSR. Hundreds of initiatives were launched in Sri Lanka alone (Fernando 2006) although many questions arose concerning the long-term engagements. For Brandix Pty Ltd, the involvement included voluntary employee time; setting up of voluntary hotlines for marooned employees, volunteer employee teams and providing tents; and providing 4000 water wells (long term). The private sector and not-for-profit organisations usually raise and donate huge sums of money for relief and recovery during disasters the world over. For example, during Hurricane Sandy, US$141 million was raised and US$1.2 billion during Hurricane Katrina (Chandra et al. 2016).

Jordan and McSwinggan (2012) portray companies as interested in furthering their lines of business through DRR. For example, the insurance and reinsurance industry educate business owners about certain disaster preparedness and to build resilience. Logistics companies find traction in assisting with managing disruptions from extreme weather events, with companies such as UPS and FedEx being specialist in this area. This is not a unique observation, as Cheng et al. (2019) also made similar findings when considering mobile CSR in the USA. Procter and Gamble (P&G) became involved in the Thailand flood of 2011 during the preparation, immediate response and reconstruction phase (Banomyong and Julagasigorn 2017). Activities supported included the distribution of water purifiers through the Princess Pa Foundation under the Thai Red Cross Society. Other revelations from India were that disaster philanthropy exercised by foreign multinational corporations (MNCs) differed from that exercised by domestic firms. From a survey of 190 MNCs and 660 domestic firms, Mitani (2017, p. 941) found that:

> The increase in MNE contributions was much larger and less strongly tied to promotional activities than the increase in contributions from domestic firms, and this difference persisted over time. Moreover, the performance implication of post-disaster philanthropy was stronger for MNEs than for domestic firms.

Hence, philanthropy in MNCs focuses more on strategic positioning. In China, Yang and Tang (2020) found that politically connected entrepreneurs were more likely to make major philanthropic donations for DRR, poverty reduction, environmental protection and charity. From such, many ended up getting political offices in government and the ruling party. Similar views are held by Gao and Hafsi (2017), based on another study on Chinese firms and their donations to the 2008 Sichuan earthquake. The authors found that firms depending more on government for support are more likely to donate towards disaster relief. Furthermore, companies under scrutiny from the government and the public were also likely to donate more.

Lastly and drawing from the Nepalese experiences, Adhikari and Matilal (2017, p. 33) argue as follows: "Disasters have long-term consequences and corporations cannot discharge their

social responsibility by simply offering immediate relief to the victims and then showcasing these efforts on their statements". Vehicles are seen loaded with items for disaster relief, communities line up to receive the goods under the corporate banner, with photographs in the newsletters and annual reports. Such actions will be heralded as valuable contributions, providing a marketing platform for future business. This remains unethical and should be stopped. A study of 84 of the Fortune 100 companies revealed that most companies engage in disaster philanthropy for ethical reasons (Johnson et al. 2011). To this end, scholars of business ethics delineate three ethical frameworks regulating disaster philanthropy, namely virtue ethics, justice ethics and an ethic of care. Some of these values were experienced in the Higherlife Foundation case study.

The next section presents the methodological design, including the main methods used to generate and analyse the data.

9.3 Methodology

As highlighted in the introduction, this chapter documents how the Higherlife Foundation became involved during Tropical Cyclone Idai, mainly during the relief and recovery stages. The objective is to demonstrate how the philanthropic engagements exceeded the usual window dressing and mileage-seeking endeavours commonly associated with such, identifying itself strongly in the ethical disaster philanthropy space. The main methods of data generation were documents, focus group and individual interviews as well as a survey that requested affected community members at the epicentre of Cyclone Idai in Chimanimani to evaluate some of Higherlife Foundation's activities. Documents containing raw data were forwarded from the official channels of the Higherlife Foundation, and such data included those gathered during impact assessment, response, recovery and the reconstruction phases. There were also observations on the ground that took place during 3 weeks of fieldwork in the study area, with photographs taken.

From the survey, a total of 153 household responses were obtained. These households were mainly drawn from the affected areas of Kopa, Ngangu, Chimanimani central business district and Machongwe. Households were also drawn from the temporary shelters. Altogether there were four temporary shelters (three around the Chimanimani business centre and the Village suburb, and one at Kopa). The respondents from the households were either household heads, or their representatives of 18 years and above. A census approach was applied to draw out respondents, with households that had no suitable respondents skipped as appropriate. Survey data were analysed through the use of QuestionPro, with interviews analysed through generic document analysis techniques. The fieldwork was undertaken in September and October 2019 for a period of 10 days with a group of 16 fieldworkers on the ground to administer the survey.

The key informants granting interviews were mainly referral, and those that were in government and traditional leadership. To this end, four chiefs were among those interviewed. In addition, representatives of relief agencies operating in the affected areas were also interviewed. To have a detailed account of what transpired, a focus group discussion involving six Higherlife Foundation representatives was organised at their offices in Harare. From this focus group discussion, additional information had to be emailed to the authors with details of donations and how they were distributed.

The methods applied are not new in disaster philanthropy studies as these have been applied elsewhere by Mitani (2017), Ouyang and Wei (2017) and McKnight and Linneluecke (2019). The next section presents the data and discusses the findings emerging from this chapter.

9.4   Presentation of Data and Discussion of Findings

The Higherlife Foundation interventions during Cyclone Idai took place mainly in the Chimanimani District and, to some extent, the Chipinge District. These interventions covered aspects in four main disaster cycle areas, namely impact assessment, response, recovery and reconstruction. The Higherlife Foundation code-

named the last phase the reboot stage. For the purposes of this chapter, the intervention areas are presented and discussed under five subsections, namely understanding the disaster magnitude and partnership with Econet; volunteerism; child protection; water, sanitation and hygiene (WASH) and education. The key activities from Higherlife Foundation included hygiene packs distribution, stakeholder engagements, procurement, assessments, food distribution, mobilisation and logistical support. Further details are deliberated in the following subsections.

9.4.1   Understanding the Disaster Magnitude and Partnership with Econet

When Cyclone Idai hit on 15 March 2019, there was a Higherlife Foundation heads-up in Harare, revealing that Tropical Cyclone Idai had hit Chimanimani and Chipinge. Those areas were sealed off due to roads and bridges being washed away and peoples' houses being destroyed. Furthermore, several deaths were reported. The two areas of Kopa and Ngangu were the epicentre of the disaster. As part of response to emergencies at Higherlife Foundation, the organisation immediately consulted with the government's Civil Protection Department to find out what had been put on the ground and where the gaps were. Given the magnitude of the disaster, there was no time for a normal rapid assessment. The Higherlife Foundation quickly mobilised through its Emergency Fund and started procuring foodstuffs. This concept of swift response from the corporate sector and related foundations emerged frequently in the literature.

After about a week from 15 March 2019, the epicentre of the disaster was still not accessible by road. As such it was through helicopters that Higherlife Foundation could go and drop off the food. Describing the challenges of access, one of the Higherlife Foundation Focus Group Discussion respondents indicated that he drove from Harare to Mutare, the capital of Manicaland Province that was hit by cyclone idai. The distance is about 270 km. The official indicated he had been overwhelmed with phone calls, including social media postings, and wanted to get first-hand information from the government officials in Mutare. On day 3 after the disaster, the Higherlife Foundation official joined hands with the Econet general manager for Manicaland Province and drove to Chimanimani. Joining hands with Econet is not surprising, given the link between the founders of the Higherlife Foundation and Econet explained in the introduction. The couple could not get to Chimanimani town. They only managed to reach Skyline some 30 km away, which was now being used as the main holding camp. The Higherlife Foundation official reported as follows: "That is where there was real disaster because we could see people with fractures who could not walk; some were being brought in by their relatives after risking crossing the flooded rivers. … Otherwise I could say Chimanimani was 100% cut-off".

The couple had to return to Mutare where they attended another meeting Monday morning. It emerged there was not much taking place as the Manicaland provincial government officials were waiting for ministers from Harare to declare the emergency. This was done around 11h00. Following the declaration of an emergency, administrative wheels started turning quicker. However, the use of the helicopters for search and rescue as well as relief was still not possible due to very bad visibility. With a helicopter, the pilot needed to see with their naked eyes the flying pathways. Hence, in such bad weather, the risk was too high. For somebody who has not been to this mountainous area it is difficult to understand. However, our fieldwork in October 2019 confirmed how bad visibility could be as it could literally go down to zero within a very short period. Figures 9.1 and 9.2 are two photos taken from the Chimanimani observatory 23 min apart; one at 14:53 and the other at 15:16 on 11 October 2019. The set of cluster buildings are Chimanimani Junior School.

On Monday evening, 15 March 2019, there was a meeting at the Mutare Bottling Company, which is part of Econet. Subsequent meetings were also held on Wednesday 17 and Thursday 18 March at the same venue. There was now a need to map out logistics as both Econet and the Higherlife Foundation were mobilising resources in Harare to be taken to Chimanimani, which was situated more than 150 km from Mutare. The World Vision warehouses were made available for all divisions in Mutare. As indicated in the introduction, World Vison is one of the partners to the Higherlife Foundation. With this offer emerged another logistics challenge, namely that all the donated goods would be in the World Vision warehouses and might not reach the

Fig. 9.1 Visibility from Chimanimani observatory at 14h53. Source: Authors, Fieldwork 2019

Fig. 9.2 Visibility from Chimanimani observatory 15h16. Source: Authors, Fieldwork 2019

intended beneficiaries. Prolonged warehousing and long chains of logistics could further lead to some items going missing. To this end, the Higherlife Foundation official was given the responsibility to be overseers for everything that was to be donated. Another trip to Chimanimani was made to a place called Silverstream, owned by the Wattle Timber company. Negotiations for a makeshift warehouse were made, resulting in their old sports club secured.

The first truckload from Mutare Bottling comprised 34 tons, offloaded at Skyline. It had mealie-meal, cooking oil, salt, sugar beans and drinking water. Silverstream was established as the official warehouse, and all the other trucks that were donating food items were directed to Silverstream. However, stock started piling up as it could not be moved due to bad roads and weather. Hence, Econet hired three helicopters from South Africa, and there were two from the Airforce of Zimbabwe in the following week. The movement of donated goods started soon after the helicopters came. The Higherlife Foundation official managed to fly to Kopa. This was the first time that relief food was delivered. The narrative of helicopters and when the first relief food reached Kopa, collaborate perfectly with what emerged from the survey and interviews with the displaced communities that were

in the Kopa tents. The Higherlife Foundation official had to instruct the helicopter crews in terms of where to drop the food items. Econet further hired tractors for other relief and recovery work that was done by the volunteers.

In terms of administration and transparency, the helicopter teams would have representatives from the Higherlife Foundation and the government, especially Social Welfare and the Civil Protection Department. These would get into the chopper to record where the food items were being dropped. Although, generally, there was order during the distribution of relief food, some areas like Mubovangwa had some indiscipline. Each time the chopper landed, the communities looted the food. More than four loads were sent down and not a single one was accounted for. As such, delivery of relief food had to be stopped until the road was accessible. The choppers worked for a week, with Econet hiring ten trucks during the second week to take over from the choppers because the roads were now accessible, with some diversions, and temporary bridges up. Areas would be surveyed for access depending on the size of the truck.

Other relief logistics were developed with members of the community in Kopa. The Higherlife Foundation official reported a certain retired air lieutenant called Kaneta. He was the first contact person in the area, and his house resembled a preschool after the disaster. His experience in the army was a welcome development in logistics and handling disaster situations.

9.4.2 Volunteerism Perspectives

Volunteerism took place under the umbrella project #ThisIsMyHOME. The project was split into two elements: #ThisIsMyHOME volunteers leaders and #ThisIsMyHOME community volunteers. From #ThisIsMyHOME volunteer leaders, 15 were involved for 6 months. Their tasks included household assessments, with a total of 5655 households assessed for the extent of damage from the cyclone and needs of interest to the Higherlife Foundation. They were further involved in the sorting of clothing donations, building equipment inventory management, assisting with food distributions, assigning of duties to builders and carpenters, as well as the disbursement of building materials for house repairs.

In terms of the #ThisIsMyHOME community volunteers element, 863 were engaged for 3 months. Their activities covered 13 wards. Specific engagements included a large-scale clean-up campaign, disbursement of building materials for house repairs, household repairs (85 houses), builder and carpenter support, as well as Blair toilets and school repairs. Higherlife Foundation work attracted other international donors. These included the Canadian friends of Zimbabwe and New Zealand friends of Zimbabwe who donated towards WASH. They gave money to build pit latrines. In contrast to lots of relief work in disaster situations that barely last a day, as documented in the literature section, the Higherlife Foundation activities lasted much longer—between 3 and 6 months.

Some of the early preliminary work that was done involved makeshift building of roads, filling of potholes and general clean-up in the areas that were affected. The many carcasses of animals that had died posed a danger. Hence some

of the work done was to burn or bury all those dead animals. The Higherlife Foundation also worked with Econet to equip the volunteers in terms of training since this was a high-risk area. The volunteers were trained in basic safety and then equipped with the tools, wheelbarrows, helmets, work suits, gumboots and everything else that would be required. At the end of the programme, Econet showed their appreciation for the volunteers by also giving them food support.

9.4.3 Child Protection Engagements

Several child protection engagements were undertaken. These were done in collaboration with IsraAid. The engagement offered psychosocial support and first aid training to the Department of Social Welfare officers. There was also a Child Friendly Corner established at the Kwirire Primary School in Ward 13 of Chimanimani, targeting 890 children. The initiative also identified and trained 20 childcare workers in child protection to support the Department of Social Welfare. Additionally, groceries were distributed to Chirinda and Houtberg children's homes.

The activity undertaken by Higherlife Foundation in terms of child protection was valid as UNICEF Childline set up even more stations in Chimanimani. At one of the stations in one of the three camps, it emerged that they would house children between the ages 0 and 17 years. The station would place them in play groups according to their age groups, namely 0–6, 7–9, 10–13 and 14–17 years. Indoor and outdoor games took place while the UNICEF officers observed the children's behaviour. Games such as volleyball, soccer and chess took place. Storytelling was also done at the station. During these engagements in this child-friendly space, problems and challenges faced by children emerged, including abuse and fees, that were referred to the Departments of Social Welfare. Issues of trauma also surfaced as some children took long to open up.

9.4.4   Water, Sanitation and Hygiene (WASH) Activities

The WASH component remains key after severe disasters such as Tropical Cyclone Idai. Without urgent action, people can die from diseases such as cholera from drinking contaminated water. The dire WASH situation became clear after a week, when the disaster area became accessible. Everything had been destroyed, and the internally displaced persons (IDP) were now put into holding camps. These included the Chimanimani Hotel, some churches and schools, including Ngangu Primary and Chimanimani Junior School. There were also IDP in holding camps in Machongwe, Skyline and the Kopa area, all in Chimanimani district. All those people needed WASH facilities.

The Higherlife Foundation distributed 5000 non-food items to 2500 households in Cashel Valley, Chikukwa and Hangani. These items included 1302 hygiene kits delivered to schools and sanitaryware distributed to 713 children at Thabanchu, Ruwedza A, Ruwedza B, Nyambeya and Chikukwa primary schools. An additional 246 units of sanitaryware were left at five schools as emergency packs. The schools were Mutambara Primary, Mutambara Secondary, Tabanchu Primary, and Hangani and Nyambeya Primary. Hygiene packs were also distributed to expecting mothers at two clinics and one hospital reaching out to 103 mothers. The packs included tissues, sanitary towels, a 10-litre bucket, Vaseline, green soap bar and bathing soap. Furthermore, cooking utensils and groceries were delivered to three schools with soup kitchens. This assisted in reaching approximately 1844 children. Medical supplies were also distributed. Some details are reflected in Fig. 9.3. Missing from Fig. 9.3 are 1500 units of disinfectant and 36×100 units of bin liners. Drip kits were also donated. Lastly, 13 disability-friendly Blair latrines were constructed in Ward 10 (Hangani, Jantia and Mura villages).

When Tropical Cyclone Idai hit, the neighbouring Mozambique had an outbreak of cholera. With the porous border and intermarriages that happen, there is a constant movement of people between Mozambique and Zimbabwe, and there was imminent danger of cholera spreading to Zimbabwe. An estimated 16 primary healthcare facilities from the Zimbabwe side and Chimanimani are situated along the border with Mozambique. Hence, the Higherlife Foundation supported the Ministry of Health and Child Welfare by providing rehydration fluids. Since the clinics are far away from the mainland, it was important that they had prepositioned stocks in the event that there was a cholera outbreak. If a patient reported diarrhoea, they could be given rehydration fluids. The kit consisted of rehydration fluids, antidote and infection control materials.

9.4.5   The Education Component

There were also hygiene packs distributed to learners (Fig. 9.4). This was done in collaboration with Unilever. These packs were distributed to 32,388 learners covering 60 schools from the worst affected areas in Chimanimani and Chipinge districts. By gender, 16,143 boys and 16,245 girls benefitted. In addition, in partnership with Hitbay Sanitation, a total of 1302 learners received hygiene packs comprising soap, sanitaryware, Colgate toothpaste, Vaseline and tissues. These were presented in Wards 1, 10 and 11 in Chimanimani. There were also 246 hygiene emergency packs (sanitaryware) given to five primary schools in Cashel Valley (Ward 1). A further 159 hand-washing kits (bucket and disinfectant) were distributed to seven schools in the same ward.

One of the most touching relief engagements was school fees assistance from the Higherlife Foundation. A total of 1639 learners were identified for educational assistance. From this number, 1463 were assisted with school fees for two school terms (terms 2 and 3, 2019) and 176 received full scholarships. Full scholarships were awarded to learners who had lost one or both parents due to Tropical Cyclone Idai, while two school terms fees assistance was granted to learners whose homes were destroyed by the cyclone. This help was aimed at providing relief to parents

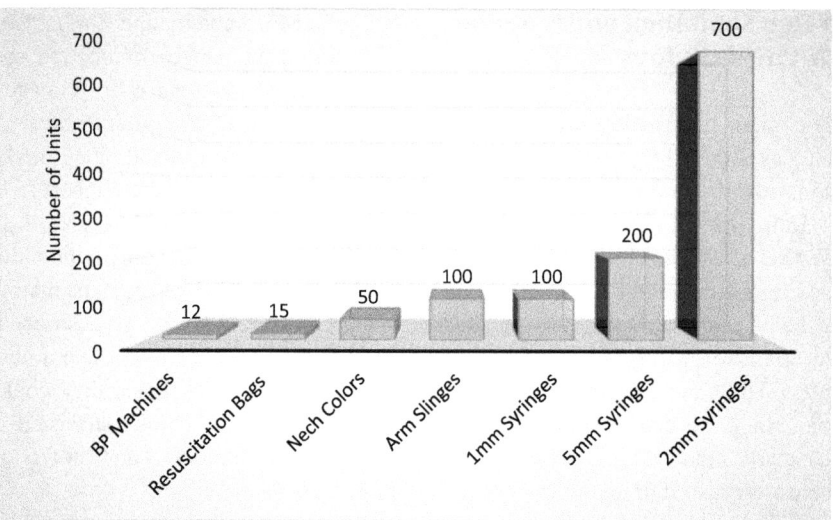

Fig. 9.3 Some medical supplies distributed. Source: Authors, Data from Higherlife Foundation

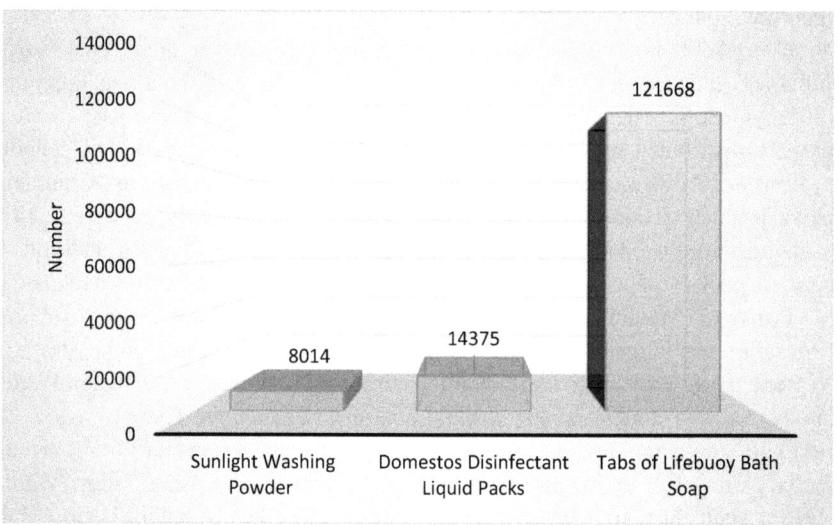

Fig. 9.4 Relief packs distributed in Chimanimani and Chipinge districts. Source: Authors, Data from Higherlife Foundation 2019

who had lost their livelihoods to be able to recover. With full scholarship, the Higherlife Foundation beneficiary is covered up to the first tertiary qualification for a degree or study at the polytechnic. Even if the beneficiary failed to get five Ordinary Level grades, the scholarship applied to vocational training, just to support them up to the first tertiary qualification. Another interesting finding from this chapter was the manner in which the Higherlife Foundation, in

partnership with the government's Tropical Cyclone Idai Education Cluster, tracked the beneficiary children across the country. In line with tradition, after both parents die, the orphans are placed with relatives wherever they might be in the country. Children were identified and tracked to the faraway provinces of Mashonaland, Midland and Harare.

The school fees issue made interesting reading prior to fieldwork. As such, there was a ques-

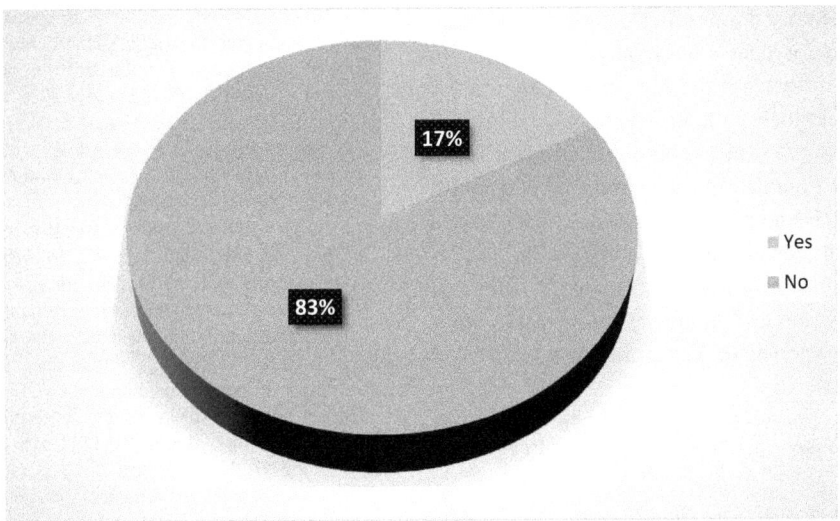

Fig. 9.5 Assistance with school fees from the Higherlife Foundation ($n = 153$). Source: Authors, data from survey 2019

tion slotted in the household survey to find out from affected communities whether there were any beneficiaries. The question asked whether the sampled household received aid towards school fees from the Higherlife Foundation. The findings are presented in Fig. 9.5. The fact that school fees were paid was further confirmed in some interviews from the Chimanimani communities. The low percentage is due to the fact that the question asked about direct benefit. This implies that from those surveyed, 17% directly benefitted. This remains a touching experience in disaster philanthropy. It emerged from the literature that the Higherlife Foundation was extending its agenda from already existing mandates of the organisation to education.

Food was also distributed in schools. Fortified rice was delivered to 13 primary schools to supplement their school feeding programmes. This activity was undertaken in collaboration with The Joseph Foundation. A total of 8 tons of rice was shared between schools and displaced people living in camps. Furthermore, the Higherlife Foundation volunteers assisted in cleaning 26 public schools and in rebuilding activities. As indicated earlier, a total of 863 volunteers participated in the programme. In addition, 600 kg of Royco Soup was distributed to villagers in tented camps in partnership with Unilever.

The work undertaken by the Higherlife Foundation during and post-Tropical Cyclone Idai was commendable. From one of the focus group respondents, it emerged that for over 23 years of its existence, the Higherlife Foundation had never been apologetic about the fact that it is a faith-driven organisation. The organisation believes in God and was happy that it did God's work.

9.5 Conclusion

The ethical work undertaken by the Higherlife Foundation cannot go unnoticed. From this work, four main disaster intervention areas are visible: impact assessment, response, recovery and reconstruction. These interventions were driven along four thematic focus areas that included volunteerism, child protection, WASH and education. This resulted in key activities focusing on hygiene packs distribution, stakeholder engagement, procurement, assessment, food distribution, mobilisation and logistical support. Given the economic situation in Zimbabwe prior to and after Tropical Cyclone Idai, the activities of the Higherlife Foundation make it difficult to classify them as anything but genuine ethical concern that led to their engagement in the disaster. The activities focused on relief and recovery and these were not a day's work only. The Higherlife Foundation does not depend on government for support, and their philanthropic engagement was never meant to please the government. The Higherlife Foundation went out of its way to fill gaps in DRR and did so without discrimination is as far as the results show. It compels one to recommend that this model be adopted by many in the same space in Zimbabwe, and possible other countries, when natural and other disasters strike.

References

Adhikari, P. & Matilal, S. (2017). Disaster Management and Corporate responsibility: An ethic of Care Approach. The Nepal Chartered Accountant, September, 32-35.

Ariyabandu, M.M. & Hulangamuwa, P. (2002). Corporate Social Responsibility and Natural Disaster Reduction in Sri Lanka. Colombo: ITDG – South Asia.

Ballesteros, L. & Useem, M. (2017). Masters of disasters? An empirical analysis of how societies benefit from corporate disaster aid. Academy of Management Journal, 60(5), 1682–1708. https://doi.org/10.5465/amj.2015.0765.

Banomyong, R. & Julagasigorn, P. 2017. The potential role of philanthropy in humanitarian supply chains delivery: the case of Thailand. Journal of Humanitarian Logistics and Supply Chain Management, 7(3), 284-303. https://doi.org/10.1108/JHLSCM-05-2017-0017.

Barnett, M.L. (2019). The Business Case for Corporate Social Responsibility: A Critique and an Indirect Path Forward. Business & Society, 58(1), 167–190. https://doi.org/10.1177/0007650316660044.

Behl, A. & Dutta, P. (2019). Social and financial aid for disaster relief operations using CSR and crowd-funding: Moderating effect of information quality. Benchmarking: An International Journal, https://doi.org/10.1108/BIJ-08-2019-0372.

Bhardwaj, P., Chatterjee, P., Demir, K.D., Turut, O. (2018). When and how is corporate social responsibility profitable? Journal of Business Research, 84, 206-219. https://doi.org/10.1016/j.jbusres.2017.11.026.

Chandra, A., Moen, S., Sellers, C. (2016). What role does the private sector have in supporting disaster recovery, and what challenges does it face in doing so? London: Rand Corporation.

Chen, J., Dong, W., Tong, Y., Zhang, F. (2020). Corporate philanthropy and corporate misconduct: Evidence from China. International Review of Economics and Finance. 65, 17-31. https://doi.org/10.1016/j.iref.2019.09.002.

Cheng, Y., Jin, Y., Hung-Baesecke, C.J.F., Chen, Y.R. (2019). Mobile Corporate Social Responsibility (mCSR): Examining Publics' Responses to CSR-Based Initiatives in Natural Disasters. International Journal of Strategic Communication, 13(1), 76-93. https://doi.org/10.1080/1553118X.2018.1524382.

Famiyeh, S. (2017). Corporate social responsibility and firm's performance: empirical evidence. Social Responsibility Journal, 13(2), 390-406. https://doi.org/10.1108/SRJ-04-2016-0049.

Fernando, M. (2006). Corporate social responsibility in the wake of the Asian tsunami: an empirical study. In J. Kennedy & L. Di Milla (Eds.), Australian and New Zealand Academy of Management Conference Queensland: Central Queensland University.

Gao, Y. & Hafsi, T. (2017). Political dependence, social scrutiny, and corporate philanthropy: Evidence from disaster relief. Business Ethics: A European Review, 26, 189–20. https://doi.org/10.1111/beer.12144.

Gao, Y., Yang, H., Hafsi, T. (2019). Corporate giving and corporate financial performance: the S-curve relationship. Asia Pacific Journal of Management, 36, 687–713. https://doi.org/10.1007/s10490-019-09668-y.

Gong, Y. & Ho, K.C. (2018). Does corporate social responsibility matter for corporate stability? Evidence from China. Qual Quant, 52, 2291–2319. https://doi.org/10.1007/s11135-017-0665-6.

Hartz, M. (2017). The Future of Disaster Philanthropy. London/New York City. The Conference Board.

Higherlife foundation. (2020). Higherlife Foundation. Retrieved from https://www.higherlifefoundation.com/ (10 March 2020).

Hotho, J. & Girschik, V. (2019). Corporate engagement in humanitarian action: Concepts, challenges, and areas for international business research. Critical Perspectives on International Business, 15(2/3): 201-218. https://doi.org/10.1108/cpoib-02-2019-0015.

Hunt, S. & Eburn, M. (2018). How Can Business Share Responsibility for Disaster Resilience? Australian Journal of Public Administration, 77(3): 482–491. https://doi.org/10.1111/1467-8500.12320.

Hwang, J., Cho, S., Kim, W. (2019). Philanthropic corporate social responsibility, consumer attitudes, brand preference, and customer citizenship behavior: Older adult employment as a moderator. Social Behavior and Personality: An international journal, 47(7), e8111.

Jain, A., Kansal, M., Joshi, M. (2020). New development: Corporate philanthropy to mandatory corporate social responsibility (CSR)—a new law for India, Public Money & Management, https://doi.org/10.1080/09540962.2020.1714280.

Johnson, B.R., Connolly, E., Carter, T.S. (2011). Corporate Social Responsibility: The Role of Fortune 100 Companies in Domestic and International Natural Disasters. Corporate Social Responsibility and Environmental Management, 18: 352–369. https://doi.org/10.1002/csr.253.

Jordan, S. & McSwinggan, G. (2012). Corporate expertise in Disasters. Washington DC: U.S. Chamber of Commerce Business Civic Leadership Center

Kanji, R. & Agrawal, R. (2019). Building a society conducive to the use of corporate social responsibility as a tool to develop disaster resilience with sustainable development as the goal: an interpretive structural modelling approach in the Indian context. Asian Journal of Sustainability and Social Responsibility, 4(5), 2-25. https://doi.org/10.1186/s41180-019-0025-7.

Liao, F.F. (2020). The Effect of Philanthropic Activities on Corporate Financial Performance: From the Perspectives of Charity Donation and Volunteer Service. Modern Economy, 11, 96-108. https://doi.org/10.4236/me.2020.111010.

Manchiraju, H. & Rajgopal, S. (2017). Does Corporate Social Responsibility (CSR) Create Shareholder Value? Evidence from the Indian Companies Act 2013. Journal of Accounting Research, 55(5), 1257-1300. https://doi.org/10.1111/1475-679X.12174.

McKnight, B. & Linneluecke, M.K. (2019). Patterns of Firm Responses to Different Types of Natural Disasters. Business & Society, 58(4), 813 –840. https://doi.org/10.1177/0007650317698946.

Mithani, M.A. (2017). Liability of foreignness, natural disasters, and corporate philanthropy. Journal of International Business Studies (2017) 48, 941–963.

Ouyang, Z. & Wei, J. (2017). Media attention and corporate disaster relief: evidence from China. Disaster Prevention and Management, 26(1): 2-12. https://doi.org/10.1108/DPM-10-2015-0247.

Owusu-Kwateng, K., Hamid, M.A., Debrah, B. (2017). Disaster relief logistics operation: an insight from Ghana. International Journal of Emergency Services, 6(1), 4-13. https://doi.org/10.1108/IJES-10-2016-0022.

Tilcsik, A. & Marquis, C. (2013). Punctuated Generosity: How Mega-events and Natural Disasters Affect Corporate Philanthropy in U.S. Communities - Working Paper 13-060. Harvard: Harvard Business School.

Wang, K., Miao, Y., Su, C.H., Chen, M.H., Wu, Z., Wang, T. (2019). Does Corporate Charitable Giving Help Sustain Corporate Performance in China? Sustainability, 11, 1491; https://doi.org/10.3390/su11051491.

Yang, Y. & Tang, M. (2020). Finding the Ethics of "Red Capitalists": Political Connection and Philanthropy of Chinese wwwwPrivate Entrepreneurs. Journal of Business Ethics. 161: 133–147. https://doi.org/10.1007/s10551-018-3934-y.

Zolotoy, L., O'Sullivan, D., Seo, M.G., Veeraraghavan, M. (2020). Journal of Business Ethics. https://doi.org/10.1007/s10551-020-04432-5.

The Chimanimani Hotel and Tropical Cyclone Idai: When Humanitarian Needs Are Priority Over Business Interests

10

Abstract

There are instances when humanitarian needs surpass business imperatives. Such was the case with the Chimanimani Hotel located at the heart of Tropical Cyclone Idai's destruction in the Chimanimani District of Zimbabwe. From the data generated from a household questionnaire survey, interviews and field observations, it emerged that following the night of severe devastation by Cyclone Idai, including the loss of lives, Chimanimani Hotel became a home for some of the internally displaced survivors. The hotel became one of the first relief centres, which accommodated, and fed approximately 600 victims in the first 2 weeks of the disaster. For a 35-room establishment, this was a phenomenal accomplishment. An estimated 200 plus internally displaced people remained accommodated in the hotel for the next 2 months. The entire hotel staff worked with one purpose, primarily to save lives from the cold and starvation. The Chimanimani hotel received recognition from the Zimbabwean Government for the role it played during the cyclone. This chapter will profile this story as an exceptional humanitarian gesture. The changing roles of the hotel from the time of search and rescue, through relief to recovery and reconstruction activities, will also be documented. The authors recommend that the Chimanimani Hotel's example be replicated across the world in the event such needs arise.

Keywords

Cyclone Idai · Chimanimani hotel · SDGs · Internally displaced people

10.1 Introduction

This chapter was inspired by an online article in The Standard of Zimbabwe by Jairos Saunyama entitled "Chimanimani Hotel turns into a sanctuary" which appeared on 26 March 2019 (Saunyama 2019). The article made reference to the 2004 real life movie, *Hotel Rwanda*. Hotel Rwanda (actual name, Hotel des Mille Collines) is well known for its humanitarian face during the worst-known global genocide that took place in Rwanda in 1994. The hotel accommodated between 1200 and 1300 refugees fleeing from the civil war (Momberg 2016; Fregonese and Ramadan 2015; Uraizee 2010). As the ethnic Tutsis and moderate Hutus fled from the ethnic Hutu militia, they sought refuge in Hotel des Mille Collines ensuring their survival. Drawing from activities following the Indian Ocean Tsunami of 2004, Henderson (2007) identifies the tensions between the business of hotels and more philanthropic activities that managers should address and reconcile.

It was, however, a war of its own kind and magnitude in Chimanimani on the night of 15 March 2019, especially in the townships of Ngangu and Kopa rural service centre as Tropical

Cyclone Idai hit and devastated the community leaving many dead, injured, traumatised and with nothing. Situated in the Manicaland Province of Zimbabwe, and in the Chimanimani Rural District, Chimanimani Hotel became a "Hotel Rwanda" as hundreds displaced people were invited to take shelter there as they fled from their homes, grieving, cold and without any hope. Figure 10.1 shows the location of both the Chimanimani district and town that were hardest hit by the impacts of Cyclone Idai. The Chimanimani Hotel has a bed capacity of 80 persons (sharing), with limited conference space. The hotel is owned by the Mataure and Makoni Families and is situated in Memorial Street, Chimanimani.

Many of the survivors' spouses, children, relatives, friends, neighbours and workers had died during the disaster. In a typical response to the United Nations 2030 Agenda for Sustainable Development's call, "Leave no one behind" (United Nations 2015), the Chimanimani Hotel's General Manager (GM), Mr Mandla Mataure, ordered that boys and girls, young and old, mothers and fathers, and those living with disabilities to take shelter in the hotel. Little did the GM knew he had embraced several provisions enshrined in the 17 Sustainable Development Goals (SDGs) including SDG 2 (Hunger and Nutrition), SDG 3 (Health), SDG 5 (Gender), SDG 6 (Water and sanitation), SDG 10 (Inclusiveness), SDG 13 (Climate Action) and SDG 17 (Partnerships). The authors further deliberate on how these SDGs were carried out in the findings section.

This chapter seeks to investigate how the Chimanimani Hotel took in and fed approximately 600 internally displaced residents of mainly the Ngangu Township and the surrounding villages. This act of Cyclone Idai kindness deserves place in this book. The objectives spelt out are as follows: (1) to investigate and document the events leading to the Chimanimani

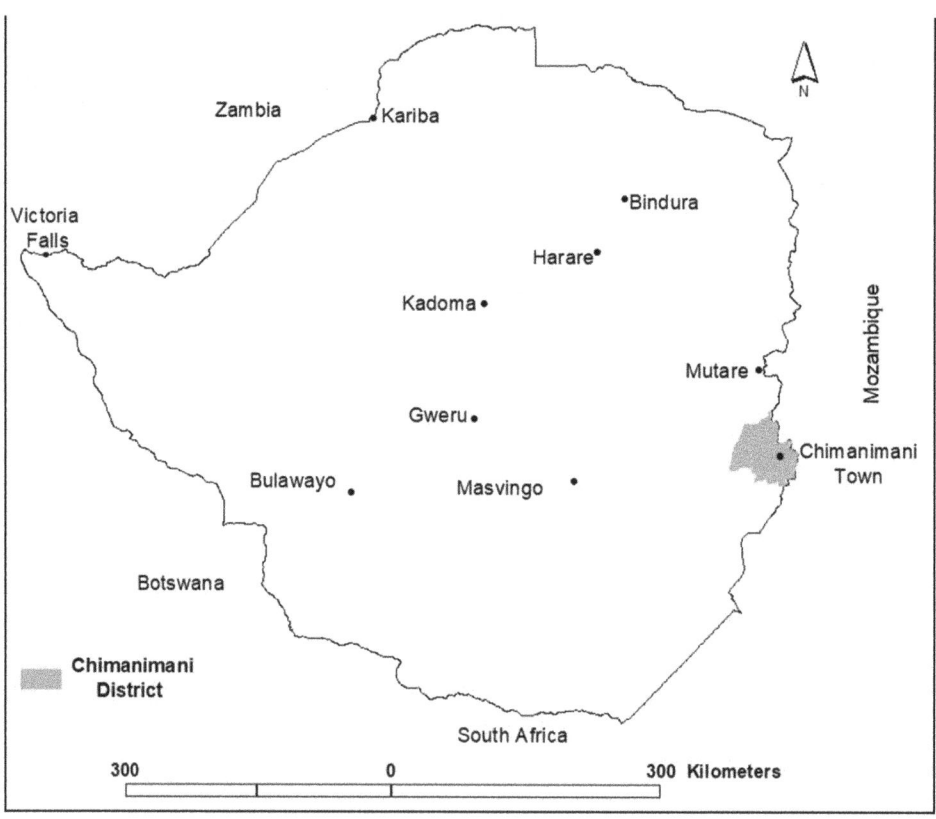

Fig. 10.1 The Chimanimani District, town and Zimbabwe. Source: Authors

Hotel inviting the fleeing residents into their premises, (2) to determine how the Chimanimani Hotel management worked out the logistics, (3) to investigate the preparedness of the Chimanimani Hotel to such disasters and how the act of kindness links to the 2030 Agenda for Sustainable Development and its development goals, and (4) to establish lessons from this event and also the role of corporate social responsibility (CSR) in disaster situations. The next section presents the relevant literature on hotels as refuge spaces during human-induced and natural disasters, thereby breaking both boundaries of CSR and the purpose of hotels.

10.2 Literature Survey

Paul Rusesabagina, a manager at Hotel Rwanda, is portrayed as a model in terms of what a single individual can accomplish when called, moved and dedicated to act (Amnesty International 2005). With the initial mind to save his Tutsi wife and family, the act expanded to the greater community residing at the hotel that he had to feed from outside sources. Following the 1994 Rwanda genocide in which the Hutu militia sought to eliminate all Tutsi in the country, Paul played a key role in protecting 1368 people (mainly Tutsi) refugees fleeing from the civil war. In fact, Ganguly (2007, p. 60) describes the act as "hospitality in the midst of terror". Eventually, an estimated 800,000 people, mainly of Tsutsi origin, died (Adhikari 2008) after being massacred by the Hutus. These events unfolded between early April and mid-July 1994. While many victims were hacked to death with machetes and spiked clubs, many more, an estimated half a million, succumbed to disease, famine and military encounters. The notion of succumbing to disease is frightening, especially under the prevailing circumstances.

Fregonese and Ramadan (2015, p. 793) present research data into the geopolitics of hotels that presents hotels as "projections of soft power, soft targets for political violence, strategic infrastructures in conflict, hosts for war reporters, providers of emergency hospitality and care, and

infrastructures of peace-building". In analysing Hotel Rwanda, Momberg (2016) portrays a picture of Paul that is intertwined with his surroundings, which makes him go beyond heroism, and into the realm of Ubuntu (the concept of "I am because you are"). In Momberg's (2016, p. 1) words, "inclusive enactment of interconnected, communal belonging opens up the possibility to understand facets of Rusesabagina's bravery as a spiritual choice". The concept of communal belonging could have played a major role for the Chimanimani Hotel's management when they brought hundreds of internally displaced persons into their hotel. Another factor of interest is that in such facilities will accommodate people filled with grief and depression; elements that require careful handling and monitoring (Mamuna et al. 2019; Sim et al. 2019).

While Hotel Rwanda's narrative centres around a genocide, some hotels in Asia were faced with similar circumstances following the Indian Ocean Tsunami of 26 December 2004 (Henderson 2005). Henderson reports the story of a general manager of a luxury hotel on the Thai island of Phuket. In the account, the priority of hotel staff was to save tourists and guests, even those that were not booked in at their hotel. This hotel remained was spared damage because it was protected by a large sand dune covered with mature vegetation. This scenario is reminiscent of Chimanimani Hotel that was also undamaged because it was located in Chimanimani town.

As the general manager arrived at the Phuket-based hotel after the waters had subsided (Henderson 2005), he witnessed piles and piles of waste, beach umbrellas, bags, cars, taxis and personal belongings scattered over the shore. Having been to Chimanimani, the scenes being described were familiar to the authors. The general manager in Thailand was then greeted by his assistant as well as several tourists in swim suits, roaming around aimlessly in shock. The hotel staff also reported their success in rescuing some of the tourists with their hotel's lifebuoys that they threw across the track into the sea. In the crowd at the reception was a father and two children who had been washed half a kilometre along the beach and rescued by the hotel staff although they were not staying in this particular hotel. From this point onwards, the hotel reception started contacting hospitals and other hotels to try and trace the mother. What is of interest and linked to the main intent of the Chimanimani Hotel's management is that the general manager provided a guestroom and free refreshments to the father and his children. This scenario played out as more guests were accommodated, with staff of the hotel requested by the general manager to "show compassion and offer the best possible service under the circumstances" (Henderson 2005, p. 91). In addition, extra staff were placed on duty in all public areas that included the lobby and main coffee shop with the sole purpose of dealing with distressed guests. All those in the hotel were treated to a buffet lunch. At about 5 pm, a report came through that the missing mother had been found and the hotel arranged transport to take the father and children to her. It is therefore noteworthy that hotel staff have key skills and competences that can be harnessed during the different phases of the disaster cycle to empower the victims to better cope with trauma and disturbance.

When management talked to the hotel staff, it emerged that families of two hotel staff members had also been swept away (Henderson 2005). However, there were no fatalities among the staff. Other measures taken for staff included allowing them a 3-day report back for duty without any penalties in the first 3 days of the aftermath of the Tsunami. Furthermore, those employees requiring leave were granted leave, and all staff had reported for work within the month. However, a new disaster was reported to the general manager by a tour operator who indicated that there was a catastrophe in Khao Lak, another island, where probably thousands of people died. The tour operator had managed to find 50 out of their 290 guests alive. Guests from Khao Lak were brought over. Although the guests were accommodated in the luxury hotel, the hotel maintained its normal rates compared to some hotels in the area that were safe and had raised their prices. Many guests were accommodated only for a night as flights were organised from around the world to take them home start-

ing on 27 December 2004. Fortunately, there were many guests who chose to stay and perform rescue and recovery work. Part of their work included visiting hospitals, embassy stations and blood banks, assisting in translations, counselling and other tasks. On 26 December 2004, the general manager called for volunteers to donate blood following an appeal by the local radio stations. Thirty guests volunteered within 15 min (Henderson 2005).

Hotels further need to address matters regarding drastic reductions in arrivals in the aftermath of major disasters such as the tsunami and in the face of increasing natural disasters linked to climate change and extreme weather events (Ivkov et al. 2019). The luxury hotel on the Thai island of Phuket later experienced the economic pinch as occupancy plummeted. While bookings for January and February 2005 stood at 85% and 92%, respectively (Henderson 2005), the realised figures were 12.5% and 15%, respectively. With such drastic reductions through cancellations, hotels with weak balance sheets can easily collapse.

Henderson (2007) follows up with a paper that focuses on the tension between hotels' commercial targets and CSR as corporate citizens in the wake of the tsunami. Managers were faced with challenges of fulfilling their role as commercial agents, even in times of disasters such as a tsunami. The work draws information from 14 hotels that completed a questionnaire about CSR activities following the tsunami where tourism infrastructure was severely damaged in Thailand, Sri Lanka and the Maldives. Dobie et al. (2018) are clear that hotels have become critical hubs for destination disaster relief. In a separate paper, Cheung and Law (2006) raise matters regarding the need to protect guests during the occurrences of natural disasters such as the tsunami. It killed between 250 and 270,000 people in South Asia, and many of those swept away were tourists from hotels. The authors observed that many hotels were built near beaches with no safety plans for saving guests in the event of disasters of the magnitude of the tsunami.

Ritchie (2008) then introduces the concept of tourism disaster planning and management focusing on response, recovery as well as reduction and readiness. This is an aspect of interest to the Chimanimani Hotel management. The authors highlighted that large resorts, including hotels, may serve as emergency shelter either prior to or immediately after a disaster. To this end, areas of the resorts that are at a lower risk must be designated as places of refuge from disasters. Linked to this, logistics on the provision of food, safe drinking water, clothing and bedding, medical care and sanitation facilities need to be identified before a disaster occurs. For this to happen smoothly, hotel or resort employees should be sanitised and prepared to carry out their roles and responsibilities. Guests look to hotel management and staff for advice, relief and help. As it becomes apparent that disaster will come, hotel management should communicate appropriately with their guests (Southon and Van der Merwe 2018). AlBattat and Som (2014) conclude that most of the issues discussed herein are lacking in the hotel industry. This view is supported by Nguyen et al. (2018) who indicate that although hotels remain active participants in disaster preparedness, very few were designated as evacuation centres during the 2011 Great East Japan Earthquake. In fact, the authors advocate for greater cooperation between hotels, resorts and local authorities in readiness for disasters, both natural and human-induced.

Poria et al. (2014) focused on Israeli hotels' social responsiveness during the 2012 fall Gaza conflict. The authors reveal that many hotels showed CSR to the communities through investing own resources, including money to serve the suffering communities. Hotels offered accommo-

dation to those affected by the conflict. The Israeli hotels' actions show that crisis situations that include war and natural disasters require non-conventional approaches to common business strategies. These situations further demand courageous leadership and innovation. The empirical evidence from 87 articles published on various media platforms established four categories of hotel CSR during the Gaza crisis. The categories were based on attributes that included (1) the level of discount offered, (2) wording of the services offered and (3) hotels' ownership and governance structure. The fourth category was the (4) business as usual approach. Details regarding these typologies are shown in Table 10.1.

The small and privately owned B&Bs establishments were ready to incur a loss as they hosted for free or agreed to forgo their profit in the event that they chose to charge at cost (Poria et al. 2014). Those advertising accommodation specials offered a large number of rooms at substantial discounts. The third category on discounts or service upgrades focused on special deals for meals and spa entrance, while the last option—business as usual offered no assistance. What emerges from this discussion is the fact that hotels offer some social and/or humanitarian services in times of crisis. The authors also discovered differences in reaction times. The B&Bs came in first, followed by the medium-sized hotels and subsequently the larger hotels.

In a study aimed at analysing hotel entities as partners in disaster management and as part of their CSR during disaster and post disaster peri-

Table 10.1 Categories of Israeli hotels' CSR during the 2012 Gaza crisis

Narrative	Response	Accommodation type	Location
Hurry up, our hotel is your shelter	39%	Bed and Breakfast (B&B)	Mainly in north of country
Special prices	54%	Mainly local hotels	Mainly in north of country and few in Jerusalem area
Discounts or service upgrading	7%	Big hotel chains	Mainly in the centre and north of country
Business as usual	No special offer	Hotels part of international chains	Mainly in big cities such as Tel Aviv and Jerusalem

Source: Poria et al. (2014, p. 85)

ods in the USA, Dobie et al. 2018 observed that 24% of the hotels made direct monetary contributions to disaster risk reduction (DRR). A further 19% donated commercial products and services for DRR, 8% of the hotels were involved in the deployment of skilled and unskilled employees both paid and those volunteering to leverage core skills and competencies of hospitality workers to contribute in DRR. About 33% contributed to activities that made the local communities proactively build capacity in DRR and improve on preparedness. Finally, close to 17% of surveyed hotels contributed to the creation of an integrated, resilient and sustainable systems for DRR. Hotels can therefore be hubs for societal resilience, providing both lifeline services and leveraging their core skills and competencies to aid in DRR. The next section presents the materials and methods applied in the research.

10.3   Materials and Methods

As outlined in the introduction, this chapter seeks to present a first-hand and reported account of how the Chimanimani Hotel took in and fed internally displaced residents of Ngangu and the surrounding villages after the devastation of Cyclone Idai. As outlined in Chap. 1, the work is informed by a snap household questionnaire survey of 219 respondents, interviews, document analysis, on the ground observations and open-ended interviews. The interviews were granted by the hotel management, workers and selected villagers that presented various testimonies regarding what transpired. The use of the methods highlighted is not new in disaster risk reduction studies having been applied by authors including the following: Hartmann (2014), Dobie et al. (2018), Ivkov et al. (2019) and Nguyen et al. (2018), to mention but a few. Among those interviewed are a woman who lost her entire family and moved into the hotel, some of the hotel employees and managers. Figure 10.2 shows the location of the Chimanimani hotel in the Village suburb of Chimanimani town.

In addition, several media reports were retrieved that enriched this work (British Broadcasting Corporation—BBC 2019; Ore 2019; Nyakunyanga 2019a, b; Saunyama 2019) and guided our 2-week-long fieldwork that followed in Chimanimani during the second and third weeks of October 2019. The analysed articles either provided fresh perspectives or supported narratives from the interviews and other sources of data on the acts of kindness from the Chimanimani Hotel. Other methods applied included the application of Geographical Information Systems (GIS) in drawing the maps.

To facilitate quick movement in the field, the research team stayed at the Chimanimani Hotel. This gave researchers additional working time and provided a meeting and interviewing place with many officials from the NGO community, development agencies, politicians, community and other stakeholders. The fieldwork involved nine fieldworkers with relevant masters and doctoral degrees qualifications. There was also a supervisory team of five doctoral qualification holders that included three professors. Prior to the fieldwork, the entire team comprising fieldworkers and the supervisory team underwent training on conducting research in disaster areas that was led by qualified psychologists. In addition, the entire research team was taken on a 2-day tour to the most affected areas and introduced to the communities by the chiefs and government officials. This process made it very easy to get an information from the communities. The fieldwork went on without any recorded incident of harm or hindrance. The next section presents data and key findings from the research.

10.4   Presentation of Data and Key Findings

Generally, the areas surrounding the Chimanimani Hotel were not severely damaged during Cyclone Idai compared to other parts of the settlement such as Ngangu and other parts of the district such as Kopa. The rest of this section is divided into three subsections. The next subsection focuses on Tropical Cyclone Idai through the eyes of the Chimanimani Hotel staff. The second subsection is dedicated to documenting the hotel's involvement during disaster relief and recovery. The third and last subsection dwells on

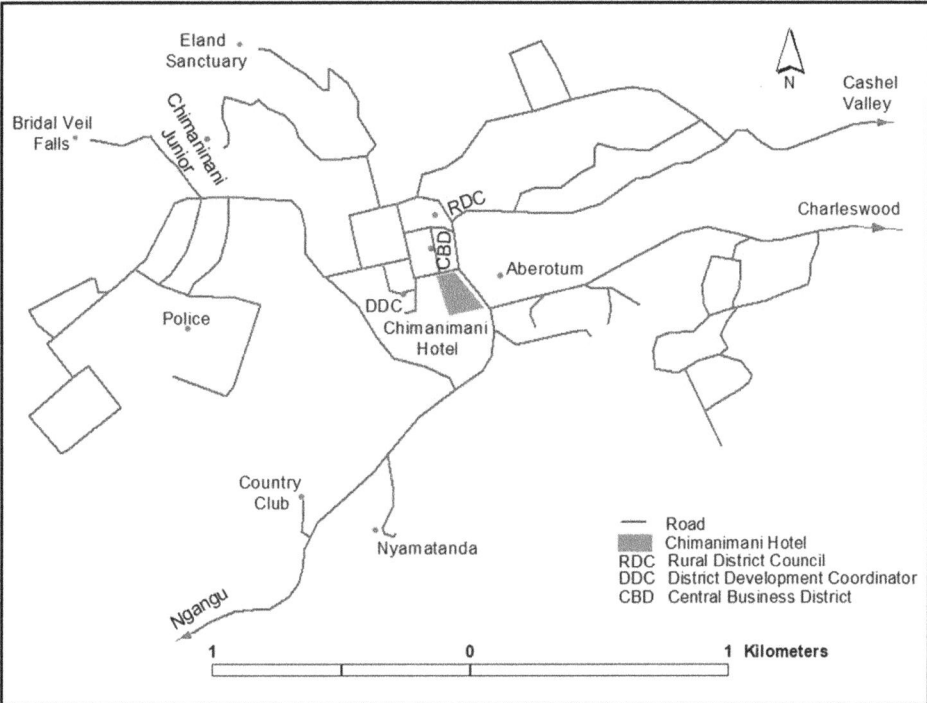

Fig. 10.2 The Village and Chimanimani Hotel. Source: Authors

outlining how the Chimanimani Hotel's engagements align with the SDGs' call for action not to leave any one behind.

10.4.1 Tropical Cyclone Idai: The Chimanimani Hotel's Staff Accounts

The development and arrival of Cyclone Idai is presented in several narratives. The General Manager (GM) of the Chimanimani Hotel, Mr Mandla Mataure gave a sobering account. He remembers watching the news on Wednesday 13 March before the cyclone hit. By then, Cyclone Idai had made landfall into Mozambique. This resulted in light rains in the Chimanimani area which started on Thursday. The hotel staff were occupied with their duties and tasks as usual. However, the GM's mother, Dr Priscilla Mataure, did research in Chimanimani at that time, and she could not travel to the field later that Thursday 14 March as rains had become heavy. By Friday, 15 March trees had started falling as it was now raining heavily. At about 21:00, it still rained heavily.

From around 21:30, the storm peaked till and lasted until approximately half past eleven. From the GM's account, this was the time the storm was most intense as it made the loudest noise. The bad weather was confirmed by Mr Michael Mataure who is the Director of the Chimanimani Hotel. He indicated that this all happened when he was in Kenya. His son, the GM sent him a text message saying, "Weather getting bad, a lot of wind".

The GM had planned to go to Harare Saturday, and when he was at the bus stop around 05:30, he was alerted that the road to Harare was no longer there. He asked the bearer of the bad news, "What do you mean there is no road? My mother was on this road just yesterday going to Harare". The GM did not understand it until he saw the extent of the damage himself. "This whole cyclone was crazy, people asked how many days it took to for that damage, but from my observation it was just a few hours. A few hours of intense rain and rock-slides and mudslides which damaged the whole area", recalls the GM. While this was happening, the GM texted the Director in Kenya and one of the messages read "Things are not okay. There is no traffic moving, the buses cannot go out, people cannot come in and there is obviously damage".

Cyclone Idai had made a massive trail in Zimbabwe, with devastating impacts in the Chimanimani and Chipinge districts and surrounding areas. The 5 days from the afternoon of Thursday it rained heavily and continuously and the weather only cleared on Wednesday, 20 March. Then the helicopters started coming in. The first helicopter managed to get it from Manyame Airbase in Harare on Sunday, and it took the crew sometime because they had to stop several times due to poor visibility. It still was not safe for the helicopter to carry people. So, there was no road access or air access until Wednesday, 20 March.

10.4.2 Involvement During Disaster Relief and Recovery Stages

While the Chimanimani Hotel was not involved in early warning (rightly so), it was hugely involved during the relief and recovery stages. When the GM could not proceed to Harare Saturday, he went back to the hotel around 11:00. He went upstairs in the hotel and looked out through the window. It was from upstairs that he saw people walking in large numbers, carrying whatever remaining belongings they could. He stood there for about 5–10 min and thought, "Where are these people going?" He then went down and found one of the hotel porters and asked, "I have seen a lot of people coming up in the direction of the Chimanimani Village[1], where are they going?" The porter told the GM that they were all going to the shops in town for cover and refuge. The GM told the porter to go there and get them into the Chimanimani Hotel. The GM also gave instruction to start fires as they were visibly cold. People started coming in slowly. According to the GM's account "20, 30, 40, 50 and 100 people came in and more were coming".

Eventually the numbers swelled to approximately 600 people. Figures 10.3 and 10.4 show the Chimanimani Hotel where these people got shelter.

Meanwhile, the GM and the Director who was in Kenya at the time of the disaster stayed in communication. In one of the texts to the Director, as it was quoted by the Director, the GM wrote, "People are here, there is no space, homes have been destroyed, etc." In typical good leadership style, the Director replied, "Do what you can to save life, life is above all the most important thing. You are on the ground, do something about it". Soon after this conversation, the communication stopped completely as the GM's mobile phone battery had run out. From the Director's account, the descriptions of what was happening was similar to what they had watched in movies and news reports from the hurricanes in the USA and other regions. Zimbabwe and southern Africa had never witnessed anything close to Cyclone Idai. From the account, the only similar experience was in 2000 in Zimbabwe when Tropical Cyclone Eline hit southern Africa. However, it is apparent that not much was learnt from that event in terms of preparedness, relief and recovery for the March 2019 catastrophe.

From the GM's account, the huge number of people in the small hotel created a logistic nightmare. There had to be separation of the people as follows: women and children had to go and sit in the lounge area (Fig. 10.5), and the men and older boys went to the conference room (Fig. 10.6). The two fireplaces were lit simultaneously to help survivors to be warm and ward off cold. The two functions rooms became their sleeping areas and temporary shelters. The day went by and there was no electricity as it went off on Thursday, 15 March, when Cyclone Idai made landfall in neighbouring Mozambique. The power lines had been destroyed. It was now the hotel's duty to look after the people and feed them. There was some initial help from church groups that gave extra plates. The hotel had some maize meal in the storeroom that was used to prepare a meal.

One safe lock had to be broken to access duplicate keys because the other of keys were with one of the staff members, who unfortu-

---

[1]Chimanimani Village is not a rural area. This is the name given to one of the upmarket suburbs in Chimanimani town.

Fig. 10.3 The front view of the Chimanimani Hotel. Source: Authors, Fieldwork 2019

Fig. 10.4 The back view of the Chimanimani Hotel. Source: Authors, Fieldwork 2019

nately passed away during the cyclone. In other instances, the hotel had to break padlocks. When the hotel ran out of food, the hotel GM started going out into the local shops to buy whatever they could find to feed people for the first 3–4 days. During this time, the hotel had to fend for itself. Contact with the outside world was difficult as the phones were down because the telephone base stations were out of diesel fuel for the stand-by generator units for powering the transmission equipment. In addition, potable water ran out on Monday, 19 March, because the main water supply pipes had also been damaged. The only water the hotel had was in the swimming pool. This was used for cleaning, laundry and bathing. On the same Monday the hotel managed to get the council to run a pipe from one of their reservoirs to the hotel. This resulted in the installation of a single tap with treated water (Fig. 10.7). This brought relief because the people now had potable drinking water.

Fig. 10.5 Residents' lounge area that was used for women and children. Source: Authors, Fieldwork 2019

Fig. 10.6 Dombi conference room that was used for men and boys. Source: Authors, Fieldwork 2029

As the food supplies were dwindle, the GM's wife, Mrs Kelly Mataure, sent a Tweet appealing for help and this went viral. People from outside started phoning on the GM's mobile to try and get information and pledging to come to the hotel and help. People also started sending money via the Ecocash platform. All the money that came in was used to buy supplies. Additional money came from the hotel's own sources. The hotel asked the shops for whatever was available rather than it being spoilt in fridges because of the power out-age. In the GM's words "Whatever it costs, let's pay for it". There were other individuals that also joined in the movement, people such as Mr Shane Kidd. Mr Kidd came to the GM and paid $2000 to a Nyamatanda Shop that the hotel could collect whatever was needed as and when it was needed. Given that food was still in short supply, the hotel devised a plan whereby people would eat at mid-day, and if possible at night to provide at least two meals a day. Women and children were priori-tised. However, it was fortuitous that additional

Fig. 10.7 The only tap the council installed at the hotel. Source: Authors, Fieldwork 2019

externally provided food supplies were sent to the hotel on Wednesday, 20 March, when the helicopters could fly in. From that time it became a coordinated effort from the District Development Coordinator's office. However, the hotel continued providing catering logistics to the entire relief programme. Because of the hotel's experience, it was easy to determine food quantities needed for other places that were hit by the disaster. The estimated number of meals prepared and served by the hotel (staff, management and volunteers) for the period (16 March–10 May 2019) is more than 51,200 meals which averages at 931 meals over 55 days. This should be compared to what had been the maximum catering capacity (pre-cyclone Idai) of just over 550 meals per day for in-residence guests when the hotel was full to capacity.

Prior to this, while he was still in Kenya, the Director was called by well-wishers who pledged foodstuffs and other goods. One group indicated they had 20 tonnes of food and were asking where to drop it as they thought the Director was in Chimanimani. The Director put them in touch with another group of volunteers he worked with

on an environmental conservation action called ZIMBOGREEN. The group started its own chain, while the GM was getting hold of his friends, stakeholders and members of the family.

The GM added more responsibilities, to his already crowded workload, by joining interim key stakeholder administrative structures for relief under the District Development Coordinator's Chimanimani Office. People had to organise themselves and work. "We had to try and come up with a way forward in terms of dealing with the crisis. The first two days council was trying to bury people because there were bodies everywhere, a lot of them being taken to the various churches and so forth and we said now we had to deal with this", explained the GM. At this point the churches got involved, and the Registrar General's office started to issue death certificates. It had to be a coordinated effort to bury people as there were so many bodies. After burials, the next priority was dealing with the injured. When the helicopters came the injured were transported to Skyline to get medical attention. Some that were critical were transferred to the Chipinge District

Hospital as there is no referral hospital in Chimanimani.

Fast forward to the second month, April, there was now more help and a better structured system in place in terms of getting rations. At the hotel, the GM had divided the cooking and other staff into two groups assigning them duties. They had to cook for both those that were coming from tents and other areas as well as those staying at the hotel. There had to be a coordinated effort to enable order and manage the hotel staff's fatigue. Hence the displaced, especially those staying at the hotel organised themselves and formed smaller teams. The teenage boys went out to fetch firewood, while men assisted in cooking, especially Sadza[2]. The women would help with the cleaning and cooking relish. In total, only 13 of the hotel staff (out of nearly 30 persons) were available at the time of the disaster as some who were on time off had been closed out as they could not access the hotel at all.

One of the survey questions sought to get information about the facilities where the displaced were accommodated during and after Cyclone Idai. Respondents were asked to identify the top three out of nine facilities where the displaced were accommodated. The findings are shown in Fig. 10.8. While the top three temporary accommodation facilities for Cyclone Idai victims were schools, churches, houses and hotels (which was only the Chimanimani Hotel), it came in fourth position with 15%. Schools were in pole position with 27%, while churches were at 20% and houses of relatives were at 17%. None were accommodated in farmhouses, and two per cent resided in community halls, 8% in shops and 10% in tents.

The question on accommodation was followed up with another that sought to determine the length of stay in whatever accommodation was granted. The details are presented in Fig. 10.9. The result of the survey is that many of those displaced by Cyclone Idai were accommodated for

more than 1 month, and some were still being accommodated. The survey responses tallied with narratives gathered from the Chimanimani Hotel and other sources that were interviewed. In fact, people stayed at the Chimanimani Hotel for nearly 2 months. This gesture is commendable by all standards of relief. An estimated 37% indicated that they stayed in the designated accommodation for a period exceeding 1 month. Together with those still being accommodated at the time of the survey in October 2019, a massive 67% of respondents fell in these two categories. This leaves an overall picture that only 33% of Cyclone Idai victims in the categories have since relocated to other areas (1%), stayed 1–5 days (6%), stayed 6–10 days (13%), stayed 11–15 days (6%) and stayed 16–30 days (6%). In addition, some of the employees affected by the cyclones were also accommodated at the hotel as well as the first relief workers from outside such as the air force and army. This is the only hotel in Chimanimani, and it could have increased its rates to capitalise on the disaster, but they did not and their rates remained the same.

The Chimanimani Hotel became a disaster shelter centre. To this end, early recovery engagements involved partnering with the organisation that included the Department of Social Development, Regional Psychosocial Support Initiative (REPSSI), Save the Children, Care International, among other for the counselling. These organisations were going around Chimanimani and other affected areas offering psychosocial support, especially to the children. However, the GM also noted that he and the hotel staff still needed to offer these services.

The hotel continued with other activities to accommodate relief and recovery teams. Figures 10.10 and 10.11 show some of the space allocations for aid agencies that included Plan International and Save the Children. People had to stay packed in an almost sardine-like style in the rooms because that was the only available place. It was traumatic. A lot of things could have gone wrong, but it all worked out because the

---

[2]Sadza is the staple food from mealie meal.

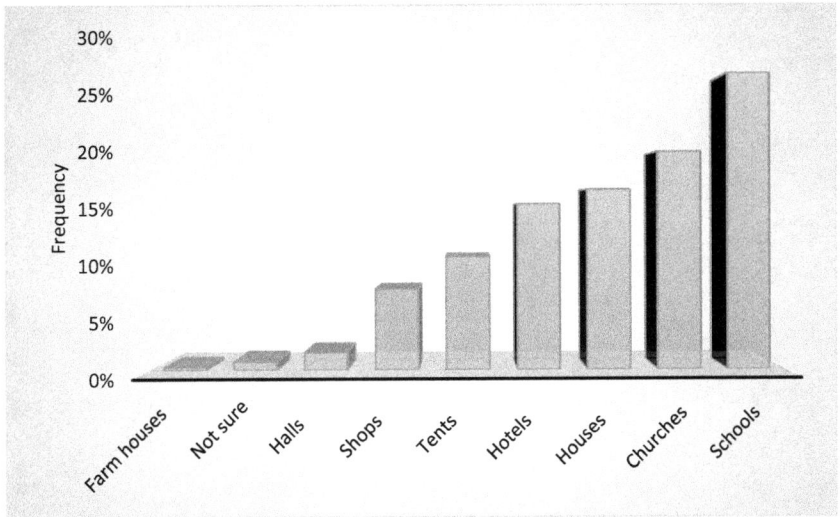

Fig. 10.8 Accommodation for displaced people ($n = 219$). Source: Fieldwork, October 2019

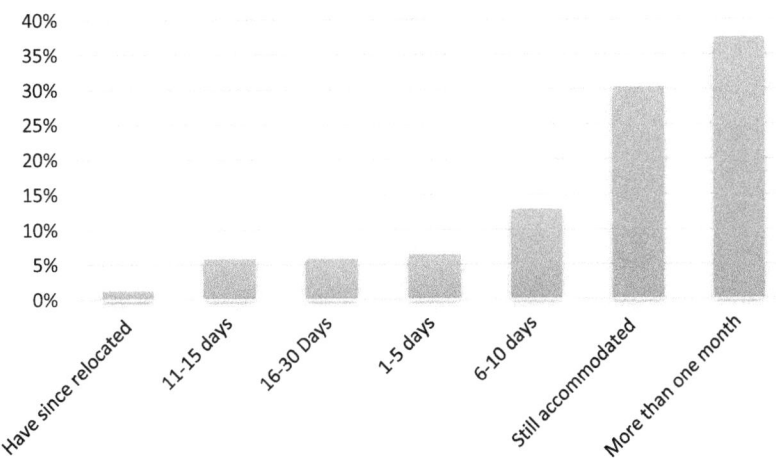

Fig. 10.9 Length of stay in accommodation facilities provided ($n = 219$). Source: Fieldwork, October 2019

community came together. As the months progressed, the hotel personnel found out that the houses of some people who had sought shelter were not completely damaged. In practice these individuals would have lived in their homes and then come to the hotel at mealtimes.

An interesting, yet sad case emerged during the fieldwork. The hotel receptionist passed away and left a 15-year-old daughter. The hotel decided to pay the girl's school fees and any other expenses. The GM also indicated that the hotel owed her mother some money in salaries which

was paid to her. The hotel was careful not to surrender the money to some family members as children usually suffer when a parent dies. The GM further revealed that they had also decided to act as foster parents. Consequently, the girl is allowed to go and spend time with them during school holidays. The front office Manager Lican Musarandega also lost a son who was doing Advanced Level studies and staying at Kopa Rural Service Centre.

The following subsection will deliberate on the GM's acts that aligned well to the SDG's call.

Fig. 10.10 Room allocated as office and storage for Plan International. Source: Authors, Fieldwork 2019

Fig. 10.11 Room allocated as office and storage for Save the Children. Source: Authors, Fieldwork 2019

10.4.3 GM's Acts Aligned with the SDG's Call for Local Action

As highlighted in the introduction, by inviting all people that were walking past the Chimanimani Hotel to the shops and government office to seek shelter, and instead come to Chimanimani Hotel, the GM displayed several provisions enshrined in the 17 SDGs including SDG 2 (Hunger and Nutrition), SDG 3 (Health), SDG 5 (Gender), SDG 6 (Water and sanitation), SDG 10 (Inclusiveness), SDG 13 (Climate Action) and SDG 17 (Partnership).

No little or no foodstuffs for many families during Cyclone Idai, it goes without saying that the assisted families would have gone hungry and possibly would have starved (SDG 2). The food shortage was acute, especially given that the area had been cut off from the rest of the world and the government reacted late—only 4 days later because of the bad weather that prohibited helicopters from landing as per the official narratives. Food insecurity are linked to health matters (SDG 3). The survey painted a picture on disease prevalence after Cyclone Idai (Fig. 10.12). Diarrhoea was identified as the most prevalent disease at 49%. In addition, malnutrition and malaria were the other diseases which made up a significant percentage. Such diseases are usually experienced in the immediate aftermath of a disaster and yet Chimanimani Hotel did not report any outbreaks of disease for the long period they had people at their premises. Other health matters related to the dire need to warm up the freezing bodies of the new hotel occupants. The extent of the cold became evident from one of the Focus Group interviews we conducted at Dombera farm some 20 kilometres from the Chimanimani Hotel. A male respondent had this to say about the extent of the cold and the treatment that followed his rescue from the flood waters. "I was given three cups of tea, but I did not feel any change in my body". Many of those that died could have died from the cold as this was a repeated refrain during our interviews.

Other SDGs addressed include one on Gender (SDG 5). As a very large number of people resided at the Chimanimani Hotel there were women and girls that could easily have fallen prey to abuse in the open spaces and during the nights. No such incident was reported during the time people were housed at Chimanimani Hotel. Women were reportedly raped during Hurricane Katrina in the USA in 2005 (Thornton and Voigt 2007). This situation was aggravated by the complete breakdown of law and order in the early phase of the disaster leaving criminals with an advantage to abuse women. In fact, eye witness accounts reveal that some were raped by those that were deployed to rescue and save them. One such victim was Neville who went public in March and April 2006 in Louisiana and Texas urging rape victims to report their rapes to victim advocates. In Japan, Yoshihama et al. (2018) reported violence against women and children following the 2011 Great East Japan Disaster. The authors recorded 82 unduplicated instances of violence against women and children that included domestic, sexual assault and unwanted sexual contact. Other findings were made on social exclusion (SDG 10) of a small group of Thai women who lived in Ishinomaki city, one of the worst hit areas by the same disaster (Pongponrat and Ishii 2018). Nguyen (2019) further observed similar trends in Eastern Visayas in the Philippines most affected by Super Typhoon Haiyan in 2013. However, the author indicates that the "heightened levels of gendered violence faced by

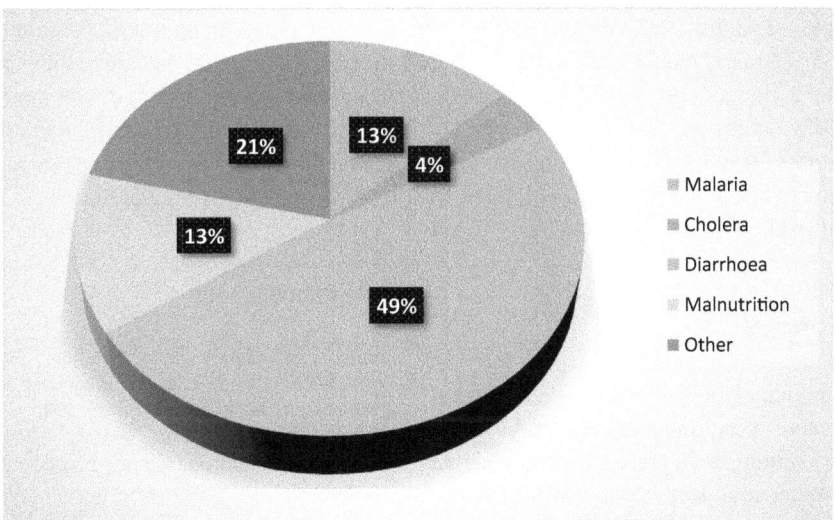

Fig. 10.12 Perceptions of disease prevalence in Chimanimani (*n* = 184). Source: Fieldwork, October 2019

women and girls are not a result of the disaster alone; rather, they are rooted in the inequalities inherent in the social construction of gender prior to the catastrophe, which then become sharpened as efforts to survive become more urgent". (Nguyen 2019, p. 421)

Although water and sanitation (SDG 6) became a challenge initially due to huge numbers at the hotel, a quick partnership (SDG 17) with the Chimanimani Rural District Council (RDC) resulted in the installation of a stand-alone pipe just within the hotel's premises. In fact, as most of the water sources had been damaged during Cyclone Idai the Chimanimani community depended on natural protected springs and piping. For example the entire Ngangu township's reticulated water and sanitation system had been damaged flying in the face of SDG 6.1 and 6.2 whose targets focus on the proportion of residents using safely managed drinking water and those using safely managed sanitation and hand-washing facilities (United Nations 2015). Other partnerships are those that developed in feeding the masses with donations that came from across the country and beyond, as promoted through the General Manager's WhatsApp platform. Lastly, the Chimanimani Hotel networked into the bigger global agenda of Climate Action (SDG 13). With extreme weather events caused by climate change increasing resulting in events such as Cyclone Idai, immediate reaction to adjust to the new reality is needed. The Chimanimani Hotel did just what the climate action doctor would have recommended. For example the SDG 13.1 target has an indicator that focuses on addressing persons directly affected as a result of a disaster such as Cyclone Idai (United Nations 2015).

10.5   Conclusion

The Chimanimani Hotel's GM, his team of managers and staff performed an act of kindness worth documenting and presenting to a global audience. What makes this story unique is the environment in which the good deeds were performed—a closed community cut off from the rest of the world for 4 days in a country with an ever-deteriorating economic situation. The initial buying of food for a crowd of approximately 600 people out of hotel and personal resources is an act of heroism. It is no wonder that many of those interviewed passed on the story of the Chimanimani Hotel. The offering of school fees scholarship to a daughter of a deceased staff member also is a true act of kindness. The Chimanimani Hotel embraced several provisions enshrined in the 17 SDGs as highlighted earlier, especially protecting the vulnerable—women, children and the old. The manner in which partnerships were built is also worth documenting, including the assistance rendered to other organisations during relief concerning logistics in feeding the hungry. As the new guests joined the hotel's relief movement, it was great to realise how the young and energetic and the mothers assisted in cooking, looking for firewood and embarked on many other tasks that would assist in improving their stay at the hotel. The areas that will require further improvements are in developing the hotel's disaster risk reduction and management plan and train all staff in how to react. The story of the Hotel Chimanimani will remain in the archives of Zimbabwean history for many generations to come. The authors would like to thank the GM for his situational leadership qualities and thank all the staff of the hotel. The hotel infrastructure and human capital make them ideal as a safe and temporary accommodation for internally displaced people. Hotels can be an important entity that can provide most of the life support services before, during and after a disaster and have key assets that allow them to enhance societal recovery from disaster.

References

Adhikari, M. (2008). Hotel Rwanda: The challenges of historicising and commercialising genocide. Development Dialogue, 50, 173–198.

AlBattat, A.R., & Som, A.P.M. (2014). Disaster Preparedness of Hotel Industry Abroad: A Comparative Analysis. SHS Web of Conferences 12, 01012, https://doi.org/10.1051/shsconf/20141201012.

Amnesty International. (2005). Hotel Rwanda: A true story. United Artists Films, Inc., New York.

BBC (British Broadcasting Corporation). (2019). Cyclone Idai: The hotel manager providing a safe haven in Zimbabwe. Available at: https://www.bbc.com/news/world-africa-47640336 (Retrieved on 25 September 2019).

Cheung, C., & Law, R. (2006). How can hotel guests be protected during the occurrence of a Tsunami? Asia Pacific Journal of Tourism Research, 11(3), 289–295, https://doi.org/10.1080/10941660600753331.

Dobie, S., Schneider, J., Kesgin, M., Lagiewski, R. (2018). Hotels as Critical Hubs for Destination Disaster Resilience: An Analysis of Hotel Corporations' CSR Activities Supporting Disaster Relief and Resilience. Infrastructures, 3, 46; doi:https://doi.org/10.3390/infrastructures3040046.

Fregonese, S. & Ramadan, A. (2015). Hotel Geopolitics: A Research Agenda. Geopolitics, 20(4), 793–813. https://doi.org/10.1080/14650045.2015.1062755.

Ganguly, D. (2007). 100 Days in Rwanda, 1994: Trauma Aesthetics and Humanist Ethics in an Age of Terror. Humanities Research, XIV(2), 49–66.

Hartmann, R. (2014) Dark tourism, thanatourism, and dissonance in heritage tourism management: new directions in contemporary tourism research, Journal of Heritage Tourism, 9(2), 166–182, https://doi.org/10.1080/1743873X.2013.807266.

Henderson, J.C. (2005). Responding to natural disasters: Managing a hotel in the aftermath of the Indian Ocean tsunami. Tourism and Hospitality Research, 6(1), 89–96.

Henderson, J.C. (2007). Corporate social responsibility and tourism: Hotel companies in Phuket, Thailand, after the Indian Ocean tsunami. Hospitality Management, 26, 228–239. https://doi.org/10.1016/j.ijhm.2006.02.001.

Ivkov, M., Blešić, I., Janićević, S., Kovačić, S., Miljković, Đ, Lukić, T., Sakulski, D. (2019). Natural Disasters vs Hotel Industry Resilience: An Exploratory Study among Hotel Managers from European Open Geoscience, 11,378–390.

Mamuna, M.A., Huq, N., Papia, Z.F., Tasfina, S., Gozal, D. (2019). Prevalence of depression among Bangladeshi village women subsequent to a natural disaster: A pilot study. Psychiatry Research 276, 124–128. https://doi.org/10.1016/j.psychres.2019.05.007.

Momberg, M. (2016). Hotel Rwanda: Individual Heroism or Interconnectedness in the Portrayal of Paul Rusesabagina? Scriptura 115(1): 1–12, https://doi.org/10.7833/115-0-1176.

Nguyen, H.T. (2019). Gendered Vulnerabilities in Times of Natural Disasters: Male-to-Female Violence in the Philippines in the Aftermath of Super Typhoon Haiyan. Violence Against Women, 25(4), 421–440. https://doi.org/10.1177/1077801218790701.

Nguyen, D.N., Imamura, F., Iuchi, K. (2018). Barriers towards hotel disaster preparedness: Case studies of post 2011 Tsunami, Japan. International Journal of Disaster Risk Reduction, 28, 585–595. https://doi.org/10.1016/j.ijdrr.2018.01.008.

Nyakunyanga, S. (2019a). On the ground in Chimanimani: System failure amid the devastating deadly force of Cyclone Idai. Available at: https://www.dailymaverick.co.za/article/2019-04-05-on-the-ground-in-chimanimani-system-failure-amid-the-devastating-deadly-force-of-cyclone-idai/ (Retrieved 25 September 2019).

Nyakunyanga, S. (2019b). Amid criticism of the government response to Cyclone Idai, Zimbabweans pitch in and self-help. Available at: https://www.thenewhumanitarian.org/news/2019/04/11/amid-criticism-government-response-cyclone-idai-zimbabweans-pitch-and-self-help (Retrieved on 25 September 2019).

Ore, J. (2019). Hotel manager in Zimbabwe shelters hundreds displaced by Cyclone Idai. Available at: https://www.cbc.ca/radio/asithappens/as-it-happens-wednesday-edition-1.5063902/hotel-manager-in-zimbabwe-shelters-hundreds-displaced-by-cyclone-idai-1.5063908 (Retrieved on 25 September 2019).

Pongponrat, K. and Ishii, K. (2018). Social vulnerability of marginalized people in times of disaster: Case of Thai women in Japan Tsunami 2011. International Journal of Disaster Risk Reduction, 27, 133–141.

Poria, Y., Singal, M., Wokutch, R.E., Hong, M. (2014). Hotels' social responsiveness toward a community in crisis. International Journal of Hospitality Management, 39, 84–86. https://doi.org/10.1016/j.ijhm.2014.02.006.

Ritchie, B. (2008) Tourism Disaster Planning and Management: From Response and Recovery to Reduction and Readiness, Current Issues in Tourism, 11(4), 315–348. https://doi.org/10.1080/13683500802140372.

Saunyama, J. (2019). Chimanimani Hotel turns into a sanctuary. Available at: https://www.thestandard.co.zw/2019/03/26/chimanimani-hotel-turns-sanctuary/ (Retrieved 25 September 2019).

Sim, T., Lau, J.L., Wei, H.H. (2019). Post-disaster Psychosocial Capacity Building for Women in a Chinese Rural Village. International Journal of Disaster Risk Science, 10:193–203. https://doi.org/10.1007/s13753-019-0221-1.

Southon, M.P. & Van der Merwe, C.D. (2018). Flooded with risks or opportunities: Exploring flooding impacts on tourist accommodation. African Journal of Hospitality, Tourism and Leisure, 7(1), Open Access - Online @ http//:www.ajhtl.com.

Thornton, W.E. & Voigt, L. (2007). Disaster Rape: Vulnerability of Women to Sexual Assaults During Hurricane Katrina. Journal of Public Management & Social Policy, Fall 2001, 23–49.

United Nations. (2015). Transforming our world: The 2030 agenda for sustainable development. New York: United Nations Secretariat.

Uraizee, J. (2010). Gazing at the Beast: Describing Mass Murder in Deepa Mehta's Earth and Terry George's Hotel Rwanda. Shofar: An Interdisciplinary Journal of Jewish Studies, 28(4), 10–27.

Yoshihama, M., Yunomae, T., Tsuge, A., Ikeda, K., Masai, R. (2018). Violence Against Women and Children Following the 2011 Great East Japan Disaster: Making the Invisible Visible Through Research. Violence Against Women, 25(7), 862–88. https://doi.org/10.1177/1077801218802642.

Part V

Looking Ahead in Order to Act Today

Ethical and Human Rights Dilemmas During Disasters: Emerging Findings from Tropical Cyclone Idai

11

Abstract

Southern Africa has witnessed increased activity in tropical cyclones and floods. However, throughout the disaster cycle, issues of ethical conduct always emerge, especially concerning survivors. Ethical conduct must be observed across all the stages of the disaster risk reduction (DRR) and management cycle, including early warning (preparedness), search and rescue, relief and recovery. This work investigates ethical dilemmas that arose when Tropical Cyclone affected Malawi, Madagascar, Mozambique and Zimbabwe in March 2019. Although some insights are drawn from the fieldwork that took place in Chimanimani, Zimbabwe, involving a household survey, interviews and on-the-ground observations, records from other countries are also included through document analysis. The work reveals that there were significant breaches in DRR ethical conduct during the cyclone. Such breaches included the distribution of expired foodstuffs, the politicisation of aid, the exclusion of those whom authorities thought had adequate resources and multiple claims for food parcels by victims, food rotting in warehouses, gender-based violence resulting from cash payments, aid shrinkage by officials and aid-related sex exploitation. Since communities in Kopa, Ngangu and other areas had to provide for the immediate needs of those directly affected—food, clothes and shelter for a week or more due to closed roads—they too were left in dire need of aid, yet they were excluded from aid lists. Moving forward, the study recommends improved beneficiary selection. This includes careful screening for food and other aid handouts for government officials as they too could be victims of a disaster as was in the case in Chimanimani. Furthermore, the work recommends that provision be made for food and other aid handouts for local communities that would not have been directly hit by the disaster but assisted with emergency food, clothes and shelter while government and development agencies made preparations. Proper protocols for documenting victims' right from the outset should be put in place, while systems for reporting sex exploitation for aid need further refinement.

Keywords

SDGs · DRR · Ethics · Cyclone Idai · Sex exploitation · Food aid · Politicisation

G. Nhamo, D. Chikodzi, *Cyclones in Southern Africa*, Sustainable Development Goals Series, https://doi.org/10.1007/978-3-030-72393-4_11

11.1 Introduction and Background

This work was triggered by many media reports of abuse related to Tropical Cyclone Idai emergency relief. The cyclone hit Malawi, Madagascar, Mozambique and Zimbabwe in March 2019, affecting an estimated 3.8 million people and leaving a trail of destruction that left more than 1000 people dead (Mushanyuri and Ngcamu 2020; Muzhinji and Ntuli 2021). From Mozambique there were reports that victims were forced to trade sex for food (Aljazeera 2019; Human Rights Watch 2019; OCHA 2019). In addition, there were also reports of dishonesty in the distribution of aid as it was diverted and/or stolen both in Mozambique and Zimbabwe (Zenda 2019). In Zimbabwe, there were also reports of food donated for Tropical Cyclone Idai victims rotting or expiring in warehouses in Chimanimani East with no clear reasons as to why this was happening while victims were starving (Maodza 2019). Drawing from Parkash's (2012, p. 383) work, "in any profession, a basic set of moral values needs to be followed to comply with what we call ethics" and the field of disaster risk reduction (DRR) and management is no exception.

Ethical conduct in DRR is also closely aligned to the 2030 Agenda on Sustainable Development and the closely linked 17 Sustainable Development Goals (SDGs). To this end, several SDGs come to mind (United Nations 2015), namely SDG 2 (ending hunger), SDG 3 (health and wellbeing), SDG 4 (quality education), SDG 5 (gender), SDG 10 (inequality and inclusiveness) and SDG 16 (peace and security). Twenty-three times, the 2030 Agenda for Sustainable Development mentions "rights". Bullet 3 specifically highlights that the SDGs will:

> End poverty and hunger everywhere; to combat inequalities within and among countries; to build peaceful, just and inclusive societies; to protect human rights and promote gender equality and the empowerment of women and girls; and to ensure the lasting protection of the planet and its natural resources (United Nations 2015, p. 3).

In addition to human rights issues, Target 5.6 of SDG 5 seeks to "[e]liminate all forms of violence against all women and girls in the public and private spheres, including trafficking and sexual and other types of exploitation" (United Nations 2015, p. 18). Intense natural disasters, including climate change-related extreme weather events such as tropical cyclones, are mentioned among the threats reversing much of the development progress from the Millennium Development Goals (MDGs) era. To this end, a holistic DRR and management approach at all spatial levels is encouraged. In earlier studies on DRR and management, Bongo et al. (2013) encouraged the Zimbabwean government to adopt an inclusive, rights-based approach. This conclusion was reached based on the fact that the DRR framework for the country did not accommodate some of the most vulnerable in DRR, especially people with disabilities, those terminally ill and the very poor.

Matters pertaining to unethical contact during disasters have also emerged during the COVID-19 relief in South Africa. Apart from massive corruption that has unfolded surrounding the procurement of personal protective equipment (PPE) involving members of the ruling party—the African National Congress (ANC)—there are also reports of unethical conduct during the distribution of food parcels meant for the poor. Such heartless acts have been condemned in the strongest of words by the head of the ruling party President Cyril Ramaphosa, in his seven-page letter to all members of the party on 23 August 2020. The party president wrote (African National Congress Office of the President 2020, p. 1) the following:

> I am sure that you are aware that across the nation there is a sense of anger and disillusionment at reports of corruption in our response to the coronavirus pandemic. This anger is understandable and justified. In recent weeks, we have heard stories of tenders for PPE that have been given to individuals associated with ANC leaders and of public servants flouting the law in issuing tenders.

The ANC and the South African president went on to indicate that the party and the government were dipping their heads in shame, especially regarding the misappropriation of food parcels meant for the families hardest hit by COVID-19. Once more the President did not mince his words:

As we have seen during the COVID-19 response, there are local ANC leaders who have used food parcels meant for the poor to buy political favours from those people in the branch or broader community whom they rely on for their positions. These practices quite literally take food out of the mouths of the poor. (African National Congress Office of the President 2020, p. 7)

Ramaphosa then outlined the immediate action that had to be taken to curb the rampant corruption taking advantage of the COVID-19 pandemic. There was a need to stop corruption within the rank and file of the ANC first. To this end, every ANC member implicated in corrupt activities pertaining to COVID-19, regardless of their position in the party had to come forward and account to the ANC Integrity Commission or face disciplinary processes. Members who failed to give an acceptable explanation or fail to step down voluntarily while their cases were being investigated were to be summarily suspended. Whether this would happen, time will tell.

Writing on ethical principles in DRR and people's resilience, Prieur (2012) identifies a range of principles, many of which are also supported by Civaner et al. (2017). Prior to disasters, the following principles should apply: prevention measures should be introduced; education, training and awareness-raising about resilience to disasters, as well as prior information should be provided; and special prevention measures for the most vulnerable groups should be instituted (Prieur 2012). Some ethical principles applied during disasters include humanitarian assistance; compulsory evacuation of populations; respect for dignity; respect for persons; emergency assistance for the most vulnerable persons; the importance of rescue workers; and the restoration of social ties (Civaner et al. 2017; Prieur 2012). Ethical principles applied after disasters include the strengthening of resilience to the effects of disasters; protection of economic, social and cultural rights and the protection of civil and political rights. In addition, the standards of medical care during a disaster remain a thorny issue and one that always raises ethical debate. Mariaselvam and Gopichandran (2016) highlight that during the 2015 Chennai floods in India, a single doctor in a camp would see between 100 and 200 patients within 3–4 h.

Another ethical perspective from the Chennai floods is the inequity in the relief and response activities, conflicts between government and non-government relief and response, focus on short-term relief rather than rehabilitation and reconstruction activities, and the lack of crisis standards of care in medical services (Mariaselvam and Gopichandran 2016). Television philanthropy, the politicisation of emergency aid, aid agencies shrinkage—stealing and holding on for too long before the distribution, cheating for more handouts from victims themselves, and the elite and/or traditional leadership capture of the humanitarian process remain matters to be investigated (Civaner et al. 2017).

Given the foregoing, this work sets the objective to investigate ethical dilemmas surrounding Tropical Cyclone Idai from the DRR and management cycle perspective. The main focus will be on search and rescue, as well as relief stages. It is hoped that this contribution will add to the volumes of knowledge in the field, and more specifically to research from southern Africa which remains devoid of such, as was revealed by the literature searches. Some of the parameters investigated include food and clothes distribution, matters of sex-for-aid, unethical contact by victims, politicisation and the capture of emergency aid logistics by ruling party members and traditional leadership.

11.2 Materials and Methods

The study focuses mainly on the relief and recovery stages of the DRR and management cycle. To address the set research objective, the work utilised a mixed methods approach that included the use a household survey administered on a QuestionPro off-line platform, key informant interviews numbering more than 25, five focus group discussions, and on-the-ground field observations that led to scenes being captured on camera. The use these methods is not new in DRR and disaster ethics. Chanza et al. (2020) used both on-the-ground observations and interviews to document the gaps in DRR and management associated with Tropical Cyclone Idai in the

Rusitu area of Chimanimani. The Global Public Policy Institute (2010) applied interviews and focus groups in researching ethics issues during the Haiti 2010 earthquake. Many households housed in tents such as those in Nyamatanda (Fig. 11.1) were visited and happily responded to questions. A total of four shelter camps were erected, namely Arboretum, Garikai, and Nyamatanda (all in Ward 15) and Kopa located in Ward 21. The QuestionPro off-line survey was administered by 16 researchers and their assistants who had undergone professional training in conducting research in disaster-affected areas. The use of surveys in DRR is also common (Mitsova et al. 2019; Keellings and Hernández-Ayala 2019).

From the household survey, the income status of the respondents was investigated and is shown in Fig. 11.2. The income status of the household is critical in terms of emergency relief and its distribution. Approximately 81.51% of the households surveyed earned US$600 or less per month. Other statistics of interest in the context of disaster relief ethics include the nature of household heads. Those households ($n = 160$) that were female-centred comprised 16.88%, followed by 29.38% of those that were male-centred, 48.12% that were nuclear, with the remaining percentage shared by those that were extended or child-headed. In terms of age groups, 1.88% of the household respondents were between 18 and 19 years, 26.88% between 20 and 29 years, while 33.12% (the bulk) were between 30 and 39. Those between 40 and 49 years comprised 22.5%, with 10% comprising household respondents between 50 and 59 years. Only 5.62% of the household respondents were aged 60 years or more.

Data were also generated regarding the size of households, which was defined by the number of people that eat from the same pot daily. The majority of households were made up of five people (26.88%). This was followed by those with four people (21.25%) and those with three people (16.25%). There were also several households with six people (11.88%) and some with seven people (9.38%). The remaining percentages were for households with one person each (4.38%) and two people (3.12%). In terms of analysis, interviews were transcribed and elements of grounded theory applied (Nhamo and Agyepong 2019). Additional analysis included importing the data from QuestionPro into MS Excel and computing graphs. Document and critical discourse analysis were also done for secondary data that included media reports (Mushanyuri and Ngcamu 2020; Aljazeera 2019; Human Rights Watch 2019; OCHA 2019).

Fig. 11.1   Some of the tents in Nyamatanda camp in Chimanimani Township. Source: Authors, Fieldwork 2019

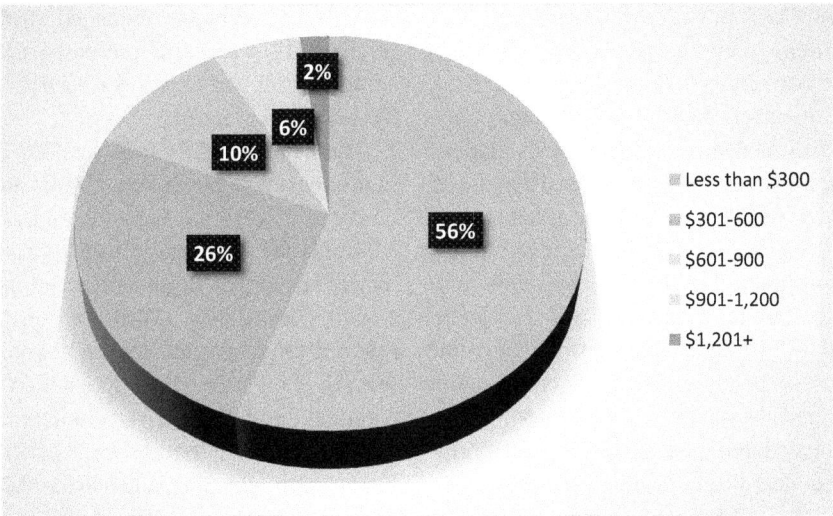

Fig. 11.2 Estimated household income per month (*n* = 146). Source: Authors, Household Survey 2019

11.3 Presentation and Discussion of Findings

This section presents and discusses the key findings of the study. This is done in three subsections, namely (1) a focus on sex-for-food (aid) and the politicisation of aid; (2) investigation into allegations of the distribution of expired and/or near expired foods, old/reject clothes and other logistics and (3) the behaviour of victims and other selected key stakeholders, including traditional leadership and allegations of hijacking emergency relief logistics. Each of these perspectives will now be considered in turn in the following paragraphs.

11.3.1 Sex-for-Food (Aid), the Politicisation of Aid and Other Dilemmas

There were widespread media reports on sexual exploitation in exchange of food aid for Tropical Cyclone Idai victims in Mozambique. Aljazeera (2019) reported that local leaders in Nhamatanda district had forced women into sex in order for them to obtain food aid. Specifically, this was done for them to get a bag of rice. This was happening as the country battled to enable more than

a million victims to resume their normal lives. The allegations also brought to the fore the politicisation of emergency relief as the local leaders concerned were known ruling Frelimo party members. There were also additional allegations of local leaders demanding money to include victims on the aid distribution lists. The lists were said to exclude households headed by women. Sex-for-food/aid is not a new phenomenon. During the Haiti 2010 earthquake, women from shanty towns were confronted with demands for sex by officials in order for them to have access to emergency aid (Global Public Policy Institute 2010). Situations occurred where people had to wait for hours in the sun to receive emergency relief.

Although not related to sex-for-food, similar reports of the politicisation of Tropical Cyclone Idai emergency relief by members of the ruling ZANU-PF party in Zimbabwe were reported in Chimanimani. Zenda (2019) highlights that in Duri village, hundreds of victims were seen jostling to receive emergency aid in May 2019 as they had been excluded from the April 2019 slot because ZANU-PF party members converted the voter registration list dominated by their members into a beneficiary list. The victims in Duri were now receiving the aid following a new, and all-inclusive list from the NGOs. The ZANU-PF

Member of Parliament for Chimanimani West and one ruling party member were identified among the perpetrators. Some of the scenes of ZANU-PF members hijacking food distribution had been recorded in the videos, with pictures also taken, making it difficult to deny the allegations as per the report. This resulted in the Minister of Local Government, Mr July Moyo, explaining to the effect that villagers may have mistaken the fact that some food was being carried in ZANU-PF branded vehicles offered by the party and not being grabbed. Another senior party official and speaker for the party, Mr Simon Moyo, indicated that the ZANU-PF youth carrying the food were genuine volunteers.

During the Tropical Cyclone Idai in Mozambique, community leaders had immense power as many areas were completely cut off as the roads were impassable (Aljazeera 2019). Three women from Mbimbir in Nhamatanda district subsequently revealed that they had been forced to have sex for food with local officials after humanitarian aid only arrived on 5 April 2019, almost a month after the cyclone. As revealed in many interviews held during fieldwork in September and October 2019, in Chimanimani (Zimbabwe), the disaster hotspots could not be reached for at least a week, either by air or road, owing to damaged roads and bad weather. This scenario presented an unwanted exploitative environment for sexual exploitation by those in power.

The issue of sex for food was also highlighted in a joint report by Oxfam, CARE and Save the Children (2019), as a recommendation that the Mozambique government needed to attend to as a matter of urgency. It also came up in the World Health Organisation (WHO) report where it is referred to as survival sex and exploitation sex from the Sofala Province (Machando 2019). Sex exploitation during crisis situations was also raised by Lafrenière et al. (2019). Knowles (2020) highlights the same case in Mozambique and several cases in which sexual favours were demanded in exchange of food aid, including in Haiti and Uganda. In Haiti, the United Nations peacekeepers bartered relief goods for sex, with more than 225 women acknowledging that they had been forced into the situation. In Kampala, Uganda, several members of the local councils were reported to be demanding sexual favours before their victims could be considered among the government COVID-19 pandemic relief food beneficiary lists. The issue of sexual abuse was also investigated by means of a household survey in Chimanimani, Zimbabwe. The focus was on

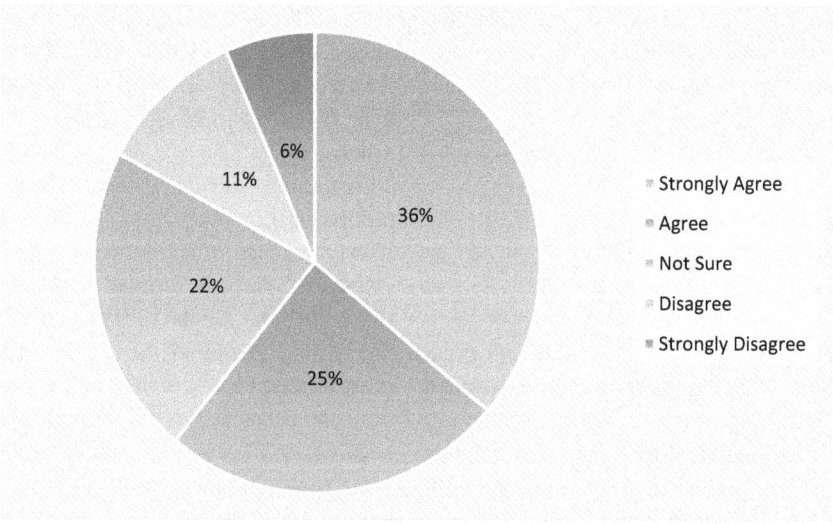

Fig. 11.3 Rescuers taking advantage of victims to gain sexual favours (*n* = 197). Source: Authors, Household Survey 2019

rescuers and how they would possibly prey on women and girls. The results are shown in Fig. 11.3. The household survey reveals perceptions confirming sexual advances by male rescuers to women and girls. An estimated 36% of the households strongly agreed that this was the case, with a further 25% agreeing that was so.

Responding to allegations from Mozambique, OCHA (2019) highlighted that it was important for stakeholders to note that aid was free. To this end, sex for aid was unacceptable. The OCHA further revealed that they were working with the government to investigate all the reported cases of abuse. The organisation indicated that while it had trained hundreds of aid workers and volunteers to prevent sex abuse, the challenge remained. The abuse of women during disasters continues to haunt humanity. During the August 2017 conflict in Myanmar, over 700,000 Rohingya refugees fled to Bangladesh, and many women and girls arrived there with unwanted pregnancies.

There were also issues regarding cash payments for emergency relief and gender-based violence. One of the key informants who is a chief in Chimanimani indicated that there were cash handouts from the World Food Programme (WFP) and CARE International. The cash payments were given in local currency (Bonds), and each family received the amount according to the number of people in the family. Many households were given money for 2 months. The reason why this was done, according to the chief, was that the two organisations had failed to get maize within the Southern African Development Community (SADC) region. Hence, households had to decide for themselves what to do with the money. Asked to explain who exactly in the household would be the person handed the money, the chief responded to the effect that:

> Where there were both father and mother, the money would be given to the mother. Where a child is heading the family, the eldest child would be given the money but under the guardianship of a close relative. Where there was a father only, the father was given the money and the same goes if there was only the mother in a family.

The chief highlighted that in September and October 2019 some people went home with 4000 Bond, with some getting up to 2500 Bond. The money was also given as a once-off payment. Previously, the victims had been given maize coupons, and five litres of cooking oil, etc. This revelation confirmed the challenge of shortages the organisations raised as the reason to move to cash payments. The cash payments were also confirmed by a focus group discussion held at Kopa, one of the townships that were the hardest hit by Tropical Cyclone Idai. From the focus group discussion, it emerged that an amount of 202 Bonds was given per person. The author also teased out whether there were any issues pertaining to gender-based and other violence associated with the cash payments. The chief gave the following response:

> There was a lot of domestic violence that was triggered by this money issue, which was given to families. Based on our culture, the men at home went and demanded part of the money and spent it on alcohol. In ward 17, there is a lady who is already in custody, she took a bayonet and cut her husband on the forehead and forearm. They were fighting over money. This money triggered a lot of domestic violence.

Although not blaming the organisations for cash payments, and also appreciating the help given, the chief was of the view that there should have been adequate research to determine the negative impacts of cash payments and to mitigate these. The chief also indicated that in a rural set-up like Chimanimani, it was rare for a family to handle something like 1000 Bond. "Then there was suddenly a mother having 2500 Bond at her disposal. This would naturally cause problems at home." Another chief from Kopa area shared his experience of dealing with cases brought before him regarding cash payments. He mentioned that the cash would be transferred to cell phones (mobile money) and often into the men's phones. Then when the wife requested a share to use the problems started. The chief gave an example of a marriage of 18 years that had been destroyed and a case he was addressing from a marriage of 13 years the day we interviewed him. It also emerged that there was another challenge associated with cash payments. The focus group at Kopa revealed that some who assisted victims during Tropical

Table 11.1 Details regarding IDP as of 20 April 2020

Emergency shelter (Camp)	Number of households	Total Individuals	Male (%)	Female (%)
Arboretum	59	309	47	53
Garikai	83	309	45	55
Kopa	53	128	48	52
Nyamatanda	29	113	50	50

Source: Authors, based on IOM Displacement Tracking Matrix (2020, pp. 1–5)

Cyclone Idai would come asking for a share of the cash, indicating that had it not been for them they would have been swept by the floodwaters.

On 27 April 2020, the IOM Displacement Tracking Matrix (2020) undertook a mission in liaison with the Zimbabwe Government and other key partners to track the mobility of the internally displaced people (IDP) in Chimanimani camps. There was a need to assess once more their living conditions, their needs and how to intervene further. The findings confirmed that the camps remained the only habitable option for the IDP, and shockingly, this was more than a year after Tropical Cyclone Idai. Furthermore, food remained an urgent need as 75% of IDP in camps were having less than three meals daily, an element that is worrying considering how women and girls are sexually exploited for food. General theft and violence were reported as worrying too. Non-food items including water containers, soap and mosquito nets were also needed. The details regarding IDP and the four camps are shown in Table 11.1.

Generally, there still remain more females in the camps. In addition, there are noticeable changes in the composition of IDP, with new births and new family member arrivals, while other family members have left the emergency shelters to look for employment and other livelihood activities (IOM Displacement Tracking Matrix 2020). In total, 224 households and 859 individuals remained in camps a year after Tropical Cyclone Idai.

11.3.2 Expired Foods, Old/Reject Clothes and Other Logistics

In Zimbabwe, there were reports of Tropical Cyclone Idai foodstuff either rotting or expiring in warehouses at the Machongwe Food Distribution Centre in Chimanimani East (Maodza 2019). Identified foodstuffs included 30 tonnes of mealie meal, 1850 kg of corn soya blend, 1150 kg of plain flour, 500 × 6 packets of sugar beans and 1000 Maheu Instant. This report was picked up by other media outlets globally and widely publicised. This alerted the research team on the ground in Chimanimani (including the author of this chapter) to be on the lookout and also to include some questions in the household survey dealing with matters of aid ethics. To verify the claims from the media, questions were inserted in the household survey concerning the crisis in standards of emergency aid. The findings are presented in Fig. 11.4.

There are several critical matters that emerge from the household survey pertaining to the issues of distributing expired and/or near expired food stuffs, old and/or reject clothes, aid agency shrinkage (stealing and holding onto aid for too long before distribution), and the elite and traditional leadership capturing the emergency aid processes. Generally, households were of the view that the matters raised indeed happened in Chimanimani, especially the distribution of expired and near expiring foodstuff. The same trend was shown in the distribution of clothes and the capture of the emergency aid logistics by the elite and traditional leadership. A focus group discussion in Kopa was one of the main sources of information to confirm what emerged from the survey. There were allegations that those in charge of clothes first selected all those that looked and/or were new for their immediate family members and friends, leaving the old clothes for the victims. Interviews with key informants further revealed that some of the newer or better clothes taken by those in charge ended up on the open markets in Kopa and Chimanimani town. This was the same with some other foodstuffs and groceries that included soap and cooking oil.

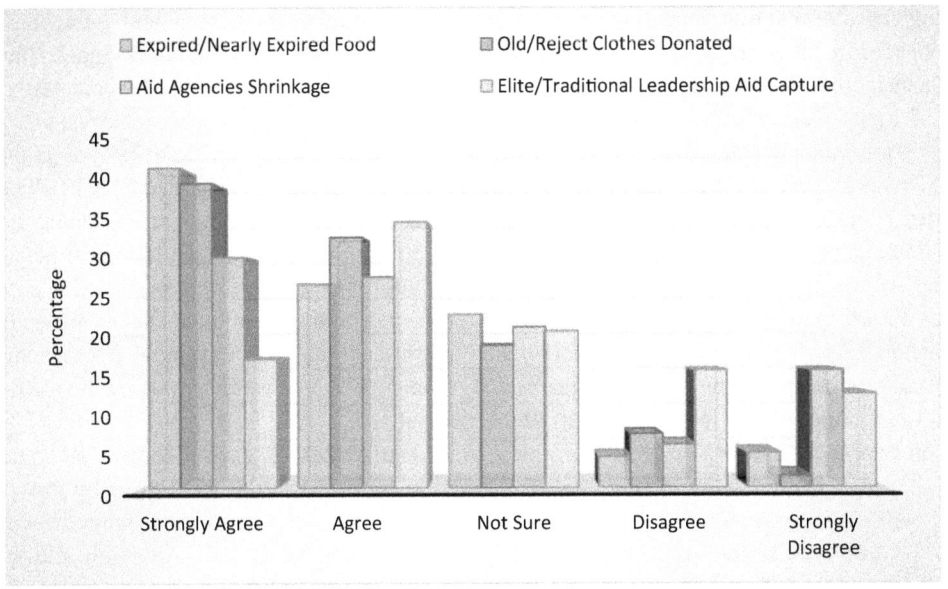

Fig. 11.4 Crisis in standards of emergency aid (*n* = 195). Source: Authors, Household Survey 2019

Traditional leadership was not spared either, with the headmen singled out as the key players in capturing the emergency aid logistics as these were mainly the men who were in charge of putting victims on lists. Yet the lists ended up with many who were not victims at all. The challenges with headmen also came up in a focus group discussion at Kopa. It emerged that the headmen would often have their secretaries list their relatives and friends first. This was done repeatedly with all the donors that came into the area and the same people ended up getting emergency aid from CARE International, World Vision, ARDRA, IRD and CCJP.

Concerning the theft of emergency relief, one case was profiled by Zenda (2019) from the Skyline warehousing and distribution point in Chimanimani. Zenda documents an incidence when a military truck arrived at the point with various food items on board. However, on inspecting the paperwork, the goods should have been a consignment of 360 L of milk and 600 kg of rice, among other things. Yet there were only 240 L of milk and 400 kg of rice, which the receiving officials did not question. Zenda concludes that it was likely that such conduct was happening every day. The shortfall could then have been explained by reports of such goods

being confiscated at roadblocks by police and some individuals getting arrested. However, on following up on issues pertaining to goods being intercepted at roadblocks, one of the key informants indicated that there were victims and government workers who were in Chimanimani without family and had to send groceries to their families as was their norm. This matter remained a thorny issue and needed further investigation for the future, with a potential to use police clearance for such instances as there would always be suspicion of aid theft.

On further follow-ups with the Chimanimani Police Station during fieldwork, the author received confirmation that there were indeed incidences of theft and conversion by some government and other officials. The cases were referred to the Magistrates Courts in both Chimanimani and Chipinge. However, one of the key informants from the security cluster, as well as another key informant from Chimanimani Hotel, were of the view that although these incidents occurred, they should not overshadow the great work that both the government and the development community were doing to assist Tropical Cyclone Idai victims. In a 10 April 2019 report by New Zimbabwe (2019), it emerged that a government employee had been arrested for stealing the aid and was appearing in court in Chipinge. From the court papers, some of the stolen goods include 2 × 5 kg of rice, 19 × 2 kg rice, 4 × 2.5 kg of baby porridge, 2 × 1 kg maheu, 11 × 500 mL mineral water, 1 classic tissue paper and 11 pairs of shoes. Three other people were reported to have been in court over stolen aid, including a deputy director from the Ministry of Women Affairs. In another separate case, a police officer was arrested for stealing Tropical Cyclone Idai aid on 4 April 2019 in Rusape Town (Ncube 2019). The policemen even had the audacity to hire a van to transport the loot. The details of the aid stolen are presented in Box 11.1.

Box 11.1 Alleged Stolen Tropical Cyclone Idai Aid by One Policeman

- 20 × 400 g Boom paste, 10 × 400 g Ideal spaghetti, 5 kg Meso Kapenta, 4 × 2 kg Maq washing powder, 20 × 500 g sugar beans, 20 × 2 kg white sugar (Goldstar), 1 × 3 kg Wave washing powder, 1 × 2 kg Surf washing powder, 4 × 2 kg Surf washing powder, 4 × 2 kg 2-in-1 Sunlight washing powder, 6 × 6400 g packs of candles, 10 × 500 g Mega rice, 6 × 2 kg Mariana rice, 1 × 2 kg Excella rice, 2 × 2 kg Red Seal rice, 1 × 2 kg Probrand rice, 1 × 2 kg Ideal rice, 2 × 12,375 mL peanut butter (Mama), 4 × 1 kg iodised Red Seal salt, 1 × 1 kg Probrand salt, 1 × 1 kg Mega iodised salt, 1 × 1 kg Royal salt, 10 Double Blade shaving machine, Smooth Air clinic set, 2 × 12 2 ℓ Zimgold cooking oil, 10 × 2 kg Probrand Rice, 12 Meiban tissues, 1 khaki shorts, 4 white drying towels, 52 assorted pairs of ladies shoes and eight Waverly blankets.
- 7 jackets, 26 pairs assorted female shoes, 16 × 200 g tablets FA bath soap, 3 dark brown shoe polish, 2 jerseys, 12 T-shirts (mixed), 1 pair of shorts, 2 ladies jacket, 1 checked blouse, 1 grey striped tie, 1 black cap, 1 × 250 mL Vaseline Blue Seal, 5 × 100 mL Vaseline bottles, 1 shoe brush, 1 × 50 mL Colgate toothpaste, and 1 × 375 mL peanut butter.

Source: Authors, based on Ncube (2019)

The Zimbabwe government later confirmed reports that some of its staff were indeed involved in stealing Tropical Cyclone Idai aid. After three of its workers had been arrested, Local Government Minister, July Moyo, confirmed this on 9 April 2019 (Munhende 2019). The confirmation came during a post-cabinet meeting briefing to journalists. One of the cases was that of the police officer who had hired a combi to transport the stolen aid. The minister further confirmed that there were also cases of staff from NGOs who were converting the aid for personal use.

The matter of expired and/or near expired foodstuffs was further confirmed by field observations, with certain already expired foodstuffs having been delivered. Although the bottled tomato sauce in Fig. 11.5 has an expiry date of July 2020, the contents had since gone bad and the cause remained unknown. As witnessed in the field, the contents turned into a dark lump surrounded by some watery stuff. Figure 11.6 shows mealie meal that expired on 17 June 2019 yet was distributed in October 2019. Once more, the reason could not be obtained but confirms the media reports.

There were further reports of dishonesty in the handing out of emergency relief in Mozambique. In the face of several allegations of stolen and/or diverted aid for Tropical Cyclone Idai, the Mozambican National Institute for Disaster Management (INGC) acknowledged that this had taken place (Frey 2019). The diversions were being facilitated by INGC staff in centres in Sofala Province. However, it was encouraging to learn that the INGC immediately established an independent commission, which subsequently led the process of aid distribution, especially food to the Idai victims. The commission was made up of an independent Beira logistics company, the World Food Programme and the INGC.

One of the key ethics matters that came up during the fieldwork, from a focus group discussion in Kopa, was the question of who should receive the Tropical Cyclone Idai aid. There were two main arguments that caused problems from an ethics perspective. The first concerned government officials, especially teachers and nurses, who were victims like any other resident. The issue was that they were paid workers and

Fig. 11.5 Expired canned tomato source (distributed in October 2019) with a 29/07/2020 date. Source: Authors, Fieldwork 2019

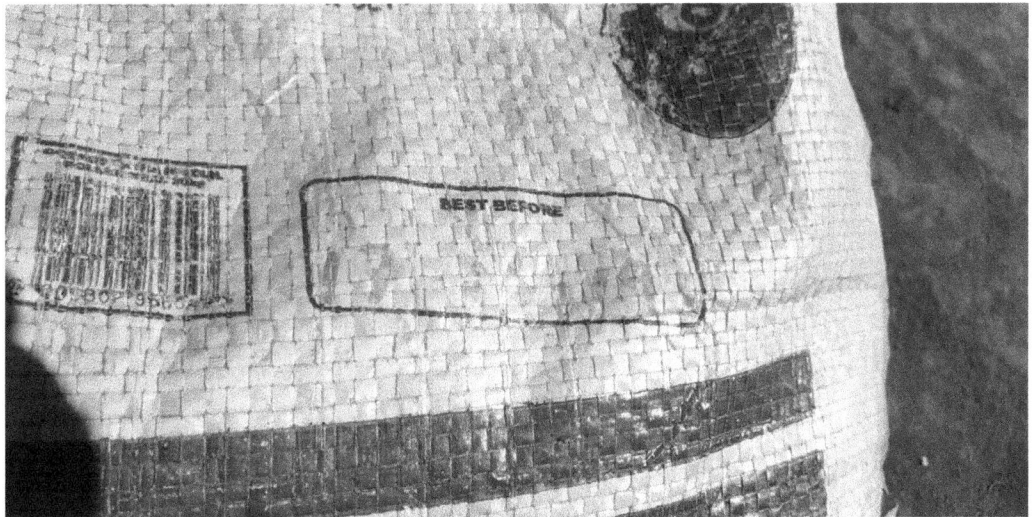

Fig. 11.6 Expired mealie meal. Source: Authors, Fieldwork 2019

should not receive emergency relief. Linked to this was the matter of whether these officials should join the normal lists or have their own express list to receive aid. The second and probably more critical was the issue of relatives and community members who housed the victims for many weeks, some of which were still being housed there rather than in tents in emergency camps. The government argued that relatives and community members that hosted victims were not beneficiaries of aid, while the relatives and community members that hosted them indicated it was proper for an agreed amount of aid and period upon this was to be put in place. This was because their food and other necessities had run out because they could not leave victims to die of hunger. This argument seems to hold water given that the government and development agencies took a long time to arrive in Chimanimani.

A summary of who accommodated the Tropical Cyclone victims in Chimanimani is presented in Fig. 11.7. From the household survey it appears that by far the majority of the victims were accommodated by their relatives (41.74%). This was followed by victims accommodated by other villagers (18.70%), bringing the total of those arguing for inclusion in the aid list to a compelling 60.44%. What was also interesting from the household survey was the fact that 93% of the respondents concurred that they had never witnessed such damage from previous cyclones that had hit the area. A further 86.71% of the households ($n = 158$) were of the view that both agricultural produce and land had been either completely destroyed (72.15%) or partially damaged (14.56%). Hence, the magnitude of the required food aid was huge and will possibly remain so for a long time.

Additional information was gathered in terms of the time it took for the victims to receive emergency relief, including food aid. The results of the household survey ($n = 158$) show that the bulk of aid came only after 1 week (Fig. 11.8). From the results it would appear that none of the victims received emergency aid within hours and none received emergency relief. In the same vein, households ($n = 178$) confirmed that the first batch of emergency aid was airlifted (84.83%), while 13.48% indicated that it came by road. The remaining small percentage indicated they were not sure or did not know (1.68%). The airlifting of food and other emergency aid was confirmed by many on the ground, including those in all the key informant interviews and focus groups. This scenario witnessed the birth of what Chanza et al. (2020, p. 1) call local heroes and good Samaritans who worked through "their social networks, norms, relationships, practices, and modest ingenuity, helped to speed up response times and minimize threats to lives and livelihoods". Some local shop owners whose shops were spared in the most damaged area of Kopa Township and their assistants were the first to respond to the needs of victims.

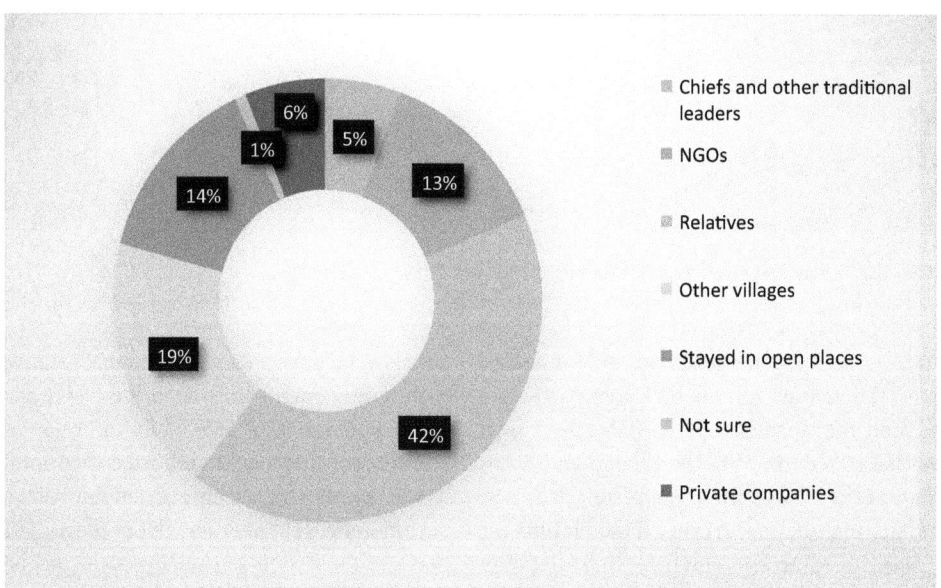

Fig. 11.7 Who mostly accommodated the victims? ($n = 230$). Source: Authors, Household Survey 2019

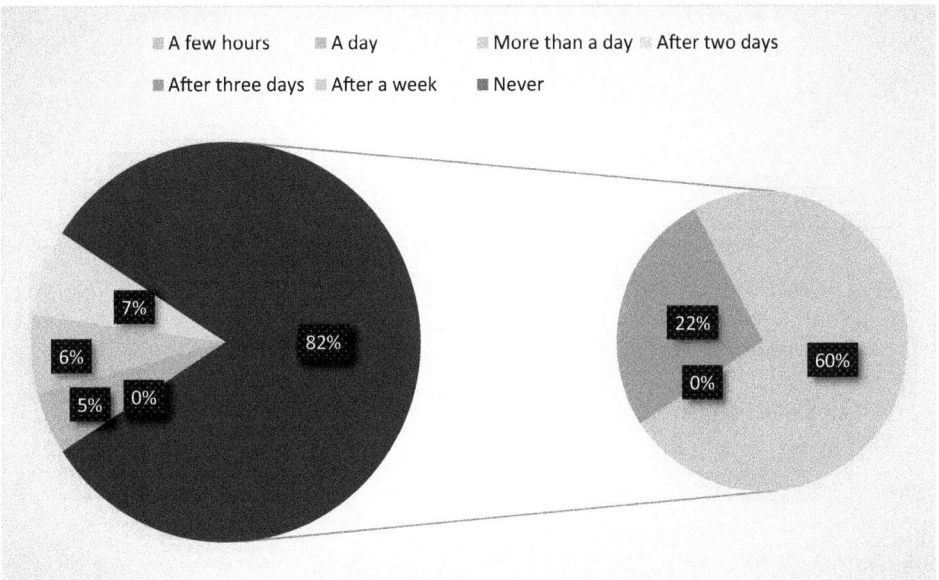

Fig. 11.8 Time it took for victims to receive emergency relief (*n* = 158). Source: Authors, Household Survey 2019

11.3.3 The Behaviour of Victims and Other Selected Key Stakeholders

In response to concerns about reports of unethical behaviour from several stakeholders that usually become involved in disaster relief, a question was included for households to give their perceptions of Tropical Cyclone Idai relief in Chimanimani district. The results are presented in Fig. 11.9. The organisations that received the most approval in terms of ethical conduct were the relief agencies, religious organisations and the defence force. The lowest score went to the municipality and political parties. The fact that the defence force got such high approval is encouraging, given that stories of sexual abuse are often reported from this sector.

One other matter pertaining to ethical conduct in disaster relief that was investigated in Chimanimani were the victims themselves. There are often reports of cheating from the victims, either through inflating the numbers of household members or the use of multiple names to obtain more than the share that would otherwise be allocated. Hence, cheating by victims was one of the household survey questions, and the responses

are shown in Fig. 11.10. An overwhelming majority of the household respondents were of the view that victims indeed cheated when receiving emergency relief. Up to 50.77% of the respondents indicated they strongly agreed with the position presented, with another 31.79% agreeing with the position. Responding to other questions on equity in relief and response activities, 22.56% of the household respondents (*n* = 195) strongly agreed that this had been observed, with an additional 36.92% indicating they agreed. Such responses give us hope in terms of what the 2030 Agenda for Sustainable Development calls for (United Nations 2015) under SDG 10. However, this is also a significant indication that equity matters were violated, with 13.85% having a strong perception and 17.95% having a negative perception. Only 8.72% of the respondents indicated they were not sure. Overall, equity matters in emergency relief seem to have been observed in Chimanimani. Matters of equity are pertinently raised in the SDGs under SDG 10 (United Nations 2015) and embedded in the overall rallying cry of letting no one be left behind in the 2030 Agenda for Sustainable Development.

Another matter concerning ethical conduct in DRR is the focus of development agencies on

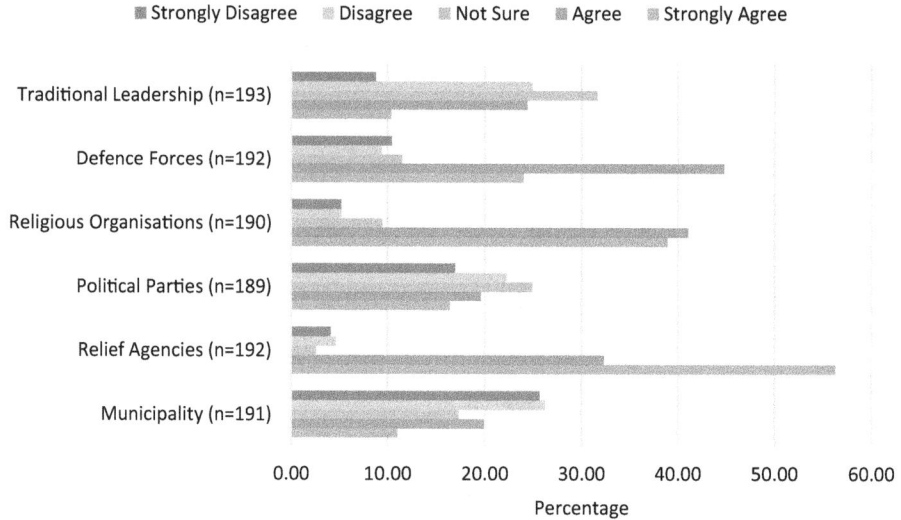

Fig. 11.9 In your view, the following acted ethically/fairly during the disaster. Source: Authors, Household Survey

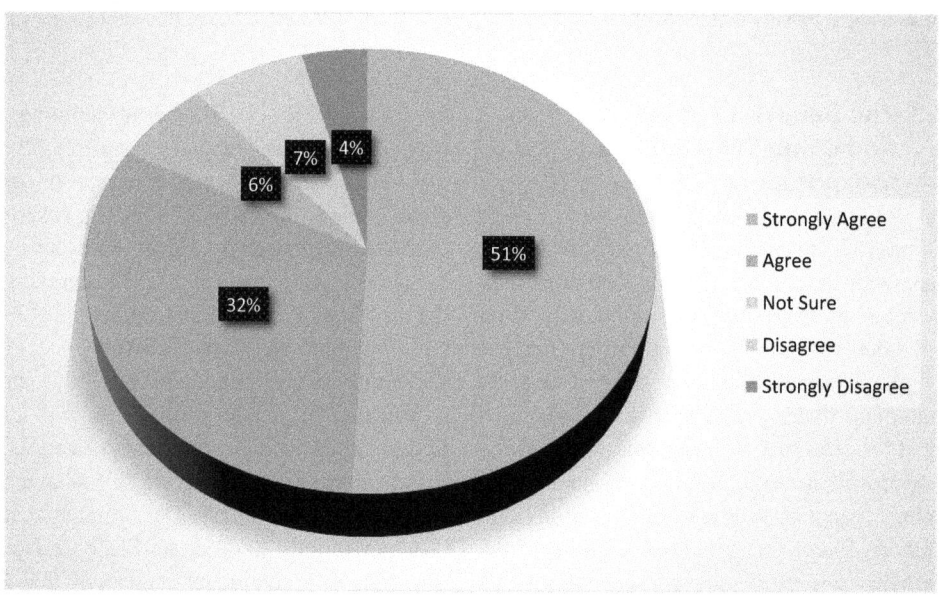

Fig. 11.10 Cheating for more handouts from victims of Cyclone Idai (*n* = 195). Source: Authors, Household Survey 2019

short-term relief, rather than medium to long-term aspects regarding rehabilitation and recovery activities. An estimated 39.09% of the household respondents strongly agreed that this was the case in Chimanimani during Tropical Cyclone Idai, with another 40.61% agreeing to the notion. Only 6.09% were of the opposing view, with another 14.21% not sure. Drawing les-

sons from the super Cyclone Orissa, which made landfall in India in 1999, and the subsequent south-eastern coast tsunami of December 2004, Mariaselvam and Gopichandran (2016) observe that short-term rescue and relief efforts usually go very well, the reason being that there will be a lot of media attention and much-needed external support. Similar scenes were also reported in

Chimanimani through two focus group discussions and several interviews with key informants. At some point there were 14 aid helicopters in the area operating from the donor and development agencies based at Skyline, which ended up being given the name Dubai by the locals as it was reminiscent of the high life of people in the United Arab Emirates. The whole place was well lit with generators and solar panels supplying electricity.

11.4   Conclusions and Recommendations

This work reveals that there were indeed significant challenges concerning ethical conduct by stakeholders during Tropical Cyclone Idai. Ethical issues pertaining to a number of scenarios arose that included the distribution of expired and/or near expired foodstuff, and the exclusion of victims (especially government employees) that authorities thought had adequate resources. There were also reports, confirmed in the literature, of abuse in the form of sex-for-food/aid and the politicisation of emergency aid by members and officials from ruling parties. Furthermore, there were incidences of multiple claims and cheating by victims to obtain more food parcels and other necessary aid. The research reveals that the traditional leadership in Chimanimani became embroiled in aid favouritism as they included their relatives and friends on the aid lists first. This allegation was made in Chimanimani and was confirmed by multiple sources during data generation, which included a household survey, interviews with key informants and focus group discussions. There was also the issue of community members and relatives who had provided emergency relief to their fellow members who could not have their food stocks replenished from incoming emergency aid. This caused tension as there was a genuine need for those affected to have at least some relief, although not for long. Given the fact that roads into Chimanimani were cut off, with the first substantial aid only getting through after a week, the local community members would have assisted as is confirmed in the write-up of this work.

In terms of general ethical conduct during the disaster, findings from Chimanimani confirm that the NGOs and relief agencies and soldiers comported themselves well, and this was appreciated by the community. However, the rural district council and traditional leadership were adjudged not to have acted ethically in the manner in which they handled emergency aid. Lastly, there emerged issues regarding aid theft and misappropriation. Although on-the-ground work revealed that the magnitude was not as wide as reported in the media, there were confirmations of court cases and one incident where a police officer was arrested transporting a combi full of allegedly stolen aid from Chimanimani. Both the governments of Mozambique and Zimbabwe confirmed there were incidences of theft from both government officials and others involved in the logistics of emergency aid. However, what was comforting was that in Mozambique, an independent commission was immediately instituted to handle the logistics of emergency aid. It also emerged that the challenges with the Tropical Cyclone Idai beneficiary list in Zimbabwe could have been due to limited experience in handling a disaster of that magnitude.

Moving forward, the work recommends that careful screening of food and other emergency relief handouts be done for government officials as they too could be victims of a disaster as was the case in Chimanimani. Furthermore, the work recommends that provision be made for food and other emergency relief for local communities that would not have been directly hit by the disaster but assisted with emergency food, clothes and shelter, while government and development agencies got themselves organised. Overall, there is a need to have proper protocols in place for documenting victims right from the outset, as doing so later leaves the aid list filled with undeserving beneficiaries. Drawing from the Mozambican case on dealing with emergency relief theft, independent institutions that include the government, NGOs, development agencies and community members may be needed to promote inclusivity when dealing with emergency relief.

References

African National Congress Office of the President. (2020). *Let this be a turning point in our fight against corruption*. Johannesburg: ANC Office of the President.

Aljazeera. (2019). Mozambique cyclone victims 'forced to trade sex for food'. https://www.aljazeera.com/news/2019/04/mozambique-cyclone-victims-forced-trade-sex-food-190425160159439.html (Accessed 23 August 2020).

Bongo, P.P., Chipangura, P., Sithole, M., Moyo, F. (2013). A rights-based analysis of disaster risk reduction framework in Zimbabwe and its implications for policy and practice. *Jàmbá: Journal of Disaster Risk Studies*, 5(2). https://doi.org/10.4102/jamba.v5i2.81

Chanza, N., Siyongwana, P.Q., Williams-Bruinders, L., Gundu-Jakarasi, V., Mudavanhu, C., Sithole, V.B., Manyani, A. (2020). Closing the gaps in disaster management and response: drawing on local experiences with Cyclone Idai in Chimanimani, Zimbabwe. *International Journal of Disaster Risk Science*. https://doi.org/10.1007/s13753-020-00290-x

Civaner, M.M., Vatansever, K., Pala, K. (2017). Ethical problems in an era where disasters have become a part of daily life: A qualitative study of healthcare workers in Turkey. *PLoS ONE* 12(3), e0174162. https://doi.org/10.1371/journal.pone.0174162

Frey, A. (2019). Mozambique: INGC acknowledges cases of dishonesty in distribution of gods Cyclone Idai victims. https://clubofmozambique.com/news/mozambique-ingc-acknowledges-cases-of-dishonesty-in-distribution-of-gods-cyclone-idai-victims/ (Accessed 23 August 2020).

Global Public Policy Institute. (2010). *Inter-agency real-time evaluation in Haiti: 3 months after the earthquake*. London: Global Public Policy Institute.

Human Rights Watch. (2019). Mozambique: Cyclone victims forced to trade sex for food – Community leaders exploit vulnerable women. https://www.hrw.org/news/2019/04/25/mozambique-cyclone-victims-forced-trade-sex-food (23 August 2020).

IOM Displacement Tracking Matrix. (2020). Tropical Cyclone Idai Response Multi-Sectoral Location Assessment (MSLA) – Round 6: Chimanimani, Manicaland. https://displacement.iom.int/system/tdf/reports/Zimbabwe_DTM_Multi-Sectorial%20Location%20Assessemnt_April%202020%20rev.pdf?file=1&type=node&id=8979 (Accessed 24 August 2020).

Keellings, D. & Hernández-Ayala, J. J. (2019). Extreme rainfall associated with Hurricane Maria over Puerto Rico and its connections to climate variability and change. *Geophysical Research Letters*, 46, 2964–2973. https://doi.org/10.1029/2019GL082077.

Knowles, I. (2020). Women, girls, children and people with disabilities face impossible choices during crises: A humanitarian mission becomes a disaster. Nairobi: Youth Agenda.

Lafrenière, J., Sweetman, C., Thylin, T. (2019). Introduction: gender, humanitarian action and crisis response. *Gender & Development*, 27(2): 187-201. https://doi.org/10.1080/13552074.2019.1634332

Maodza, T. (2019). Cyclone Idai food donations rotting in warehouses. https://www.herald.co.zw/cyclone-idai-food-donations-rotting-in-warehouses/ (Accessed 23 August 2020).

Machando, Z. (2019). Investigating 'Sex for Food' Allegations in Mozambique. Retrieved from https://www.hrw.org/news/2019/05/08/investigating-sex-food-allegations-mozambique (Accessed 6 May 2021).

Mariaselvam, S. & Gopichandran, V. (2016). The Chennai floods of 2015: urgent need for ethical disaster management guidelines. *Indian Journal of Medical Ethics*, 1(2), 91-5. https://doi.org/10.20529/IJME.2016.025

Mitsova, D., Escaleras, M., Sapat, A., Esnard, A.M., Lamadrid, A.J. (2019). The effects of infrastructure service disruptions and socio-economic vulnerability on hurricane recovery. *Sustainability*, 11: 516. https://doi.org/10.3390/su11020516

Munhende, L. (2019). Government confirms own officials stealing Cyclone Idai aid. Retrieved from https://www.newzimbabwe.com/government-confirms-own-officials-stealing-cyclone-idai-aid/ (Accessed 6 May 2021).

Mushanyuri, B.E. & Ngcamu, B.S. (2020). The effectiveness of humanitarian supply chain management in Zimbabwe. *Journal of Transport and Supply Chain Management*, 14(0): a505. https://doi.org/10.4102/jtscm.v14i0.505

Muzhinji, N. & Ntuli, V. (2021). Genetically modified organisms and food security in Southern Africa: conundrum and discourse. *GM Crops & Food*, 12(1): 25-35. https://doi.org/10.1080/21645698.2020.1794489

Ncube, M. (2019). Cop hires kombi to ferry stolen Cyclone Idai donations. https://bulawayo24.com/index-id-news-sc-national-byo-160060.html (Accessed on 24 August 2010).

New Zimbabwe. (2019). Another government official in court over theft of cyclone donations. https://www.newzimbabwe.com/another-gvt-official-in-court-over-theft-of-cyclone-donations/ (24 August 2020).

Nhamo, G. & Agyepong, A.O. (2019). Climate change adaptation and local government: Institutional complexities surrounding Cape Town's Day Zero. *Jàmbá: Journal of Disaster Risk Studies*, 11(3): a717. https://doi.org/10.4102/jamba.v11i3.717

OCHA (United Nations Office for the Coordination of Humanitarian Affairs). (2019). Mozambique: "Aid is free and sexual exploitation and abuse are unacceptable". Retrieved from https://www.unocha.org/story/mozambique-aid-free-and-sexual-exploitation-and-abuse-are-unacceptable (23 August 2020).

Oxfam, CARE, Save the Children. (2019). From cyclone to food crisis: Ensuring the needs of women and girls are prioritized in the Cyclone Idai and Kenneth responses. https://doi.org/10.21201/2019.4634.

Parkash, S. (2012). Ethics in disaster management. *Annals of Geophysics*, 55(3): 383-387. https://doi.org/10.4401/ag-5633.

Prieur, M. (2012). *Ethical principles on disaster risk reduction and people's resilience*. Brussels: Council of Europe.

United Nations. (2015). *Transforming our World: The 2030 Agenda for Sustainable Development*. New York: United Nations Secretariat.

Zenda, C. (2019). Politicisation, theft of Cyclone Idai aid worsen survivors' anguish in Zimbabwe. https://www.fairplanet.org/story/politicisation-theft-of-cyclone-idai-aid-worsen-survivors%E2%80%99-anguish-in-zimbabwe/ (Accessed 24 August 2020).

Religious Engagements with Tropical Cyclone Idai and Implications for Building Back Better

12

Abstract

In March 2019, Tropical Cyclone Idai hit Malawi, Mozambique, Madagascar and Zimbabwe. For most indigenous African people, there is a very thin line between the sacred, secular and scientific, hence interpretations of natural disasters are informed by religious belief systems. Through a household questionnaire survey, face-to-face interviews, focus group discussions and document analysis, this chapter documents and analysis how local people in Chimanimani (Zimbabwe) linked the cyclone to their religious beliefs. It emerged that the destruction caused by Cyclone Idai in the village of Kopa was believed to have been caused by mermaids tracking ill-gotten gains of locals who had travelled to Mozambique and used *muti* (traditional charms) to amass their wealth. Another credence was linked to clay pots that had been broken by the Apostolic Faith Christians in the Ngangu Mountain that could have angered the ancestors. It also emerged that some chiefs had kept their areas safe by performing rituals through their spirit mediums who asked *Musikavanhu* (God) to protect them. Furthermore, cleansing ceremonies (*Mapira eChenuro* in the Shona language) were conducted by chiefs and the government, with another great push to rebury bodies placed more than one in a grave at the Chimanimani district heroes' acre at Ngangu. Bearing in mind that the United Nations 2030 Agenda for Sustainable Development clarion call of "Leaving no one behind", we recommend that religious belief systems be an integral part of managing and raising awareness of natural hazards such as tropical cyclones, including building back better.

Keywords

SDGs · Cyclone Idai · Religion · Beliefs · Spirit mediums · Culture · Rituals

12.1 Introduction

Ayeb-Karlsson et al. (2019) acknowledge the splendid work that has been done in addressing the economic and technical basis for disaster risk reduction (DRR) and management. However, they note a gap in directing efforts towards understanding the cultural and social roadblocks in DRR and management, an argument advanced earlier by Ha (2016). Studies on stress and religious coping among flood victims in Malaysia found that religion was the most apparent coping skill among victims (Sipon et al. 2014). In New Zealand, Sibley and Bulbulia (2012) found that believe in God increased in Christchurch in the earthquake-affected communities in 2011. This was despite an overall decline in religious faith elsewhere. From studying Hurricanes Sandy (2012) and Katrina (2005) in the USA and an 8.9 magnitude in Japan in 2011, Hirono and Blake (2017) observe that the clergy were sometimes the missing link because they were needed 2–3 weeks after the disasters, a period critical in post-traumatic stress disorder. Without this engagement, building back better (BBB) communities after disasters is a challenge.

Religion and belief systems are usually contested. Shoko (2007) identifies three common religions among the Shona[1] people of Karanga tribes in the Masvingo Province of Zimbabwe. These religions are as following: (1) traditional religion—commonly referred to as African Traditional Religion, (2) mainline Christianity and (3) independent churches. This division is common across the country (Kazemba 2009). Clarke and Parris (2019) highlight that in the past decade, humanitarian events affected an average of 120 million people every year. From this number, up to 40 million were affected by natural hazards. Lives are lost, injuries occur, survivors get maimed and many are left traumatised, requiring the intervention of governments as well as religious organisations in DRR and management. Hence, apart from search and rescue, one of the immediate requirements is multi-stakeholder psychosocial support. However, Kulatunga (2010) notes that culture is a double-edged sword when it comes to DRR and management. On the one hand, culture has become a factor of survival from disasters in communities, and, on the other hand, culture remains a barrier to effective DRR.

Twice, the United Nations 2030 Agenda for Sustainable Development and its 17 Sustainable Development Goals (SDGs) make reference to religion, five times to culture and once to traditional knowledge (United Nations 2015). The highlighted perspectives are a testimony that

matters considered in this chapter are not only of local importance but also considered at the global level. With reference to the Charter of the United Nations, the 2030 Agenda for Sustainable Development calls for respect and the protection and promotion of fundamental freedoms for all, without distinction of any kind of language, religion, and national or social origin. Specifically, Target 10.2 of SDG 10 (reducing inequality) calls for stakeholders to "empower and promote the social, economic and political inclusion of all, irrespective of age, sex, disability, race, ethnicity, origin, religion or economic or another status" (United Nations 2015, p. 21). In Bullet 36, the United Nations unequivocally pledges:

> To foster intercultural understanding, tolerance, mutual respect and an ethic of global citizenship and shared responsibility. We acknowledge the natural and cultural diversity of the world and recognize that all cultures and civilizations can contribute to, and are crucial enablers of, sustainable development. (United Nations 2015, p. 10)

Target 4.7 of SDG 4 (equitable and quality education) calls for the promotion of a culture of peace and non-violence, as well as the appreciation of cultural diversity and the valuable contribution of cultures to sustainable development (United Nations 2015). Target 8.9 of SDG 8 (decent jobs) encourages communities to devise and implement policies that promote sustainable tourism and local culture. The same focus is raised in Target 12.b of SDG 12 (sustainable production and consumption). Researching disaster preparedness in Bangladesh and Nepal, Ayeb-Karlsson et al. (2019, p. 752) found that "a deeper cultural and religious reasoning serves to explain disasters, and how to prevent them or find safety when they strike".

From the foregoing discussion, Tropical Cyclone Idai presents a contested platform in terms of what happened, especially why it happened. Did people die because they had sinned or because some ancestors had been angered? These are questions that the author seeks to answer in this chapter. The chapter, therefore, has an objective to investigate religious engagements surrounding Tropical Cyclone Idai, with the view to document and tease out how such impact on BBB communities.

---

[1]There are also other Shona-speaking people with similar religious beliefs including Zezurus (mainly from Mashonaland Central, East, and West Provinces); as well as Manyikas and Ndaus from Manicaland (although there has been some contestation as to whether Ndau is a Shona dialect or not).

12.2 Literature Review

Given the contested space of religion in Zimbabwe, some paragraphs of this chapter are reserved to discuss some of the common practices and belief systems. The local tradition will mainly be that of the Shona-speaking people, with some examples drawn from Manicaland Province where Tropical Cyclone Idai hit. From the Shona group of languages, there is Zezuru (mainly spoken in the three provinces of Mashonaland Central, East and West[2]), Chikaranga (Masvingo and some parts of Midlands Provinces) and ChiManyika (Manicaland Province). There is also ChiNdau, which is spoken in the Chimanimani and Chipinge areas that were significantly damaged by Tropical Cyclone Idai. Other minor languages spoken in Manicaland Province include Shangani, Venda and Ndebele.

Many Christians believe in the creation of the earth in seven literal days as per the account of Genesis Chapters 1 and 2. According to this account, God created the heavens, the earth and humankind. Genesis 1 verses 26 and 27 indicate that God said, "Let us make mankind in our image, in our likeness, so that they may rule over the fish in the sea and the birds in the sky, over the livestock and all the wild animals, and over all the creatures that move along the ground. So, God created mankind in His own image, in the image of God he created them; male and female he created them" (the Bible, New International Version—NIV n.d.). Once that happened, God saw all that he had made and said it was indeed very good. The fact that God says "Let us" is interpreted by many Christians to imply the trinity, a notion of three in one as God the Father, God the Son and God the Holy Spirit. The son is Jesus Christ, who came to earth and lived for about 33 years, died at the cross of Calvary on a Friday and rose again on the third day, early Sunday morning. To this end, many Christians are looking forward to the second coming of Jesus Christ when the earth shall be cleansed of

its sins and the devil cast into the lake of fire and brimstone.

From the Biblical account, the God spoken about has the power to cause damage through the use of nature and natural disasters to those who do not follow his law. God's moral law is summarised in Exodus 20, written by Moses. From the Ten Commandments given to Moses by God as originally written on two tablets of stone, the first four commandments are obeying God (on one tablet of stone), and the other six are about harmonious living with fellow human beings (on the other tablet of stone). From the Ten Commandments, the fourth commandment requires Christians to observe the Sabbath Day, and this has caused commotion between traditional set-ups such as *Chisi* (day of rest from all work in the Shona culture) and the Sabbath Day of rest from all forms of work in the Bible. *Chisi* is usually observed on a Tuesday, Wednesday or Thursday, depending on the area. Hence, from both the fourth commandment and *Chisi* (known as *Magarai* or *Makarani* in the Chimanimani and Chipinge areas), there is agreement that humanity must have a day of rest from the busy week. From Exodus 20 verses 8 to 11, the Bible calls upon humanity to:

> Remember the Sabbath day by keeping it holy. Six days you shall labour and do all your work, but the seventh day is a Sabbath to the Lord your God. On it you shall not do any work, neither you, nor your son or daughter, nor your male or female servant, nor your animals, nor any foreigner residing in your towns. For in six days the Lord made the heavens and the earth, the sea, and all that is in them, but he rested on the seventh day. Therefore, the Lord blessed the Sabbath day and made it holy.

This set-up remains contested as the Christian belief system stands in opposition to the Shona *Chisi* tradition. Those who observe the Sabbath as the day of rest from all work go to work on *Chisi* days, usually with accusations of disobedience from either side. What is of interest is that within the two major belief systems, God is acknowledged as the ultimate power. In the Shona culture, many traditional ceremonies like *Bira reMukwerera* or *Doro reMusoso* (rain-making ceremony) and *Bira reMatakapona* (ceremony to celebrate deliverance from a disaster

---

[2]ChiKorekore is also spoken widely in some parts of Mashonaland West and Central Provinces

such as Tropical Cyclone Idai) or *Bira reChenuro* (cleansing ceremony) use a hierarchy that ends with God as *Musikavanhu* (the one who created humankind).

Therefore, the main starting point in understanding Shona belief systems is ancestral spirits (*vadzimu/midzimu*), believed to be the guardians of the people and the land (Shoko 2007). However, the matter of ancestral spirits is not exclusive to the Shona culture. In Italy, there is a record of a participant who indicated he had been "awakened by the spirit of his deceased mother seconds before the earthquake and saving his entire family" (Massazza et al. 2019, p. 301). From the Shona culture, ancestral spirits are those who died but still exist in the spiritual form and world called *Nyikadzimu* (Shoko 2007). The ancestral spirits include important spirits of chiefs that are called *Mhondoro* (lion spirits).

As indicated earlier, the Shona people believe there is a God somewhere, and this God is called by different names that include *Musikavanhu* (as indicated before), *Mwari* (God), *Nyadenga* (the owner of skies or one who resides in the sky), *Mutangakugara* (the one who originated before anyone else) and *Chirozvamavi* (the one who blesses and withholds blessings) (Tatira 2014, p. 108). Since God is so important, the Shona people believe that they can only communicate with him through the ancestral spirits. Hence, it is improper to by-pass the ancestral spirits and go directly to God, an element that places them in opposition to Christian believers. From the lived experiences of the author, the concept of hierarchy is so strong in the Shona culture that even picking pieces of meat when eating from a communal and/or family plate takes place in a specific order, thus from the oldest to the youngest. This arrangement may not make much sense in modern set-ups where people eat from their own separate plates. In addition, as observed by Tatira (2014), the hierarchy is also so strong that greeting the elders in some set-ups must be done from the youngest to the oldest, until it gets to a father or uncle. Moreover, it is patriarchal. At times, some matters must be told to the mother to pass on to the father. Other matters such as intent to marry or get married have their protocol, as these cannot simply be told to the father or mother. Such is the arrangement.

Drawing from the story of the two great ancient cities of Sodom and Gomorrah recorded in the Biblical historical account of Genesis 19, many Christians believe that God can destroy communities through calamities—as was the case with Sodom and Gomorrah. Genesis 19 verses 24–26 states, "Then the Lord rained down burning sulphur on Sodom and Gomorrah—from the Lord out of the heavens. Thus, he overthrew those cities and the entire plain, destroying all those living in the cities—and also the vegetation in the land. But Lot's wife looked back, and she became a pillar of salt" (NIV). Other stories of God taking difficult decisions for wrongdoing by the society are also recorded, including Noah's ark and the great flood that lasted 40 days and 40 nights (Genesis 7:12). From Noah's story, water destroyed the bad people, leaving behind those who were good and obeyed God's commandments. There is also the story of the 10 plagues of Egypt to force Pharaoh to let the Israelites go. The plagues are found in Exodus 7 verse 14 to Exodus 12 verse 36. These include blood, frogs, lice, flies, boils, hail, locust, darkness and the death of the Egyptian firstborns. However, the Bible also has stories of deliverance about those who trust the Lord, including many healings performed by Jesus and his disciples in the New Testament. One of the key deliverance events in the Bible is the Israelites' crossing of the Red Sea and the swallowing of Pharaoh's armies recorded in Exodus 14. This deliverance of God's people is at times set against the delivery of some Shona people by their ancestors in difficult times (provided the protocols and rituals are followed).

Kazemba (2009) investigated the relationship between God and people who follow the Shona traditional religion. He found that the clients of leading spirit mediums called maGombwe were from many provinces (including Mashonaland Central, Mashonaland East, Manicaland and Harare Provinces) and even outside the borders of Zimbabwe. The maGombwe resided in Mashonaland Central and had clients who included some of the maGombwe themselves, ancestral spirit mediums, spiritual healers and herbalists. They also serviced other patients, including church leaders, businesspeople, politicians and ordinary people. The rituals were performed at gatherings at maGombwes' residences, in mountains, by rivers and in forests, depending on the specific purpose. The food and beer consumed during these rituals had to be of African origin and had to be prepared following strict protocols under customary laws. Common healings included chasing away evil spirits that could be from the dead, business rituals gone wrong, witchcraft and other sources.

Tatira (2014) highlights that not all Shona cultural belief systems are useful to the modern generation as some go against human rights. This perspective came out clearly during the fieldwork conducted by the author in 2019. One chief from Chimanimani indicated that during the installation of a chief, the chief-elect had to sleep with his sister in front of people. Of course, this practice has since been stopped. The enthronement ceremony is now done with the chief presiding together with his sister. Tatira (2014) is of the view that many of the Kalanga traditional belief systems are for the protection of human rights, and they are mainly driven from the interaction with the supernatural. Summarising the practices, Tatira writes the following:

> The breach of a belief mostly triggers a supernatural punishment, not only for the offender but for the whole community where the offender resides. This makes the observance of beliefs self-mandatory where beliefs touch on human rights, environment, justice, morality and a host of other issues that regulate human relations. The beliefs become effective in checking against untoward behaviour. (Tatira 2014, p. 106)

Another common Shona traditional ceremony linked to the ancestors is the rainmaking rituals known as *Mukwerera* (Risiro et al. 2013). This practice is also common in other African countries, including Kenya and Nigeria (Ombati 2017). From the author's view, the practice may be recorded as one of the early acknowledgements of climate change through indigenous knowledge systems in Africa. The *Njelele* (a mystical and highly revered rainmaking shrine in Zimbabwe's Matobo National Park in Matabeleland South Province, close to Bulawayo) is highlighted as one of the common places where rainmaking ceremonies are performed. During these ceremonies, dancers from the Shona people imitate certain animals and birds associated with rain: *Nyenze* (icada), *Fudzamombe* (stork birds) and *Nyenganyenga* (swallows). The lack of rain was associated with community sins and such sins had to be confessed and presented to *Musikavanhu* through the chiefs and spirit mediums. Other interesting matters surrounded the participants in the rainmaking ceremony, including that only married women could join in, with sex workers and others as designated excluded. From the authors' experiences and indigenous knowledge, the handling of beer brewing and carrying it to the sites were done by post-menopausal women. Part of the beer had to be poured on the ground accompanied by rituals, while some food would be left there for the ancestors. All the other food for the day would be shared among the participants following strict protocols. Should things be done properly, rain would start to fall immediately, and no one would be allowed to cover themselves with an umbrella.

In the Ndau and Shona cultures, water bodies remain sacred places. This is the concept of *kuyera*, according to Muyambo and Maposa (2014). Hence, such water bodies should not be trivialised. The author concluded that (1) water bodies are highly revered, (2) are ritualised places and (3) are governed by taboos. Two sacred water bodies are well-known in Chipinge, namely Kubiri and Chisurudza (Box 12.1). The value of water is embedded in the Ndau culture, to the extent that most of the clans in Chimanimani and Chipinge have a *dziva* (water pool) embedded in their totems, including the Musikavanhu chieftaincy as well as Muyambo and Dziya. *Madziva* (many pools) are also associated with mysterious creatures (such as mermaids and snakes), spiritual singing and many more. The tempering of water points (mainly in the form of springs) also emerged during Tropical Cyclone Idai fieldwork in October 2019. Many springs were destroyed, and rituals had to be performed before building them back better using cement. However, the fear that the water points could dry up sometime in the future remained. This will be discussed in the findings section.

Sacred places are dotted across Zimbabwe. For the Masvingo area, Risiro et al. (2013) note the following: Chikona hills in Zaka, Bvuma Mountain and sacred wells around Bvuma Mountain, Chikona and Runinga hills. As with the case in Chipinge, metal objects and blackened clay pots are prohibited from being used in collecting water from wells. The belief is that the wells have mermaids that if offended will make the wells dry or the water muddy. Some three wells were reported dried up due to the failure to observe these protocols, as people used soap, and because one ancestral water snake had been killed. Another reason, as in Chipinge, was associated with fencing the well. All the narratives

show that many areas in Zimbabwe are intertwined with traditional belief systems. Given the many cultural set-ups, even farmers in the Gaza Province in southern Mozambique believe the droughts are caused by spiritual and supernatural forces that include angry ancestors or God (Salite 2019). Drought punishment could be from both known and unknown wrongdoings. Cox et al. (2018) looked at Christian interpretations of Tropical Cyclone Winston and climate change in Fiji, and came to similar conclusions.

In the next section, the key methods used in generating and analysing the data are provided.

12.3 Materials and Methods

Although Tropical Cyclone Idai hit Malawi, Mozambique, Madagascar and Zimbabwe, this chapter focused on Zimbabwe. In Zimbabwe, the study further concentrated on the areas of Chimanimani that were the hardest hit, including Ngangu, Kopa, Machongwe, Peacock and Rathmore Farm. The main research methods were key informant interviews, focus group discussions, a household questionnaire survey administered on an offline QuestionPro platform, field observations and literature review. Details on the methods were alluded to earlier in Chap. 1. The use of surveys, interviews and focus group discussions is common in DRR and management as was used by Tammar et al. (2020) in Jeddah City (Saudi Arabia) after the 2009 and 2011 flash floods and on earthquake preparedness among religious minority groups in Israel's Jewish ultra-orthodox society by Erblich et al. (2020). Massazza et al. (2019) used a survey and interviews in researching the nature of natural disasters from survivors' explanations of earthquake damage in Italy. Interviews were also used in studying the role of faith-based institutions in urban DRR for immigrant Cambodian and Thai communities New Zealand (Ngin et al. 2020) and on Vanuatu Island when three stories were investigated on navigating religious climate change narratives (Fair 2018). To analyse the data, elements of grounded theory and document analysis were applied (Tammar et al. 2020). Data from the household survey were imported into MS Excel with graphs plotted. The next section presents and discusses the findings.

12.4 Presentation and Discussion of Findings

This section deliberates in depth on some of the findings pertaining to religious belief systems and how these related to Tropical Cyclone Idai, particularly as they emerged from open-ended key informant interviews and focus group discussions. The three sub-sections focus on: (1) the causes of Tropical Cyclone Idai and excessive floods from a religious point of view, (2) community warning of Tropical Cyclone Idai by the spirit medium, and (3) religious belief system and their implications for BBB after Tropical Cyclone Idai.

12.4.1 Causes of Tropical Cyclone Idai and Excessive Flooding

Clarke and Parris (2019) are of the view that disasters can be affected by religious beliefs and

given that up to 85% of global citizenry self-profess religious beliefs; the understanding of disaster causes is enveloped within such belief systems. In many instances, these belief systems embrace sacred texts, religious teachings and sectarian practices. Hence, those affected by such natural disasters will easily relate to "acts of God", meaning God could be punishing them somehow. Therefore, to effectively address needs after disasters, the local contexts should be fully embraced.

To start with, the respondents in the household survey were asked to identify what they perceived as the main cause and source of the floods from Tropical Cyclone Idai. They were given a choice to select up to five causes (Fig. 12.1). What is of interest in the context of this chapter is that angry ancestors ranked 4th (9.64%) out of the 11 possible options provided. Excessive rains and storm surge were ranked 1st at 25.07%, followed by low-lying and flood-prone areas (20.94%) and poor settlement designs (13.77%). Another interesting point was that witchcraft ranked 10th (2.2%). However, from one of the key interviews with a chief whose area includes Kopa, it was clear that Kopa Township encroached into the main flood plain and confluence of three

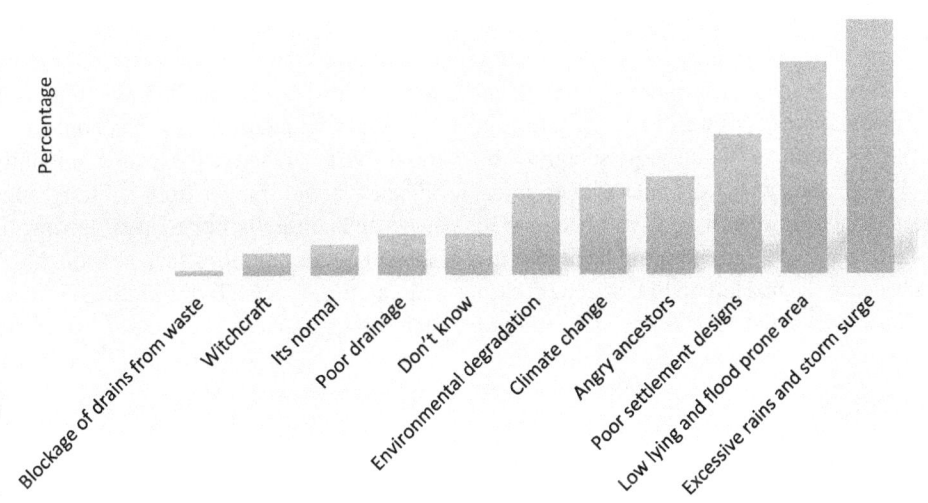

Fig. 12.1 Perceptions on the leading causes of flooding ($n = 363$). Source: Authors

rivers. Just before independence in 1980, the area was still reserved for growing crops related to the flood plain. However, somehow the council decided to allow the growth point to go into the flood plain, as the former rich Kopa Township residents seem to have bought their way onto the flood plain from the council. The chief recalled one of the headmen, Mr Dzavanda Dzingire, in the mid-1970s fighting the colonial Local Board Authority and indicating that they should not settle people in the Kopa area as this was reserved for the growing of traditional crops that required almost waterlogged areas. Crops like *madhumbe* (scientific name *Colocasia esculenta*) were given as examples. These crops did not require daily tendering, allowing for no frequent visitation of the sites, giving the mermaids some peace and quietness. What was interesting is that a good number of household respondents based their responses on modern science, with 84.57% falling into this category. Hence, the consideration of religious belief systems and Tropical Cyclone Idai is done in the interest of the United Nations 2030 Agenda for Sustainable Development's call to leave no one behind (United Nations 2015).

There emerged a cluster of believers convinced that Kopa was destroyed because of the nature of business conduct by its residents. The first narrative that came to the fore was that business people from Kopa Township got their wealth in dodgy ways involving *Kuromba* from Mozambique (as the situation where a business-person uses dubious traditional means or lucky charms also known as *muti* in the local ChiManyika language, as well as goblins, to gain wealth). This narrative was strongly supported by some eyewitnesses who were interviewed referring to a deceased woman who had been saved from the floodwaters, only to jump back into the waters because she had left up to US$25,000 cash in the house. Given the rural set-up and nature of businesses usually undertaken by the community, the respondents strongly believed that the money had been from *Kuromba*. Another story linked to the issues of *Kuromba* was that of a woman who rescuers could not reach because of a snake that kept raising its head from a water pool. However, one of the chiefs indicated that they had to join hands with another chief (as they shared the water pool) and performed rituals, resulting in the woman being rescued after the snake disappeared.

On investigating further, another story was told that two ladies who looked like *njuzu* (mermaids) had been seen in the area just before Tropical Cyclone Idai hit Kopa. The issue of mermaids was confirmed by the chief whose area included Kopa. In his account, they had to inquire from their spirit mediums and local diviners, who confirmed that the mermaids had indeed caused the destruction in Kopa. The issue of *njuzu* confirms the Ndau beliefs documented by Muyambo and Maposa (2014) when they researched the link between culture and water technology in the Chipinge area. The chief also confirmed the matters of ill-gotten wealth and prostitution as the ultimate causes of the destruction. Another key informant who was a guide in the area also confirmed that certain young people in Chimanimani had suddenly become very rich in a way that was suspicious. Five prostitutes were known to have been swept away in Kopa. The chief went further and indicated that during the night of the great flood from Tropical Cyclone Idai (known locally as *Dutumupengo*—a violent and insane storm), prostitutes from Chipinge had been swept away by the floodwaters. The areas of Ngangu and Machongwe that were also hit hard by Tropical Cyclone Idai were included as having ill-gotten wealth and high prostitution levels. The account of the chief confirmed Tatira's (2014) findings that offended ancestors (the guardians of the people and land) could cause damage to the local community. The offences could range from spilling blood to incest and culturally bizarre acts, leading to punishment of communities such as droughts, floods and cattle diseases. Tatira highlights that when such things happened, the chief had to find the culprits and publicly punish them (including chasing them away from the area in the case of proven witchcraft). If the offence included spilling blood, the chief had to arrange ceremonies to ask for forgiveness through *Birai reChenuro* (ceremony to cleanse), linking the living with the ancestors and appeasing them.

Although Christians' beliefs that God could have destroyed Kopa for its sins were raised, this narrative was not as pronounced. Contrary to God permitting calamity on people over sin, Massazza et al. (2019) report the story of Sant'Emidio, a local patron saint of protection from earthquakes, who protected religious and faithful communities from the fury of God's wrath. God's existence and power to save were recounted by survivors from an earthquake, one indicating that they had been able to see in the dark, resulting in a nephew being saved from a collapsing ceiling, while another reported they were led by the light from the heavens out of the debris. However, during the 2010 Ms 7.1 Yushu earthquake in the Qinghai Province of China, the Tibetans (informed by their Tibetan Buddhist faith) believed that it was a form of punishment from God (Sun et al. 2019). In Samoa, the Christian understanding of two communities impacted by the December 2004 tsunami resulted in them attributing the event to divine punishment and/or an end time sign of Jesus Christ's second coming. Similar sentiments were expressed during the Haiti 2010 earthquake (Abbott 2019). The challenge presented is that this positioning undermines disaster response, including readiness. Hence, if help is to come, it must be channelled through the temples and their leadership.

In one of the interviews with a chief knowledgeable about the happenings at Ngangu Mountain, a narrative of three big clay pots came up. From the chief's perspective, these three pots had been filled with rapoko (*rukweza* in ChiManyiaka), which never rot over the years. However, there was a group of Christian Apostolic believers who went into the mountain and broke the pots associated with the world of the dead and their appeasements. Hence, Tropical Cyclone Idai was the result of angry ancestors, an aspect that came up earlier in the household survey. The issue of the clay pots came up in two other separate conversations with key informants from Chimanimani. A question was asked "We gather that there were clay pots that a group of Christian Apostolic believers broke from the Ngangu Mountain alleging that the clay pots were

associated with the world of the dead and their appeasements. Is there any truth in this?" Both responses were in the affirmative. One of the respondents wrote the following on 19 October 2020, "One of the survivors from Tropical Cyclone Idai told me of the issue of how some religious group tampered with a revered site in Ngangu and broke artifacts, which is why the storm was took harsh on Ngangu". The respondent went further to impose his own analysis of the harshness of the storm adding that "It is also surprising that despite the eye of the storm coming onto land on 16 March 2019, its strength did not reduce even when it crossed four ranges of mountains from the sea".

Spiritualism is a major issue in Chimanimani. In the second response to the same question posed on the clay ports, the respondent gave a historical perspective to the sacredness of the clay pots. In two WhatsApp voice recordings of 15 October 2020, the respondent spoke of a certain farmer (names supplied) in the early 1970s (prior to Zimbabwean independence in 1980) who owned a farm including where modern-day Ngangu Township stands. The farm also included the Ngangu Mountain, which had a cave where the three ports were placed. It then happened that the white farmer short the three pots destroying one and making holes in the other two that remained. After the incident, the farmer is said to have been bitten to death out of nowhere by a certain local man (names supplied). The respondent went further to confirm that indeed there was a group of Christian Apostolic believers who broke the two remaining clay pots and even removed them from the cave. However, on the day of Tropical Cyclone Idai calamity, the respondent indicated that up to 16 people, including those that had come to pay lobola (bride price) to a relative of those involved in the act of breaking the pots died in their house. The respondent was of the view that this unfortunate incident was caused by the spirits that were showing their displeasure in what had taken place concerning clay ports. The respondent described the Tropical Cyclone Idai flood waters to have gashed out of that particular cave direct to the house where the victims were. This account was traced further, and no any other confirmations from additional sources and/or eye-witnesses could be gathered.

In one of the interviews with a chief, he revealed that near Mount Binga, the highest peak of the Chimanimani Mountain (Mawenje) range on the way to Mozambique, there is a bubbling spring called *Chiserosero*. People could speak to this spiritual spring and it responded. Hence, the chiefs would go there to consult when trying to find out if someone from their community was or had done bad things. In speaking to the spring, the words "*Maseruseru ndisekerero*" (spiritual spring, can you smile at me) would be mentioned by the concerned person. Should the person be bad, the spring died down and even threw out muddy water.

12.4.2 Community Warned of Tropical Cyclone Idai by the Spirit Medium

The examination of religious beliefs and customs for disaster preparedness is not new. Drawing from the Jewish ultra-orthodox society in Israel, Erblich et al. (2020) found that the majority of community members had low levels of hazard knowledge and high levels of disbelief that an earthquake would occur in their area in the future.

An interview with one of the chiefs responsible for the Kopa area revealed that there was indeed early warning regarding Tropical Cyclone Idai. A female spirit medium from an area called Mutambara, specifically by the name Mupati, is said to have sounded the early warning. The spirit medium is said to have instructed that there had to be a *Bira* to protect the area from some form of pending disaster. The message was also taken to the Chimanimani District officials, according to the key informant's recollection. However, the message was not believed by many. It is said that the spirit medium was on point, indicating that there would be a great storm. Part of the *Bira* solution was to slaughter a *Ngogoni* (wildebeest). Wildebeests are plenty in the Chimanimani area and specifically in Chimanimani National Park (Fig. 12.2). The reason why people did not believe the spirit medium was also given—in the past, what it said did not come to pass.

Apart from the spirit medium's early warnings, one chief and a focus group discussion at Rusitu Power Station confirmed that the community had been forewarned by the extremely hot months of September to December 2018. According to one of the respondents in the focus group discussion, banana plants could not grow—something they had never seen in years (Fig. 12.3). In line with the local belief system,

Fig. 12.2 A wildebeest. Source: Pixabay (2020)

Fig. 12.3 Visible lack of growth of bananas in the Nyabamba area. Source: Authors, fieldwork 2019

this was a *shura* or *maninji* (a unique and extraordinary happening that predicts a bad omen in the near future). When there is *shura*, the chiefs gather their kraal heads and headmen and arrange to perform rituals to prevent the predicted future mishappening. To this end, other chiefs did the rituals to safeguard their areas from Tropical Cyclone Idai. Two examples were given by chiefs in the Muusha and Ngorima areas whose kraal heads had undertaken to perform these rituals, and their areas were spared from extensive damage with no deaths reported.

Talking of other bad omens or signs, Chief Saurombe related how wild pigs had caused great trouble in his area in February 2020, pointing to something gone wrong in the community. The extent of the damage to crops caused by the wild pigs, especially maize, had never been reported earlier. Hence, the chief gave permission to one of his aides to kill one wild pig from the fields and performed their rituals. Since that day, the problem went away. The chief further spoke of places that are sacred and to which people cannot just go without consulting chiefs as they ended up getting lost or even disappeared forever. An example is a mountain in the Chimanimani mountain range called Maweje. The chief indicated that there is an area that people are not permitted to visit, and some tourists who disobeyed this ended up getting lost there.

12.4.3 Religious Belief System and the Implications for BBB

Religious belief systems have major implications for how communities can BBB after disasters such as Tropical Cyclone Idai. One of the key issues that came up during the fieldwork was the lack of closure concerning the burial of loved ones. Given that there was limited time to bury the dead, with bodies decomposing, the authorities decided to bury more than one body in one grave in the local heroes' acre at Ngangu in Chimanimani Town. According to the local tradition and belief system, this was not supposed to be. The living wished to perform rituals, and this could not be done as totems had to be used to wake up the spirit or rest the dead. The challenge was that there are three major tribes in the area, namely the Manyika, Ndau and Changani. Hence, the traditional practices differed, and if a family wished to put up a tombstone, it would be difficult. Furthermore, among the dead were visitors (including sex workers) to Kopa (a popular commercial centre). Thus, quick physiological recovery remained elusive. The suggestions from some key informants in the interviews and from the focus group discussions were that there had to be some reburials.

It also emerged that many bodies were still missing, and some who had been buried elsewhere in Mozambique were from Kopa Township. All these matters remained unfinished business in terms of closure for families and the area. By the time of the fieldwork, the families of those who had been swept away were still awaiting government confirmation and declaration of those missing as dead. From one of the focus group discussions at Kopa, this was likely to take place only after 5 years due to the Zimbabwean laws governing missing persons. However, further discussions with the acting district development coordinator of Chimanimani revealed that the government, through the Ministry of Local Government, was working on fast-tracking the process.

The other matter concerning religious belief systems and BBB was the performance of traditional ceremonies (*Mapira*). This celebration includes alerting and/or appeasing the dead in the land of souls (*Nyikadzimu* in Shona/ChiManyika). The belief system has it that when the living die, they go to this land where their spirit would come back to protect the living if appeased or to destroy if the opposite. After Tropical Cyclone Idai, those in the land of the dead had to be informed of the trouble. Hence, all six chiefs had to undertake *Mapira*. Those who were done at of the time of the fieldwork in October 2019 included one for Chief Saurombe on 5 June 2019, one for Chief Muusha on 13 June 2019 and one for Chief Ngorima on 16 July 2019. Such *Mapira* also had to ask for future protection from disasters. Following the December 2004 tsunami in Indonesia, Adiyoso and Kanegae (2013) explored the role of Islamic teaching in DRR from Banda Aceh. The findings were as follows:

> In school, the religion–natural disaster relationship messages are focused on the stories of natural disaster in the Holy Quran, accepting natural disaster as God's will and the importance to ask (*Doa*) protection to Allah. Such themes can be found in the textbooks, school walls and teaching processes in school. (Adiyoso and Kanegae 2013, p. 918)

Furthermore, in collaboration with the Ministry of Local Government, five chiefs from the affected Chimanimani area were to hold one huge *Bira* code named *Bira reChenuro yaIdai* (translated a celebration to cleanse away the spir-

its of the victims of Tropical Cyclone Idai). Such traditional ceremonies are held to prevent similar occurrences in the future. After being delayed and postponed many times, the last time being in March 2020 (due to COVID-19), the cleansing ceremony finally took place on 7th October 2020 at the Machongwe/Peacock area. However, while acknowledging the good intention of the ceremony, one of the key informants present during the event bemoaned the limited attendance of people from Peacock who experienced the damage from Tropical cyclone Idai. Those from Machongwe were present in good numbers, according to a WhatsApp communication response following our probing of the event on 8 October 2020. A cow, some goats and chicken were slaughtered during the celebration. The celebration was further confirmed by another key informant who indicated that there were five chiefs from Chimanimani and one from Chipinge in attendance.

While awaiting for the *Bira reChenuro yaIdai*, other village heads from Chimanimani organised *Bira reMatakapona Idai* (literally translated as a ceremony to acknowledge that we survived Tropical Cyclone Idai). One such village was that under Headmen Bunza and Musiyandaka. In this area, not even a single person died due to Tropical Cyclone Idai. The traditional authorities there considered it a miracle owing to one key informant from the area. The chiefs were once again advised to slaughter a wildebeest to block future tropical cyclones. One key informant hinted that the *Bira reMatakapona* had to go ahead as the Chimanimani area remained a pathway for tropical cyclones into Zimbabwe. He indicated that "growing in Chimanimani, we know that it is just a matter of a few years before another tropical cyclone or flood comes taking away roads and bridges". Christian faith-based organisations were also encouraged by the chiefs and government to perform their own ceremonies about Tropical Cyclone Idai.

From a Christian perspective, there is usually an immediate passion to assist those in need. As such, there were many Christian faith-based relief agencies involved, from search and rescue, to relief and recovery that included the Catholic

Church and the Seventh Day Adventist Church. An interview with a Catholic Church bishop revealed that the Catholic Church made a call to immediately contribute towards relief for the people of Chimanimani and Chipinge. In response, "every Catholic brought whatever they had. They brought pots, plates, clothes, food, and they were collected at every centre and every branch". However, as these goods could not reach Chimanimani for 3 weeks due to damaged roads, they were left at Caritas in Mutare some 250 km or so away from Chimanimani centre. Some of the donations came all the way from South Africa. From the interviews, it emerged that the relief had not stopped as donations were still being received at the time of the fieldwork in October 2019. After the relief phase, the Catholic Church initiated the 5-year Beyond Cyclone Idai Programme (BCIP).

The BCIP was aimed at raising US$3 million towards reconstruction, recovery and restoration of livelihoods in Malawi, Mozambique and Zimbabwe, with US$1 million raised by 13 October 2019. The idea was to refrain from giving people food and clothes. Instead, the victims had to get their livelihoods back. Hence, the BCIP focuses on the construction of health, education and faith-based centres. Three religions (Christianity, Buddhism and Confucianism) were found to play a role in supplementing care after disasters (Ha 2016). The BCIP's reconstruction would not look only at the Christian faith; if a mosque had been destroyed, the programme would assist to bring it back and build it better. The bishop further revealed how the BCIP money would be spent. The money would be distributed as per the damage, with Mozambique getting the bigger share, followed by Zimbabwe and Malawi getting the smallest portion. In the interview, the bishop also touched on some belief systems of interest:

> Now, what is the problem? The problem is we don't know how to educate our people. The natural disasters are real and natural disasters are not sent by a *N'anga* (witchdoctor) and it is not a bad omen from God. And it is not a bad omen from nature. So, we have to educate. Part of our educational system has to be part and parcel of basic sciences that make us be who we are, what we are. We are one and the same with Mother Nature and if we don't listen to nature, nature will deal with us.

The involvement of religious organisation as part of social capital in post-disaster recovery was also documented by Tammar et al. (2020) in Jeddah City (Saudi Arabia) after the 2009 and 2011 flash floods. Since many religious institutions were involved in evacuating people and providing emergency relief, their involvement in recovery also continued. Ngin et al. (2020) examined the role that the Buddhist temples could play in DRR from preparedness and relief to recovery. The temples were found to be centres of community organisation and information sharing. However, language barriers, generational divides, member exclusivity and personal conflicts were observed as having the potential to limit the temples' full involvement. This observation is critical and probably explains why the Catholic Church's BCIP had to be all-inclusive.

12.5 Conclusions

There are many religious beliefs across the world, and all seem to have a bearing on the manner in which DRR and management are handled. While there are arguments about religion being a double-edged sword in DRR and management, strong positioning emerged on beliefs pointing to Tropical Cyclone Idai's causes from both the Christian and local ChiManyika and Ndau cultures. Allegations pointing to wrongdoing that included ill-gotten wealth and prostitution in the most affected areas of Kopa, Ngangu and Machongwe in Chimanimani (Zimbabwe) arose from the investigations. Angry ancestors were said to have led to the flooding from Tropical Cyclone Idai. Sighting of mermaids, snakes and early warnings from spirit mediums to perform *mapira* to appease the ancestors and avert the pending danger were among some of the key findings. It also emerged that the chiefs and their kraal heads and headmen played a critical role as the intercessor between the ancestors that reside in the supernatural *nyikadzimu* and the living. Those who survived Tropical Cyclone Idai were asked to organise one collective *Bira reChenuro* to cleanse the evil spirits from Tropical Cyclone Idai and also appease the ancestors leading to future protection from incidences such as *Dutumpengo*. It also emerged that some headmen from some villages had gone ahead and performed their *Matakapona*. It was also found that some chiefs performed *mapira* to prevent Tropical Cyclone Idai and had their areas spared. However, from one account by somebody who grew up in the Chimanimani area, *maipira* were part and parcel of their lives as Chimanimani sits along the tropical cyclones' route. In this way, every few years, it is known that there will be such and the chiefs will always perform *mapira*.

Moving forward, the chapter recommends that the matters of religion and DRR be strongly considered in designing DRR policies at national and sub-national levels. For example, there are cultural setting that must be observed. However, as was the case elsewhere and in Chimanimani, both the traditional leadership (as custodian of cultural values) and organisations of Christian-based faith played fundamental roles during Tropical Cyclone Idai, from search and rescue, relief and recovery (including BBB). There is still a lot of work that will require all religious organisations to come together (SDG 17) to rebury the victims of Tropical Cyclone Idai that were laid to rest in the Chimanimani district heroes' acre. There is also work that remained to assist many families that were still in emergency shelters (camps) for the second year in such conditions, and the third rainy season. Furthermore, religious leaders will be needed to track bodies and perform rituals in Mozambique, where some bodies were buried. There will also be a need to join hands to undertake rituals to rest the cases of bodies that will never be recovered. After all, the leadership of the regions will be needed to continue with the psychosocial support for the victims. Lastly, the rich experiences gained by all the religious leaders in dealing with Tropical Cyclone Idai will be needed in the future should similar or other destructive disasters hit Zimbabwe.

References

Abbott, R.P. (2019). "I Will Show You My Faith by My Works": Addressing the Nexus between Philosophical Theodicy and Human Suffering and Loss in Contexts of 'Natural' Disaster. Religions, 10, 213. https://doi.org/10.3390/rel10030213.

Adiyoso, W. & Kanegae, H. (2013). The preliminary study of the role of Islamic teaching in the disaster risk reduction (a qualitative case study of Banda Aceh, Indonesia). Procedia Environmental Sciences, 17: 918–927.

Ayeb-Karlsson, S., Kniventon, D., Cannon, T., Van der Geest, K., Ahmend, I., Derrington, E.M., Florano, E., Opondo, D.O. (2019). I will not go, I cannot go: cultural and social limitations of disaster preparedness in Asia, Africa, and Oceania. Disasters, 43(4), 752–770. https://doi.org/10.1111/disa.12404.

Clarke, M. & Parris, B.W. (2019). Understanding disasters: managing and accommodating different worldviews in humanitarian response. Journal of International Humanitarian Action, 4:19. https://doi.org/10.1186/s41018-019-0066-7.

Cox, J., Finau, G., Ksnt, R., Trai, J., Titifanue, J. (2018). Disaster, Divine Judgment, and Original Sin: Christian Interpretations of Tropical Cyclone Winston and Climate Change in Fiji. The Contemporary Pacific, 30(2), 380–411.

Erblich, T., Orr, Z., Gottlieb, S., Barnea, O., Weinstein, M., Agnon, A. (2020). Earthquake preparedness among religious minority groups: Natural Hazards and Earth System Sciences. The case of the Jewish ultra-orthodox society in Israel. https://doi.org/10.5194/nhess-2019-387.

Fair, H. (2018). Three stories of Noah: Navigating religious climate change narratives in the Pacific Island region. Geo: Geography and Environment, 2018: e00068. https://doi.org/10.1002/geo2.68.

Ha, K.M. (2016). The Role of Religious Beliefs and Institutions in Disaster Management: A Case Study. Religious, 6, 1314–1328.

Hirono, T. & Blake, M.E. (2017). The Role of Religious Leaders in the Restoration of Hope Following Natural Disasters. SAGE Open, 1–15. https://doi.org/10.1177/2158244017707003.

Kazemba, T. (2009). The Relationship between God and People in Shona Traditional Religion. The Rose+Croix Journal, 6, 51–79.

Kulatunga, U. (2010). Impact of Culture towards Disaster Risk Reduction, International Journal of Strategic Property Management, 14(4), 304–313. https://doi.org/10.3846/ijspm.2010.23.

Massazza, A., Brewin, C.R., Joffe, H. (2019). The Nature of "Natural Disasters": Survivors' Explanations of Earthquake Damage. International Journal of Disaster Risk Sciences, 10: 293–305. https://doi.org/10.1007/s13753-019-0223-z.

Muyambo, T. & Maposa, R. (2014). Linking culture and water technology in Zimbabwe: Reflections on Ndau experiences and implications for climate change. Journal of African Studies and Development, 6(2), 22–28. https://doi.org/10.5897/JASD2013.0266.

Ngin, C., Grayman, J.H., Neef, A., Sanunsipl, N. (2020). The role of faith-based institutions in urban disaster risk reduction for immigrant communities. Natural Hazards, 103, 299–316. https://doi.org/10.1007/s11069-020-03988-9.

Ombati, M. (2017). Rainmaking rituals: Song and dance for climate change in the making of livelihoods in Africa. Journal of Modern Anthropology, 10, 74–96. https://doi.org/10.4314/ijma.v1i10.3.

Pixabay. (2020). Wildebeest. Retrieved from https://pixabay.com/photos/black-wildebeest-wild-horns-grumpy-4538695/ (Accessed 30 August 2020).

Risiro, J., Tshuma, D.T., Basikiti, A. (2013). Indigenous Knowledge Systems and Environmental Management: A Case Study of Zaka District, Masvingo Province, Zimbabwe. International Journal of Academic Research in Progressive Education and Development, 2(1), 19–38.

Salite, D. (2019). Explaining the uncertainty: understanding small-scale farmers' cultural beliefs and reasoning of drought causes in Gaza Province, Southern Mozambique. Agriculture and Human Values, 36, 427–441. https://doi.org/10.1007/s10460-019-09928-z.

Shoko, T. (2007). Karanga Indigenous Religion in Zimbabwe: Health and Well-Being. Hampshire: Ashagate Publishing Ltd.

Sibley, C.G. & Bulbulia, J. (2012). Faith after an Earthquake: A Longitudinal Study of Religion and Perceived Health before and after the 2011 Christchurch New Zealand Earthquake. PLoS ONE 7(12): e49648. https://doi.org/10.1371/journal.pone.0049648

Sipon, S., Narrah, S.K., Nazli, N.N.N.N., Abudullah, S., Othman, K. (2014). Stress and Religious Coping among Flood Victims. Procedia - Social and Behavioral Sciences, 140, 605–608.

Sun, L., Su, G., Tian, Q., Qi, W., Liu, F., Qi, M., Li., R. (2019). Religious belief and Tibetans' response to earthquake disaster: a case study of the 2010 Ms 7.1 Yushu earthquake, Qinghai Province, China. Natural Hazards, 99, 141–159. https://doi.org/10.1007/s11069-019-03733-x.

Tammar, A., Abosuliman, S.S., Rahaman, K.R. (2020). Social Capital and Disaster Resilience Nexus: Study of Flash Flood Recovery in Jeddah City. Sustainability, 12, 4668. https://doi.org/10.3390/su12114668.

Tatira, L. (2014). Shona Belief Systems: Finding Relevancy for a New Generation. The Journal of Pan African Studies, 6(8), 106–118.

The Bible (New International Version). (n.d.). Nashville: Tommy Nelson.

United Nations. (2015). Transforming our World: The 2030 Agenda for Sustainable Development. New York: United Nations Secretariat.

Exploring the Potential of Dark Tourism in the Aftermath of Tropical Cyclone Idai in Chimanimani

Abstract

Although places of human-induced death, suffering and torture have aroused great interest in terms of their local, national and global following, a new wave of natural disasters has fuelled and re-ignited research interest in dark tourism. The chapter investigates the potential of dark tourism at the epicentre of Tropical Cyclone Idai that devastated Zimbabwe in March 2019. A household questionnaire survey, interviews and field observations were the main methods applied to generate data. The GIS and document analysis were methods applied to analyse data. Findings are that the dark tourism idea found resonance in government as Cyclone Idai was the worst natural disaster the country ever witnessed. The authorities have decided to put up memorial sites in the three most affected areas of Ngangu, Machongwe and Kopa, all of them in the Chimanimani district. This new development has the potential to attract special tourists in the niche of dark tourism, although some of the artefacts of interest like the ropes that were used to save lives in Kopa were removed from the site. Furthermore, communities that had struggled for years to get quarry stone for construction were mining these at potential dark tourism sites. Hence, the value of the sites was constantly being degraded. Among some of the stones that were at risk from quarrying were those marked as potential sites for the missing dead. The paper concludes that there is a potential for dark tourism in Zimbabwe and steps need to be taken to realise and promote it.

Keywords

Dark tourism · Cyclone Idai · Chimanimani · Stakeholders · Community

13.1 Introduction

Dark tourism, also known as tourism of desolation, thanatourism or catastrophic tourism, was introduced in the 1990s (Molokáčová and Molokáč 2011; Stone and Sharpley 2008) and remains a diverse and multifaceted phenomenon even today (Kunwar and Karki 2019; Becker 2019). From the developed northern hemisphere, dark tourism comes up as embedded in history and heritage as well as tourism and catastrophe like the holocaust (Kidron 2013). To this end, a niche has been identified with dark tourism as special interest tourism (Stone 2006). Among the range of typologies of dark tourism, the following can be included: blackspot tourism, morbid tourism, natural disaster tourism, conflict tourism, dissonant heritage tourism and so forth (Seaton 2019a). Given the forgone, Tropical

Cyclone Idai's devastating impact that left thousands dead, many more homeless, injured and traumatised in Malawi, Mozambique and Zimbabwe qualify as a dark tourism phenomenon following a natural disaster.

From the literature, one may not separate dark tourism from aspects such as repressed sadism, commercialisation of grief, commoditisation of death and death seekers capitalism (Kunwar and Karki 2019). Several dark tourism sites have emerged in the literature. These include the Concentration and Extermination Camp in Auschwitz (Poland), Ground Zero—The National September 11 Memorial in New York (USA), Hiroshima Peace Memorial Park in Hiroshima (Japan); and Hector Pieterson Memorial Museum in Soweto (South Africa) (Stone 2006; Seaton 2019b; Nhlabathi and Maharaj 2019). There is also the Apartheid Museum (Johannesburg, South Africa) and the Robben Island Museum (Cape Town, South Africa), Chapel of Bones Évora (Portugal); Island of the Dolls Mexico City (Mexico); The London Dungeon (London, UK); Chernobyl Disaster Pripyat (Ukraine); Kigali Genocide Memorial Centre (Rwanda) and various 2004 Tsunami sites (Pratt et al. 2019; McDaniel 2019).

This paper explores the potential for dark tourism in the aftermath of Tropical Cyclone Idai, mainly in Zimbabwe. Two research questions were raised: (1) What plans were afoot from the Zimbabwe government to work towards establishing dark tourism sites? (2) Which sites were identified as having potential for dark tourism in Zimbabwe in the aftermath of Tropical Cyclone Idai?

13.2 Literature Survey

Kumar et al. (2018) bring the conventional understanding that tourism is associated with, thus, leisure, happiness and enjoyment. However, dark tourism, falling in the category of alternative or niche tourism, is a product originally thought of as specific to a tourist on a spiritual journey to gaze upon real and recreated death (Stone 2006). Although "it may be categorically unpleasant to visit cemeteries, crash sites and death camps, tourists queue up to see such places" (Hartmann et al. 2018, p. 269). After all, even the place where the assassination of former American President John F. Kennedy took place and the Holocaust happened still draw multitudes of people (Foley and Lennon 1996). Whereas the world could loosely share particular product features, perceptions and characteristics translating into various "shades of darkness" (Stone 2006), dark tourism products lie along a spectrum of intensity. Hence some sites tend to be "darker" than others. Their shades of darkness are a function of certain characteristics, perceptions and product traits (Fig. 13.1). Sites of death and suffering have higher political influence and ideology, while sites associated with death and suffering are the opposite. Furthermore, sites of the darkest shades focus on education orientation, they are history focused, are perceived as authentic and attract tourists in the short term of the aftermath

Fig. 13.1 Dark tourism continuum. Source: Authors, modified after Stone (2006), p. 151)

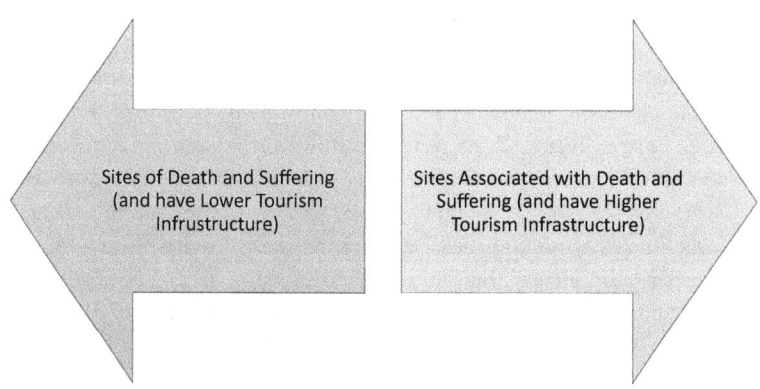

Sites of Death and Suffering (and have Lower Tourism Infrustructure)

Sites Associated with Death and Suffering (and have Higher Tourism Infrastructure)

of the event. On the other hand, sites of lighter shades are entertainment-oriented, heritage-centred and attract visitors on the long-term time horizon.

Stone (2006) goes further to identify seven dark suppliers, namely fun factories, dungeons, conflict sites, resting places, exhibitions, genocide camps and shrines. Since Stone's work came prior to the world gaining consensus on the impact of natural disasters related to climate change, an eight-supplier code-named disaster tourism now exists (Kunwar et al. 2019). Dark fun factories are sites providing entertainment that include fictional death and macabre events. Dark exhibitions may include museums that showcase death and associated suffering with a commemorative focus (Kim and Butler 2015). The Smithsonian Museum of American History exhibited images and artefacts of the 11 September 2001 terrorist attacks. However, to manage emotions, the exhibit displayed only 45 objects (Stone 2006; Potts 2012). Dungeons revolve around past prisons and courthouses. The Robben Island Museum off Cape Town (South Africa) remains one such popular dungeon where many South African leaders from the National African Congress (ANC) were imprisoned. The iconic Nelson Mandela spent 27 years there. Dark resting places revolve on cemeteries and graves (Nhlabathi and Maharaj 2019). Several tour operators now offer packages to various battlefields, either specifically or as part of a comprehensive holiday itinerary. Sites of interest include the battlefields of the 1st and 2nd World Wars (Miles 2014). Although not common, genocide sites attracting tourist are found in Rwanda, Cambodia and Kosovo.

From Nhlabathi and Maharaj's (2019) work that surveyed 100 tourists seeking to understand what value dark tourism brought to them as they came to South Africa, it emerged that most international travellers saw value in the sites in order to learn history and pay respects to victims of the past atrocities. In this case, Nelson Mandela became part of the attraction. Visitors further indicated that they also went to concentrations camps like Auschwitz (Poland) and others in the Netherlands and in Germany to have a global perspective regarding issues. Some revealed that they had visited these dark tourism sites as part of either school or family heritage tours, concurring with Fonseca et al. (2016) who indicate dark tourism visits purposefully or otherwise. The perspective of heritage is also experienced by African Americans that trace their slave trade routes and holding cells (Parry 2018; Bright et al. 2018), especially the Cape Coast Castle in Ghana (Boateng et al. 2018). Originally built by the Swedes in 1653 and later acquired by the British in 1663, The Cape Coast Castle was used as a centre of transatlantic slave trade. Similar, The Gorée Island in Senegal provided the shortest slave transportation route to North America (Small 2013) and is the country's first site on the UNESCO World Heritage List (Thiaw and Wait 2018). The Gorée Island has become the prime destination for global dark tourism, particularly for the African diaspora (Everill 2017). When Nelson Mandela visited The Gorée Island, he cried and indicated that the slave house he got into reminded him of his experience at The Robben Island. Hence, emotions can run amok in dark tourism experiences.

From Kunwar et al. (2019), disaster tourism emerges when historical and cultural identity is destroyed, and violent death takes place in large numbers of people from even seismic events like the 2015 Earthquake in Nepal. To this end, the Barpak and Langtang Seismic Memorial Sites in Nepal have started attracting tourists. The earthquake's landslides and avalanches destroyed the settlements. Among several reasons given by tourist visiting the sites were: black spot, history and heritage, cultural values, heritage and identity, survivors' guilt, acts of memory, empathy, remembrance, education and entertainment. Similar studies focusing on earthquake sites have been done (Wang and Luo 2018; Wright and Sharpley 2018). The studies include the Beichuan County in Sichuan Province of China after the 2008 and the L'Aquila in Italy following the 2009 earthquake. A total of 516 respondents were involved in a Beichuan County survey that revealed that the residents tended to form positive attitudes towards dark tourism development after such events (Wang and Luo 2018).

From the L'Aquila experience, it emerged that appropriate disaster tourism responses support the disaster recovery process and ability to host tourists from the local communities (Wright and Sharpley 2018). However, depending on the set-up, some communities may need to be re-oriented towards dark tourism as they both host and may have been directly affected by death. Dark tourism sites have attracted huge volumes of tourists. Figures compiled by Fonseca et al. (2016) reveal some sites have attracted millions annually. Figure 13.2 presents selected examples in terms of numbers they have attracted over time. Not included in Fig. 13.2 is Hiroshima Peace Memorial Park, Japan, that has been visited by over 53 million people since it opened its doors.

Reasons people visit dark tourism sites and harness their (negative) emotions are well documented in psychology and consumer behaviour (Nawijn and Biran 2019). However, González-Tennant (2013) considers these sites a "difficult heritage". As such, the concept of "new heritage" is brought in for social justice, particularly with local communities that are both directly and indirectly affected. "New heritage is the intersection of new media technologies and cultural heritage" (González-Tennant 2013, p. 62) that could emerge in instances of the aftermath of Cyclone Idai. The next section focuses on the materials, methods and the study site.

13.3 Materials, Methods and Study Site

Two research objectives were stipulated: (1) to document plans put in place by the Zimbabwe government in working towards the establishment of the dark tourism phenomenon and dark tourism sites, and (2) to identify and document potential dark tourism sites in Zimbabwe in the aftermath of Tropical Cyclone Idai. The methodology involved gathering data and information from government sources through interviews and written communications and pronouncements. Authors further engaged with communities that were impacted during fieldwork that took 3 weeks with a group of 16 fieldworkers. Given that dark tourism is under-researched in many African countries, including Zimbabwe (Nhlabathi and Maharaj 2019), the authors preferred an innovative and exploratory approach. The exploratory

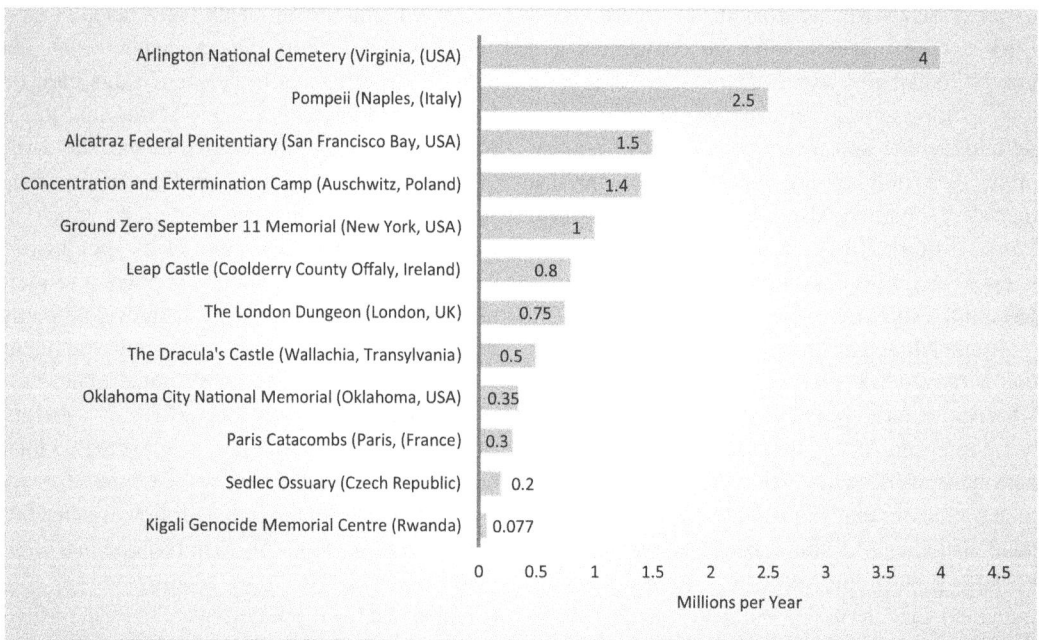

Fig. 13.2 Dark tourism sites in visitor numbers. Source: Authors, Data from Fonseca et al. (2016, pp. 4–5)

approach (Nhlabathi and Maharaj 2019, p. 5) "is a useful means to qualitatively explore experiences, attitudes, behaviours and beliefs where little is known about a phenomenon". In this case, it becomes the potential of dark tourism resulting from Tropical Cyclone Idai.

Not only did the authors walk the routes and recorded detailed accounts of what took place, but the authors further started plotting potential dark tourism sites and routes based on the narratives and the artefacts found on the sites. In addition to familiarisation and acclimatisation tours led by government officials and community leaders to the most affected areas, five tour guides further took us to various other sites. Four main areas (as ranged from worst hit) emerged, and these included Kopa, Ngangu, Machongwe and Rathmore farm areas. However, part of our methodology was to identify other sites that add value to the narratives and development of the dark tourism products and include them in the proposed dark tourism map. The map of the study area is shown in Fig. 13.3.

Three questions related to international relations and tourism as well as deaths of popular religious leaders were slotted in the overall household survey instrument uploaded onto the QuestionPro survey tool. The use of a survey is not new in dark tourism studies (Nhlabathi and Maharaj 2019; Wang and Luo 2018; Wright and Sharpley 2018). Our survey questions were as follows: (1) In your view, did Cyclone Idai led to improved diplomacy and international relations in general? (2) In your view, did Cyclone Idai led to the creation of opportunities in tourism, especially accommodation in bed and breakfast establishments (B&Bs), hotels and homes? (3) Please indicate if there were deaths associated with popular religious leaders. The realised samples (n) were 203, 206 and 195 respondents for the three questions in that order. Furthermore, one focus group discussion (FGD) was held with the internally displaced persons (IDPs) living in the tents at Kopa. The other method involved direct observation to artefacts and capturing thousands of scenes on camera. Although not deliberately

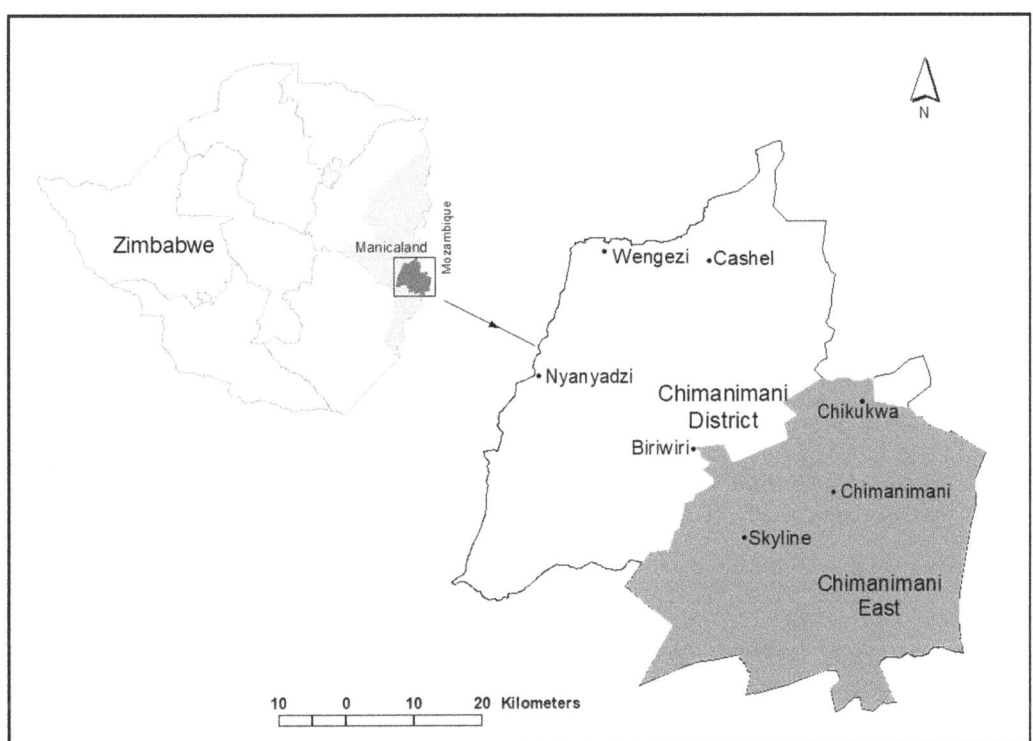

Fig. 13.3 Location of the study area—Chimanimani East. Source: Authors

setting up the journey on a dark tourism pathway as we were on research fieldwork, the experiences of two of the authors were that from the dark tourism space, which emerged from the literature at times as forced by circumstances. Document reviews, interviews and direct observations are common methods used in dark tourism research (Kunwar et al. 2019; Wright and Sharpley 2018), having been applied in Nepal to study the 2015 earthquake.

Data analysis was done using QuestionPro. The FGD and interviews were transcribed first and themes concerning dark tourism mined during analysis. The findings are considered in the results and discussion section that follow.

13.4 Presentation of Data and Discussion of Findings

This section comes in four sub-sections. The first sub-section analyses survey results on perceptions regarding improvement in international relations and diplomacy following the disaster, while the second sub-section is dedicated to understanding what the Zimbabwe government was doing regarding the need to promote dark tourism. The third sub-section narrows down to presenting the authors proposals for dark tourism sites and routes as they emerged during the fieldwork; and the last sub-section discusses proposals for establishing multiple museums in key attraction and more accessible cities like Harare the capital city, Mutare, the provincial capital for Manicaland Province, and Chimanimani town as the main host centre of the disaster.

13.4.1 General Overview of Tourism Attractions

As indicated earlier, two questions were included in the household survey to gauge if Tropical Cyclone Idai led to improved diplomacy and international relations in general and to the creation of opportunities in tourism, especially accommodation in B&Bs, hotels and homes. The responses for both questions are presented in Figs. 13.4 and 13.5.

What emerges from Figs. 13.4 and 13.5 is that the Tropical Cyclone Idai disaster brought some positivity in terms of international relations and general diplomacy (at least from the perceptions of those that were impacted). This concurs with the observation of Fonseca et al. (2016) that dark tourism plays a crucial role in both a country's economy and its image. A total of 52% of those surveyed were affirmative there was a change and only 13% thought Cyclone Idai did not result in improvement in diplomacy and international relations in general. The buttered international image of Zimbabwe resulted in bad international relations with many countries, among them the USA, Europe and the UK, all major sources of tourists. This scenario led to the late President Robert Gabriel Mugabe strongly driving the "Look East" policy, where China could be the main source of tourism and investment. As for the creation of "dark tourism" opportunities, 38% of those surveyed were positive, while 31% responded on the contrary. Overall, there were shades of positivity in both the improvement of international relations and diplomacy in general and the creation of tourism opportunities, at least in the short term. Two of the authors of this chapter witnessed the challenge of accommodation in Chimanimani due to the relief and recovery work that was going on. Most places were fully booked, many for months in advance. The next section turns the focus to government commitments towards establishing facilities promoting dark tourism in the aftermath of Cyclone Idai.

13.4.2 Government's Commitment to Develop Dark Tourism Sites

The first time the researchers heard from the Zimbabwean government regarding the concept of dark tourism development in Chimanimani was from an interview with an official from the Zimbabwe Tourism Authority during fieldwork in Harare at the end of September 2019. The respondent indicated that steps were underway to declare selected sites for erecting memorial monuments, thereby creating heritage sites (heritagization) according to Becker (2019). The same message was conveyed by the Chimanimani

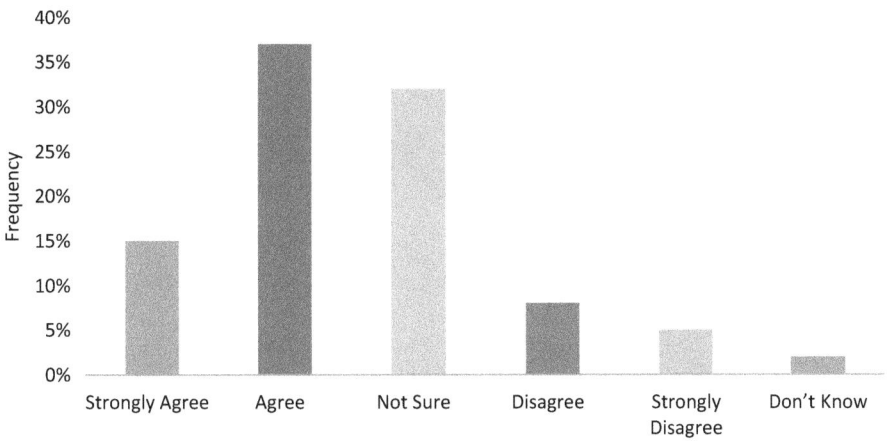

Fig. 13.4 Cyclone Idai led to improved diplomacy and international relations (*n* = 206). Source: Fieldwork 2019

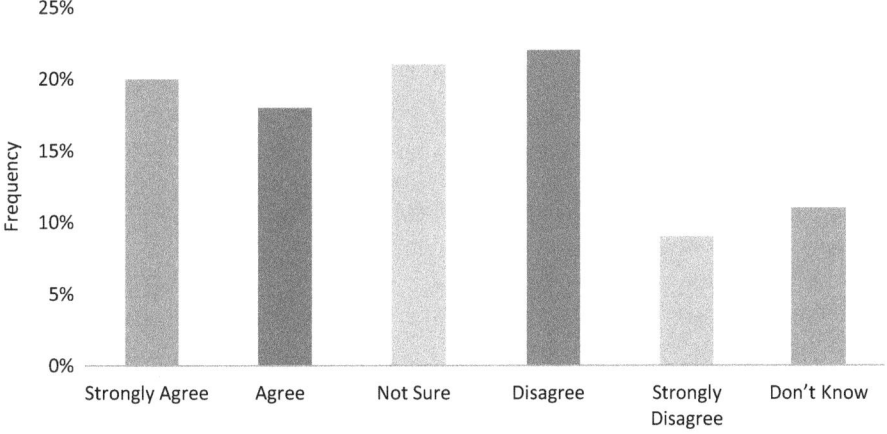

Fig. 13.5 Cyclone Idai led to the creation of opportunities in tourism (*n* = 203). Source: Fieldwork 2019

district development coordinator and some officials from the Chimanimani Rural District Council.

During the subsequent engagements with data and information provided by the Manicaland Provincial Development Coordinator's (PDC) Office, it came to light that indeed there was a memorialisation sector that regularly gave progress updates. In a 21 June 2019, the Manicaland Provincial Disaster Management Committee reporting cycle, the memorialisation sector revealed that it had commenced preparatory work to enable setting up of the Cyclone Idai temporary exhibition at Mutare Museum offices. The target date was September 2019. It also emerged that the provisional designs for the memorial monuments that were to be erected at Ngangu,

Kopa and Machongwe were still with the Ministry of Home Affairs for approval. At the same reporting meeting, there was the information to the effect that the quantity surveyors from the Public Works Department are working on the bill of quantities.

The picture regarding deaths and the injured from Chimanimani and those that were swept away from Kopa with bodies recovered in neighbouring Mozambique is presented in Fig. 13.6. The Cabinet also approved to have a team of pathologists to visit Mozambique to identify the bodies of Zimbabweans who are buried there. However, one focus group discussion at Kopa revealed that the process was taking too long, and the affected families were desperate for closure. From a total of 341 reported deaths in all eight

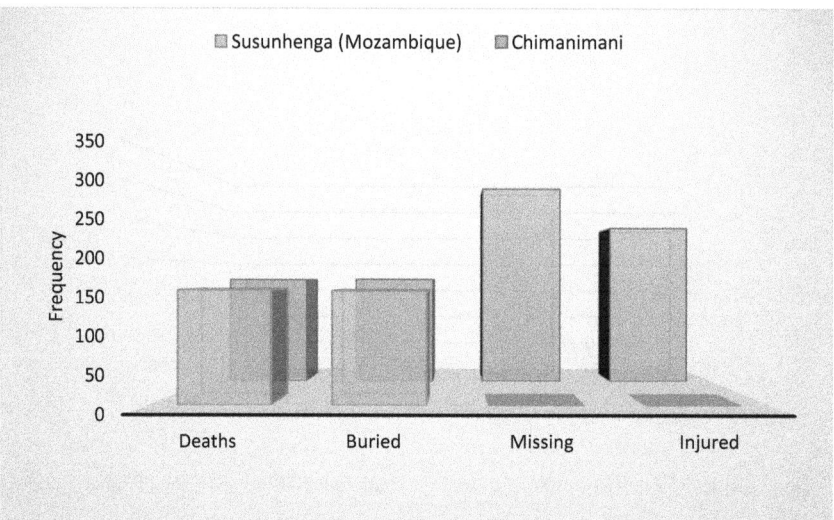

Fig. 13.6 Statistics regarding deaths, the missing and those injured. Source: Authors, Data from Manicaland PDC's Office

districts of Manicaland Province, 329 were from Chimanimani. From a total of 344 reported missing persons, 325 were from Chimanimani. From 295 injured, 259 were from Chimanimani. Clearly, Chimanimani was the epicentre of the Cyclone Idai disaster. Not reflected in Fig. 13.6 are the following statistics: one death in Buhera district, six from Chipinge district, one from Makoni district and four from Mutare rural district.

Additional statistics concerning the affected households, moderately affected households, internally displaced persons and internally displaced persons still in camps as of 31 September 2019 are shown in Fig. 13.7. From a total of 43,883 affected households, close to half (20,021) were from Chimanimani district alone. From the 26,682 households moderately affected, close to half (11,216) were in Chimanimani. Of the 17,201 internally displaced persons (IDPs), more than half (8805) were from Chimanimani. Lastly, from 2326 that were still in camps almost all (2251) were from Chimanimani. The other districts that had significant numbers were Chipinge and Buhera.

In his speech of November 2019 when commissioning the two Bailey bridges at Kopa in Chimanimani, Vice President of Zimbabwe. Mr Chiwenga reiterated the government's commit-

ment to erect monuments in honour of those killed by Cyclone Idai. Three places were identified, namely Ngangu, Machongwe and Kopa (Maodza 2019). The Vice President indicated that:

> In line with our culture and tradition, as we bring closure to this painful cyclone disaster, the Government is going to erect monuments at Ngangu, Machongwe and Kopa where names of all those who perished will be inscribed on. The chiefs were consulted and have since performed their rites as per tradition. Furthermore, Government will declare all missing persons dead before the expiry of the usual stipulated five years as provided by our laws.

Zavar and Schumann III (2019) focus on post-disaster landscapes that they claim remain replete with memorials that assist communities to remember such destructive events collectively and also recover psychologically. To this end, dark tourism sites will attract both local and international visitors. Among the physical artefacts that tourists may wish to see as they visit the sites in Chimanimani are residual houses, newly created landscapes now filled with rock boulders and gullies, marked rocks from search and rescues, trees with a history of search and rescue, especially those at Kopa, ripped off and washed away bridges. Figure 13.8 shows a site in Ngangu where houses were swept away and where lives

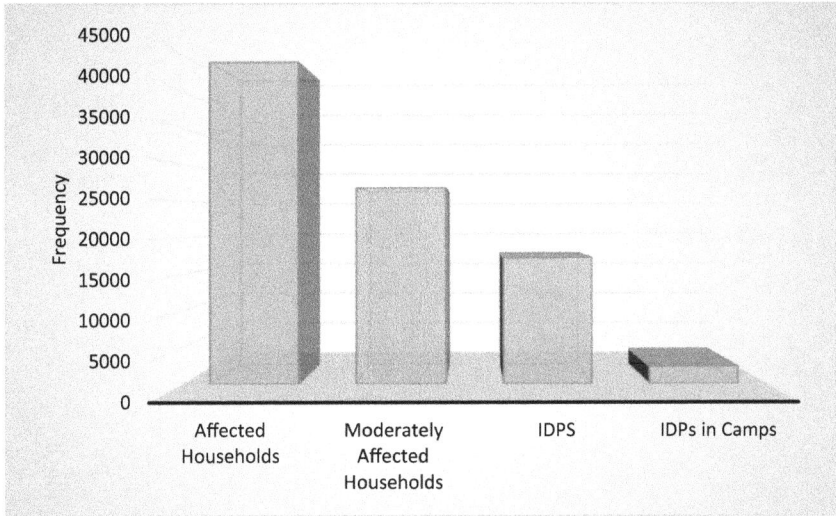

Fig. 13.7 Statistics on affected households and internally displacements. Source: Authors, Data from Manicaland PDC's Office

Fig. 13.8 Massive destruction that took place in some sites in Ngangu. Source: Authors, Fieldwork October 2019

were lost. Figure 13.9 depicts the massive destruction that took place when the water knocked down houses creating a completely new feature leaving few houses in ruins either sides. The seemingly open pathways and galleys on mountain slopes show areas of land and mudslides that may never be seen again after the terrain recovers, with new vegetation growing.

Figure 13.10 shows an open area where many houses in Kopa were located. On the ground nar-

ratives from two eyewitnesses that were involved in the search and rescue revealed that a rescue rope was tied to the mango tree on the right to the other big tree right ahead. As the waters were rising, rescue efforts were taking place. This mission was cut when the volume of water increased and swept away a lorry that hit the rope and cut it. This was indeed a painful narrative that will be told for many years to come to those that will visit. Also linked to this is the idea of a need to

Fig. 13.9 New feature created from the destruction. Source: Authors, Fieldwork October 2019

Fig. 13.10 The ruins of Kopa housing settlement. Source: Authors, Fieldwork October 2019

identify authentic community guides that will assist in telling many stories from Cyclone Idai across many settings and communities. Figure 13.11 reveals the extent of the damage with huge boulders, some that were rolled over from long distances and some potentially exposed from under the ground. From Fig. 13.12, some search and rescue markers on some stones can be seen. The markings show where search teams with sniffer dogs imported from South Africa suspected there could be something trapped and dead, while a house left with minimal damage still stands (Fig. 13.13).

In as much as the government identified three areas broadly, other incidences that are documented in this work will add value to the tourism product being discussed herein. These areas specifically include St Patrick Luanga Boarding School, Rathmore Estate and a site where a Chimanimani-bound long-distance bus from Harare was swept away by landslides and mud killing some passengers. To this end, as authors who have been in the field in the affected areas and who also spoke to people on the ground at length, we have added these and other areas of interest. The proposed potential dark tourism

Fig. 13.11 Huge boulders at Kopa. Source: Authors, Fieldwork October 2019

Fig. 13.12 Boulders with markings inserted during search and rescue. Source: Authors, Fieldwork October 2019

map and routes are the subject matter for the next section.

13.4.3 Proposed Dark Tourism Map: From the Authors' Field Experiences

Based on the evidence gathered on the ground during the October 2019 fieldwork, as authors, our suggested dark tourism map is shown in Fig. 13.14. The detailed explanations and

justifications for our choice are presented in the paragraphs that follow. The bus destined for Chimanimani was swept away by a mud and landslide, killing some of its occupants. By the time of finalising this article, the wreckage of the bus was still on site with no chance it will be recovered anytime soon. The bus disaster was captured in an interview with the Chimanimani East Member of Parliament. He described the incident as follows:

> We drove for about a kilometre and met the Trip Trans Bus, which was swept away. It was coming

Fig. 13.13 One of the houses that remained standing with minimal damages. Source: Authors, Fieldwork October 2019

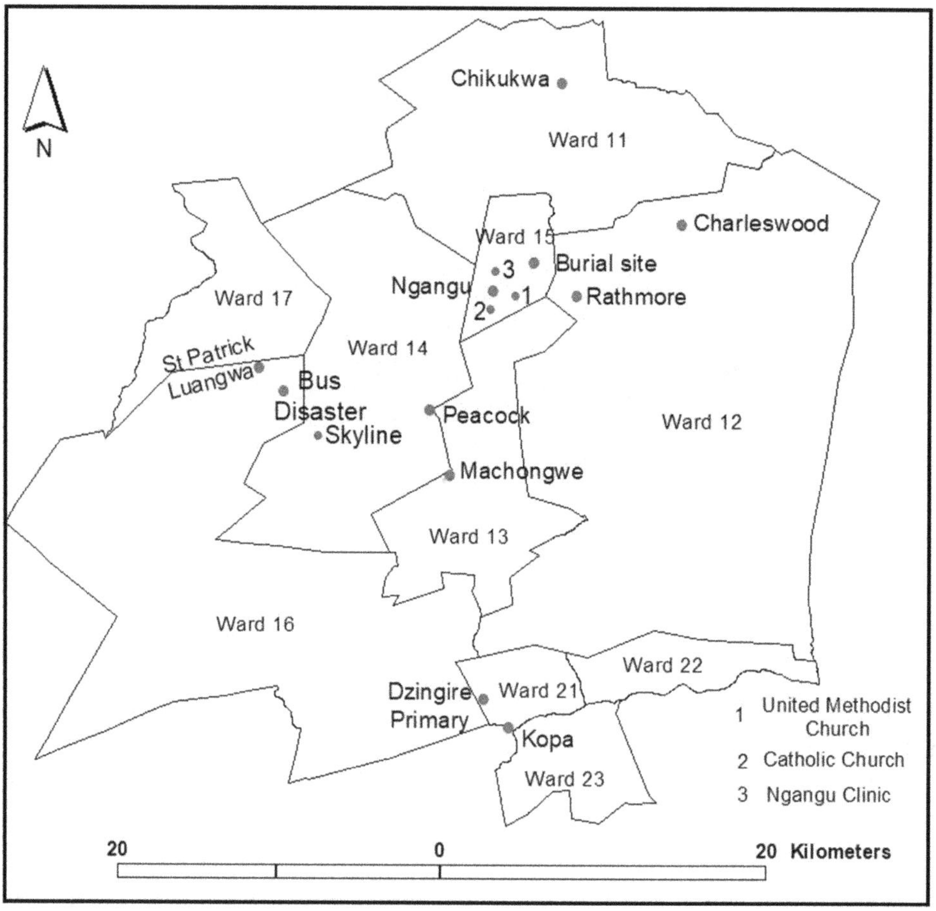

Fig. 13.14 Proposed dark tourism sites after Cyclone Idai. Source: Authors

up and we were going down. From what we gathered, the bus got to where the road was blocked, it could not pass because of a fallen tree. As passengers were trying to get axes to cut the tree so that the bus could pass, then there was an avalanche of mudslide coming from above and literally swept the bus down into the valley below. Unfortunately, we also had a loss of life at that point.

Numbers 1–3 are sites that played the role of mortuaries. The sites include the United Methodist Church, the Catholic Church and Ngangu Rural District Clinic. There was also a make-shift mortuary at Skyline. Apart from being the temporary command centre and relief hub, Skyline was popularly known as "Dubai", the Emirate from the United Arab Emirates, given that it remained the only place well light from generators that provided reasonable comfort to many development agencies staff. Skyline also played a role as a transitional health post for complicated cases referrals to Mutambara and Chipinge District Hospitals as well as Mutare General Hospital. Therefore, Skyline remains one such value-added dark tourism site.

One of the less spoken events about the area we discovered from an interview with the farm manager at Rathmore farm is depicted in Fig. 13.15. At this farm, 28 people lost their lives when Cyclone Idai hit. Most of them were swept away in the dark from their married quarters while they were asleep. Also swept away was the second largest operational timber mill in Zimbabwe on the same farm. Another site includes St Patrick Luangwa. The damage at St Patrick Luangwa school, where two pupils lost their lives after mudslides that resulted in huge boulders rolling over the dining room and into one of the dormitories, is also shown in Fig. 13.16. The sad story is that the bodies had to be transported to the next nearest accessible point, the Skyline, which was 15 km away. This narrative is one that will remain to be told to future generations and dark tourism tourists. From Skyline, the bodies were transported by road to Chipinge District Hospital outside Chimanimani by the Zimbabwe National Army. This is the story we got from the St Patrick Luangwa Boarding School in an interview. However, Chimanimani East Member of Parliament gave a supporting account of what transpired. In the Member of Parliament's words, he said:

We also had children from St Charles Luangwa who had been trapped in the boarding school. Unfortunately, we lost 3 lives as well, the security guard and 2 school children. Boulders tumbled down from the mountain and went into the dormitory, passing through the guard's house and caused the death of these people.

Fig. 13.15 Destruction at Rathmore farm. Source: Authors, Fieldwork October 2019

Fig. 13.16 Damage to St Patrick Luangwa infrastructure. Source: Authors, Fieldwork October 2019

At Dzingire Primary School, some teachers and pupils lost their lives. At Peacock, 17 people died, and the entire settlement was washed away. Only four people survived as per interview data with the Member of Parliament for Chimanimani East. Further up, several cars and their owners, drivers and passengers were swept away at Machongwe. These people had slept while awaiting the river to recede. The Chimanimani East Member of Parliament painted the picture of the event as follows:

> I drove from Chimanimani to Machangwe Growth Point where I stay, 20 km from Chimanimani. I managed to cross over a bridge along Mutandagwari River. It was in flooding but still possible. ... By the time we drove past, the river had risen quite substantially. As we passed the river, there were a number of cars parked by the river, 3 or 4 cars, and we actually had conversations with the drivers that the water level was rising and that they should move away from the bridge. Incidentally, this is where the Nyahode and Mutandagwari rivers meet. The Nyahode River flooded, and it was getting dark. The four cars were literally washed away after we had left and there was a loss of life.

The other dark spot we picked is the Chimanimani District Heroes Acre that remained the burial place for many that lost their lives from Ngangu. Although loved ones were buried, there are still contestations from the community regarding the fact that there are some graves with more than one body. The matters of concern surround the fact that people not related to their totems could be rested in the same grave. Other community members were also complaining of shallow graves. However, an interview granted from the District Development Coordinator's office and other government officials gave clarity as to why some graves contain more than one body. Additional clarity was also given to the effect that the graves were not shallow. The burials took such a heavy toll on the community in that they had to bury over 40 bodies in a day at one point. The bodies were decomposing and given the lack of experience to handle such disasters, the community was also traumatised. There became a point when people became afraid to recover the bodies scattered, with some partially buried in mud and others in debris.

There are, however, additional attractions in Chimanimani that could be marketed under the dark tourism scenario. These include the Bridal Veil Waterfalls, Tarsus Pool and Waterfalls, and Mawenje Mountains Range with rock paintings and curves. The Mawenje Mountain Range further links Chimanimani with Mozambique, and a tourism route can be developed networking both dark tourism cites and other regular attractions. Figures 13.17 and 13.18 present some of the attractions highlighted herein.

Fig. 13.17 Bridal Veil Waterfalls. Source: Authors, Fieldwork October 2019

From the proposed sites are many that could be later linked to pilgrimage (Collins-Kreiner 2016, p. 1186). The author reveals that, indeed, "both dark tourism and pilgrimage emerge from the same milieu to include the sites of dramatic historic events that bear the extra meaning". However, pilgrimage is normally portrayed as a holistic phenomenon with religious and secular foundations. This understanding is of interest to this work as our work revealed that there were indeed some religious leaders that died in Chimanimani. Asked to indicate if Cyclone Idai led to the death of popular religious leaders, 25% of the respondents (n = 195) indicated this had happened. Follow-up interviews revealed that there was a popular religious leader who had even travelled from outside Kopa, who was washed away by the flood waters. The next subsection focuses on the call to solidify the establishment of Cyclone Idai museums.

13.4.4 Beyond the Monuments: Call for Tropical Cyclone Idai Museums

The idea of a monument is noble. However, there may be many tourists that will not be able to drive all the way to Chimanimani, more than 400 km away from the capital city of Harare. As such,

creating a dedicated Tropical Cyclone Idai museum replicated in Chimanimani town, Bulawayo, Harare, Mutare and Victoria Falls help in both preserving the history and also attracting more tourist volumes. This idea is one that even require exclusive government rights as it can be taken even across borders and abroad. In the museum, photos gathered from the Chimanimani community, rescue work and any other sources including satellite images could be included. Artists and other stakeholders could also bring their impressions and artefacts to the galleries of such a museum. Stories not told and some already told can find their way into these artistic impressions and serve the purpose to preserve the history and the event for a long time. In the museum, pictures and/or paintings could be placed depicting how Cyclone Idai caused landslides from the mountains (Fig. 13.17). The photo in Fig. 13.17 was taken on 13 October 2019 from the Chimanimani observatory.

In the museums, President Emerson Dambudzo Mnwangagwa's Tropical Cyclone Idai's awards citations could be included. Several awards were conveyed on 9 August 2019 during the National Heroes Day public holiday at the National Shrine. The President honoured individuals, corporates, religious organisations and diplomats in recognition of their efforts and sacrifices in providing humanitarian support to

Fig. 13.17 A bird's eye view from the Chimanimani observatory. Source: Authors, Fieldwork October 2019

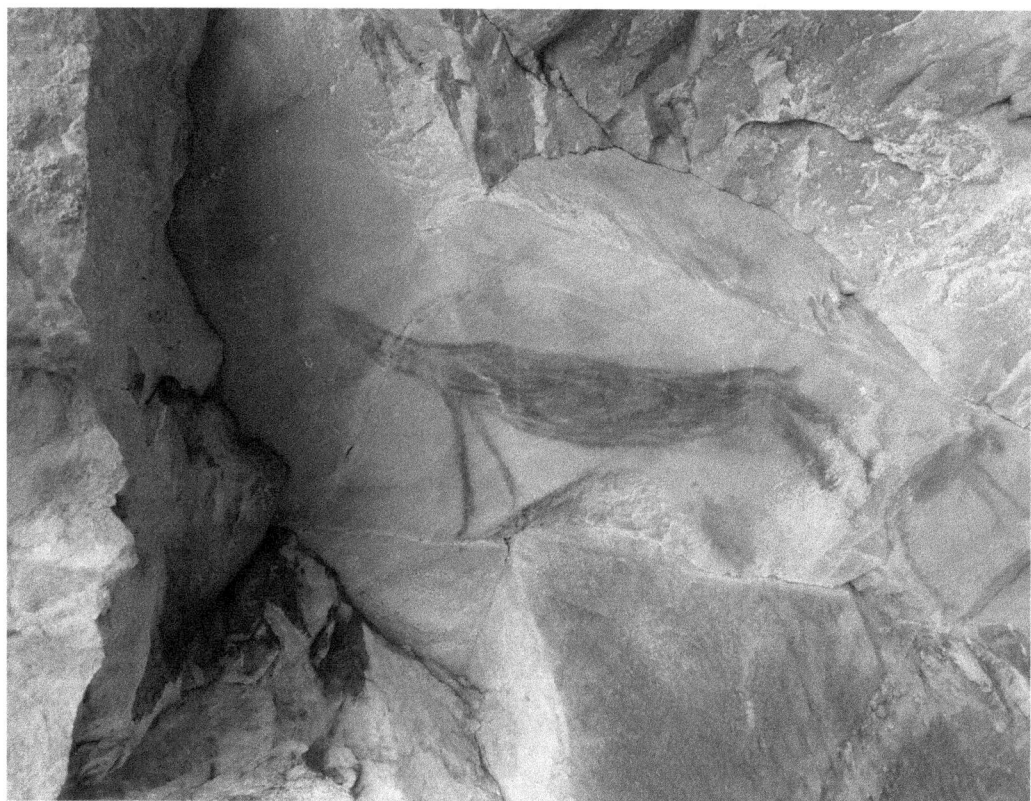

Fig. 13.18 Rock paintings in Mawenje Mountain Range. Source: Authors, Fieldwork October 2019

lessen the burden from the effects of Cyclone Idai (Murwira 2019). Among the individuals and organisations honoured were the following: United Nations represented by Ambassador Bishow Parajuli, Dean of African Ambassadors accredited to Zimbabwe and Democratic

Republic of Congo Ambassador, Mwawapanga Mwanananga, Bitumen World, represented by Mr Andre Zietsman and Fossil Contracting represented by Mr Ronald Mashura. The Awards categories included Zimbabwe International Friendship Award in Gold, Jairos Jiri Humanitarian Award and Zimbabwe Bravery Award. Extracts from the citations for the different categories are presented in Box 13.1.

Box 13.1 Awards Citations Extracts

Zimbabwe Bravery Award in the Gold Cross: They rescued people trapped by mudflows and rockfalls in the Ngangu extension, ferrying of cyclone survivors to safe shelter for first-aid treatment, venturing into the epicentre of the disaster, crossing flooded rivers to rescue marooned victims and establish contact with the outside world and, above all, in the process sacrificing their lives in rescuing operations.

Jairos Jiri Humanitarian Award: The remarkable actions they did include: transporting desperate cyclone victims to hospitals under difficult weather conditions and, at times, using rudimentary means, extracting the critically injured and deceased under massive boulders that had crushed them, conveying dead bodies for burial, providing emergency shelter, first aid and water to cyclone victims, coordinating with district authorities concerning the disaster thereby triggering rescue operations, coordinating the district civil protection committee and other agencies on the ground in rescue operations, sourcing and distributing necessities, mobilising resources for the burial of deceased persons, providing their own resources such as blankets and curtains to wrap the deceased and those rendered naked by the devastating cyclone. Other actions included mobilising the youth for the digging of graves and for burials for Cyclone Idai victims, donating cash, fuel and shelter to victims, and using

air transport to facilitate transport for the first doctors for cyclone victims.

Zimbabwe International Friendship Award in Gold: Ambassador Parajuli exhibited outstanding leadership and diplomatic skills in fostering constructive, positive engagement between the United Nations and Government of Zimbabwe in areas of collaboration such as electoral support, human rights, inter-political party dialogue, peace building, gender equality and public sector capacity building.

Source: Murwira (2019, Online)

While the potential and business case for dark tourism has been presented, the major challenge is that many artefacts have been removed from many sites. For example the rope that was used to rescue people at Kopa was taken away by some individuals from the account on the ground. Thieves were also grabbing whatever could be of value to them. In addition, given that construction material has always been a perennial problem in Chimanimani, the exposed rock boulders and other building material and aggregates were being harvested on a semi-commercial basis. Tractors full of these aggregates were witnessed, many as the night fell.

13.5 Conclusions

While the authors are in full agreement with the Zimbabwean government's decision to erect monuments at Kopa, Ngangu and Machongwe in memory of many people that lost their lives during Tropical Cyclone Idai, there are other sites that will enhance the dark tourism product. Additional sites included in the proposed and developed dark tourism map include the main burial site at the Chimanimani District Heroes Acre, Rathmore, the facilities that acted as mortuaries, the bus disaster site, Skyline (Dubai), Dzingire Primary School, St Patricks Luangwa boarding school, Chikukwa and Charleswood. It must also be noted that all the highlighted sites may not be all the dark tourism sites there could be in and around Chimanimani. Without doubt, Chimanimani presents a potential for both ends of the dark tourism spectrum (sites of death and suffering as well as sites associated with death and suffering). Among some of the dark tourism consumers could be environmentalists, hydrologists, academics and climate change scientists.

We are also of the view that Cyclone Idai-related museums can be opened in more accessible tourist areas like the capital city Harare and Mutare, the provincial capital for Manicaland Province where the Chimanimani district is. It also emerged that the concept of dark tourism has just begun filtering through the local level although still not a popular engagement. The dark tourism route from Kopa could also be extended to Susunhenga in Mozambique where some bodies washed away into rivers and later buried by traditional leaders there. A key recommendation that can be taken to similar sites and incidences in Zimbabwe and elsewhere is to have local governments moving in quicker to protect the potential dark tourism sites and their artefacts through temporary by-laws. For example a temporary by-law could have prohibited unrestricted access to the sites, harvesting of building material and other aggregates as well as removing of artefacts in an ad hoc manner. Some of the artefacts could have found their way to the national museums for safe keeping in preparation for a local museum or any such similar arrangement.

References

Becker, A. (2019). Dark tourism: The "heritagization" of sites of suffering, with an emphasis on memorials of the genocide perpetrated against the Tutsi of Rwanda. International Review of the Red Cross. 101(1), 317–331. https://doi.org/10.1017/S181638311900016X.

Boateng, H., Okoe, A.F., Hinson, R.E. (2018). Dark tourism: Exploring tourist's experience at the Cape Coast Castle, Ghana. Tourism Management Perspectives, 27, 104-110. https://doi.org/10.1016/j.tmp.2018.05.004.

Bright, C.F., Alderman, D.H., Butler, D.L. (2018). Tourist plantation owners and slavery: a complex relationship, Current Issues in Tourism, 21(15), 1743-1760, https://doi.org/10.1080/13683500.2016.1190692.

Collins-Kreiner, N. (2016). Dark tourism as/is pilgrimage, Current Issues in Tourism, 19(12), 1185-1189, https://doi.org/10.1080/13683500.2015.1078299.

Everill, B. (2017). All the baubles that they needed: Industriousness and slavery in Saint-Louis and Goree. Early American Studies, 15(4), 714–739.

Foley, M. & Lennon, J.J. (1996). JFK and dark tourism: A fascination with assassination, International Journal of Heritage Studies, 2(4), 198-211. https://doi.org/10.1080/13527259608722175.

Fonseca, A.P., Seabra, C., Silva, C. (2016). Dark Tourism: Concepts, Typologies and Sites. Journal of Tourism Research & Hospitality, S2, 2-6. https://doi.org/10.4172/2324-8807.S2-002.

González-Tennant, E. (2013). New Heritage and Dark Tourism: A Mixed Methods Approach to Social Justice in Rosewood, Florida. Heritage & society, 6(1), 62–88.

Hartmann, R., Lennon, J., Reynolds, D.P., Rice, A., Rosenbaum, A.T., Stone, P.R. (2018). The history of dark tourism, Journal of Tourism History, 10(3), 269–295, https://doi.org/10.1080/1755182X.2018.1545394.

Kidron, C.A. (2013). Being there together: Dark family tourism and the emotive experience of co-presence in the Holocaust past. Annals of Tourism Research, 41, 175–194. https://doi.org/10.1016/j.annals.2012.12.009.

Kim, S. & Butler, G. (2015). Local community perspectives towards dark tourism development: The case of Snowtown, South Australia, Journal of Tourism and Cultural Change, 13(1), 78-89, https://doi.org/10.1080/14766825.2014.918621.

Kumar, S., Shekhar, R., Attri, K. (2018). Exploration of Potential for Development of Dark Tourism in India. International Journal of 360 Management Review, 6(2), 2320–7132.

Kunwar, R.R. & Karki, N. (2019). Dark tourism: understanding the concept and recognising the values. Journal of APF Command and Staff College, 42–49.

Kunwar, R.R., Aryal, D.R., Karki, N. (2019). Dark Tourism: A Preliminary Study of Barpak and Langtang as Seismic Memorial Sites of Nepal. Journal of Tourism & Hospitality Education, 9, 88–136.

Maodza, T. (2019). VP Chiwenga commissions Bailey bridges at Kopa. Available at: https://www.herald.co.zw/vp-chiwenga-commissions-bailey-bridges-at-kopa/ (Retrieved 9 December 2019)

McDaniel, K.N. (2019). The Palgrave handbook of dark tourism studies, Journal of Tourism History, 11(1), 93-95, https://doi.org/10.1080/1755182X.2019.1585510.

Miles, S. (2014). Battlefield sites as dark tourism attractions: an analysis of experience, Journal of Heritage Tourism, 9(2), 134-147, https://doi.org/10.1080/1743873X.2013.871017.

Molokáčová, L. & Molokáč, Š. (2011). New phenomenon – Dark tourism. Acta Geoturistica, 2(1), 1–7.

Murwira, Z. (2019). President honours Cyclone Idai heroes. Retrieved at https://www.herald.co.zw/president-honours-cyclone-idai-heroes/ (Accessed 20 December 2019).

Nawijn, J. & Biran, A. (2019) Negative emotions in tourism: a meaningful analysis, Current Issues in Tourism, 22(19): 2386–2398, https://doi.org/10.1080/13683500.2018.1451495.

Nhlabathi, S.S. & Maharaj, B. (2019). The dark tourism discipline: a creative brand in a competitive academic environment? Current Issues in Tourism, https://doi.org/10.1080/13683500.2019.1636770.

Parry, T.D. (2018). What is Africa to me' now? African-American heritage tourism in Senegambia, Journal of Contemporary African Studies, 36(2), 245–263, https://doi.org/10.1080/02589001.2017.1387236.

Potts, T. J. (2012). 'Dark tourism' and the 'kitschification' of 9/11. Tourist Studies, 12(3), 232–249. https://doi.org/10.1177/1468797612461083.

Pratt, S., Tolkach, D., Kirillova, K. (2019). Torusim & death. Annals of Tourism Research 78: 102758. https://doi.org/10.1016/j.annals.2019.102758.

Seaton, P. (2019a). Islands of "Dark" and "Light/Lite". Japan Review, 33 (Special Issue: War, Tourism, and Modern Japan), 299-328. : https://www.jstor.org/stable/10.2307/26652985.

Small, S. (2013). Still Back of the Big House: Slave Cabins and Slavery in Southern Heritage Tourism, Tourism Geographies, 15(3), 405–423, https://doi.org/10.1080/14616688.2012.723042.

Stone, P.R. (2006). A dark tourism spectrum: Towards a typology of death and macabre related tourist sites, attractions and exhibitions. Tourism, 54(2), 145–160.

Stone, P.R. & Sharpley, R. (2008). Consuming dark tourism: A thanatological perspective. Annals of Tourism Research, 35(2), 574–595. https://doi.org/10.1016/j.annals.2008.02.003.

Thiaw, I. & Wait, G. (2018). Presenting Archaeology and Heritage at a UNESCO World Heritage Site: Gorée Island, Senegal. Advances in Archaeological Practice 6(3), 238–247, https://doi.org/10.1017/aap.2018.19.

Wang, J. & Luo, X. (2018). Resident perception of dark tourism impact: the case of Beichuan County, China, Journal of Tourism and Cultural Change, 16(5), 463–481, https://doi.org/10.1080/14766825.2017.1345918.

Wright, D. & Sharpley, R. (2018). Local community perceptions of disaster tourism: the case of L'Aquila, Italy, Current Issues in Tourism, 21(14), 1569–1585, https://doi.org/10.1080/13683500.2016.1157141.

Zavar, E.M. & Schumann III, R.L. (2019). Patterns of disaster commemoration in long-term recovery. Geographical Review, 109(2), 157–179, https://doi.org/10.1111/gere.12316.

Tropical Cyclone Idai and Flood Hazard Modelling in the Eastern Parts of the Save Catchment, Zimbabwe

14

Abstract

Floods are the most frequent and destructive natural disaster affecting human settlements, livelihoods and the environment. Over the past two decades, economic losses due to hydro-climatic disasters such as flooding have risen more than threefold. Although floods cannot be prevented completely, their impact can be reduced through appropriate preparation and organisation in areas that are at risk. This research aimed at modelling the flood hazard occurrence in the eastern parts of the Save Catchment of Zimbabwe. The height above channel base for the study catchments, together with the flood presence and absence data were run in a logistic regression and in a Geographic Information System (GIS) to determine the probability of flood occurrence at specific places. The results show a significant negative relationship between the height above channel base and flood-risk probability. Over 20% of the studied catchment area is noted to be flood safe, and 62% is shown to be moderately vulnerable to flooding. A further 18% of the area is zoned to be at high risk of flooding. Well-engineered institutions and stakeholders were found to be key in the effective utilisation of flood-risk maps generated from GIS projects to enhance community resilience to flooding. Furthermore, a contingent flood hazard management plan has been produced for the area.

Keywords

Flood hazard; flood modelling · Flood hazard contingent plan · Save Catchment · Cyclone Idai · Logistic regression

14.1 Introduction and Background

Floods are one of the most frequently occurring and distressing natural hazards that the global community faces (Twumasi et al. 2019). They are responsible for the most damage and destruction compared to other extreme hydro-meteorological occurrences (NOAA 2009). In the tropics, cyclones are a major cause of flooding and rank high among the most distressing natural disasters causing human fatalities and destruction to livelihoods, infrastructure and the environment (Sidek et al. 2016). Moe et al. (2018) highlight that, throughout the last century, floods have been one of the costliest hazards in both the value of destroyed properties and human casualties. The destructive characteristics of land-falling cyclones include persistent high-speed winds and intensive rainfall, and generally their trail of destruction is enormous at any level of society when compared to other hazards (Nelson 2008). Under most climate change scenarios, it is projected that the occurrence and magnitude of cyclones are likely to increase the world over (Deo and Ganer 2014).

In African countries tropical cyclones and the consequent flooding have a very high socio-economic and ecological impact because most of these countries do not have adequate disaster preparedness, they lack real-time forecasting technology or resources for post-disaster recovery are limited (Nelson 2008). The impact of tropical cyclones on people, their livelihoods and the environment can be reduced by using appropriate structural and non-structural management approaches (Sidek et al. 2016). An efficient tropical cyclone hazard intervention strategy is organised into four main segments: reduction risk, risk preparedness, response and recovery (Nelson 2008). The least costly approach that can save lives and reduce damage caused by tropical cyclones is having appropriate disaster risk reduction and preparedness plans in place (Chadburn et al. 2010). Predicting flood intensity, frequency and hotspots as non-structural measures are ideal for reducing the exposure of populations and property to floods by informing appropriate land use planning and management approaches (Sidek et al. 2016). This approach includes discouraging settlement growth in areas that are prone to flooding and may include moving values at risk from the floodplains and ensuring that structures which are permanently located within floodplains are resilient to potential threats. These elements can only be informed by accurate flood-risk maps of an area (Kwang and Osei 2017).

At global level, the Sendai Framework encourages countries to "urgently anticipate, plan for and reduce disaster risk in order to effectively protect persons, communities and their livelihoods, health, cultural heritage, socioeconomic assets and ecosystems, and thus strengthen their resilience" (United Nations 2015, p. 10). The Sendai Framework also articulates the necessity for countries to occasionally generate new location-based data for disaster risk reduction (DRR). This new geospatial data may include disaster risk maps that need to be communicated to intervening authorities, interested parties and the at-risk society.

For effective communication, the GIS-generated risk maps must be packaged in a format easily understood by non-experts. Remote sensing and spatial analysis techniques produce data and products that support effective decision-making in reducing the exposure of communities to cyclone-induced flooding (Twumasi et al. 2019). Together with GIS and satellite remote sensing, the techniques offer reliable and effective tools for supporting effective decision-making during all the phases of the DRR cycle. This is so because most of the data required for flood disaster management has spatial components and also changes over time; remote sensing and GIS are therefore very useful for analysing these various components of risk (Santillan et al. 2016). Remote sensing methods are less expensive and provide in quick turnaround time the situational awareness of a flood disaster even for the hard-to-reach areas. GIS methods provide the visual, analytical and modelling capabilities to help DRR managers and interested parties make effective decisions (Franci et al. 2015).

Global warming and related climate change have stimulated increases in incidences and magnitude of weather extremes such as tropical cyclones. In the past, landlocked countries such as Zimbabwe were relatively safe from cyclone-induced flooding. Tropical cyclone Eline was the first major cyclone to reach Zimbabwe in 2000 (FEWSNET 2000). Since then, cyclones Japhet in 2003, Dineo in 2017 and Idai in 2019 have hit the country. Despite efforts to manage floods in Zimbabwe, they still result in loss of life, destruction of livestock, crop, and properties, and trigger outbreaks of diseases such as cholera and malaria in Zimbabwe (Madamombe 2004). The response from the government has mainly been reactionary rather than preventive or mitigatory of settlement vulnerability to flooding. Using lessons from tropical cyclone Idai, which hit Zimbabwe in March 2019, the aim in this chapter is to model the probability of flood occurrence in the eastern parts of the Save Catchment in Zimbabwe. A contingent flood hazard management plan is also produced for the area.

14.2 Literature Survey

Floods are the most recurrent natural hazards that lead to serious damage to structures and human casualties (Twumasi et al. 2019). Sidek et al. (2016) highlight that floods were responsible for close to 40% of the total financial losses induced by all forms of natural hazards. These financial losses caused by flooding run into billions of dollars globally. From 1998 to 2017, for example, US$2908 billion worth of losses occurred due to disasters (UNISDR and CRED 2018). Losses that were due to hydro-climatic extremes for the same period amounted to over US$2245 billion, which translates to about 77% of the total economic losses. Over the past 20 years, economic disasters caused by hydro-climatic disasters have risen by almost 151% (UNISDR and CRED 2018). In Indonesia, for example, flooding is the most frequent disaster with at least 8498 flood events being recorded from 1980 to 2017 (Moe et al. 2018). Damage from floods is noted to be more intensive in countries of the Global South due to limited financial resources and technical know-how and a lack of political will to implement long-lasting interventions (Mahmood 2017).

Although floods cannot be prevented entirely, Twumasi et al. (2019) observe that their impact can be reduced through appropriate preparation and organisation in areas that are most vulnerable. The destructive impact of floods on livelihoods, economic assets and property can be mitigated by both structural and non-structural interventions. Keys et al. (1996) argue that floods are among the most manageable of all disasters, unlike rapid impact disasters such as wildfires and earthquakes. The reason is that in most cases their area of occurrence is predictable and, as such, can be planned for. Proactive measures to reduce flood risk involve identifying places at the highest risk of flood occurrence, gauging the probability for hazard occurrence in the at-risk places, modelling the impact of hazards, approximating the possible damage that will occur in communities and gathering information during a disaster in order to help build resilience to future incidences (Pawaringira 2008).

A battery of factors influences stream flow and, by default, the prospect for flooding in an area. Among many, the most significant are the amount and type of precipitation received at a place, the condition and structure of the catchment area and antecedent soil moisture conditions, land cover and climatic swings. Climatic swings may include the occurrence of tropical and extra-tropical cyclones, intense thunderstorms and tornadoes, all of which can be exacerbated by the El Nino phenomenon (Pawaringira 2008). Other causes may include monsoons, glaciers and structural failure of reservoirs which can flood downstream areas even under dry weather conditions. Although most flooding occurs during or after intense rainfall activity which exceeds the capacity of streams to convey excess water, floods are not always caused by heavy rainfall. They also occur as a consequence of backflow of water from a flooded river without any rains falling, which normally occurs in the Mazarabani area of Zimbabwe (Chingombe et al. 2015).

Integrated flood-risk management requires both structural and non-structural interventions. Structural interventions for flood control have been noted to be very expensive for most countries of the Global South and therefore not always feasible to implement (Sidek et al. 2016). This leaves non-structural interventions to mitigate flood risk as the most feasible option for them. Twumasi et al. (2019) posit that countries at risk of flooding in the Global South need mitigation measures that are financially viable, technically affordable yet also socially and ecologically acceptable. Such measures need to be implemented based on the need and affordability of the user community. Flood hazard mapping has been shown to be the first and most integral part of land use and emergency planning (Mahmood 2017). It provides the basis for informed decisions because it provides data essential to understand the form, risk and characteristics of floods that are likely to occur at a place. These maps can function as a key source of data informing decision makers, engineers and planners to come up with balanced solutions in dealing with flood hazards (Kwang and Osei 2017). Flood hazard mapping can also function as part of spatial decision support systems in locating critical infrastructure and evacuation zones or as part of an evacuation manual, and supports search and rescue operations. It is also key for insurance experts and investors as it helps them to calculate the premium an entity has to pay based on their risk level to the flood hazard (Mahmood 2017). Flood hazard mapping is therefore not only cost efficient, but it can be used as a tool for social resilience (Khaing et al. 2019).

Sidek et al. (2016) highlight that the foremost considerations for executing integrated flood-risk management in countries of the Global South are effective institutions and stakeholders, local community participation, funding of flood-risk management measures, sustainable maintenance and continuous monitoring and evaluation to make the system responsive to new and evolving issues. Effective institutions and stakeholders aid in connecting flood hazard management with governance and help allocate stakeholders' roles and responsibilities. Engaging the local community will assist in assessing the flood hazard at grassroots level and in appreciating local capacity and knowledge. Adopting community-based methods where applicable is strongly advised. This will ensure that the development and implementation of the integrated flood-risk management is based on community knowledge and experience (Jaha et al. 2012; Khaing et al. 2019).

GIS and remote sensing (RS) have become critical tools for the effective assessment of hazards and risks associated with flood disasters. These tools can perform simulations of flood characteristics such as flood extent in order to assess the socio-economic and environmental implications (Albano et al. 2015). The applications of GIS and RS are also now very widespread in almost all disaster-related issues. These include applications in multiple hazard assessment such as modelling of floods, landslides and land degradation, earthquakes, wildfires and general DRR (Twumasi et al. 2019). GIS and RS permit faster and easier handling and integration of numerous spatial and non-spatial datasets, automate the development of the flood-risk models through the use of algorithms and facilitate flood-risk analy-

ses through the identification of settlements, utilities, assets and land cover that may be at risk under several flood scenarios. The tools can also facilitate efficient generation of inundation maps and statistics of utilities damaged or at risk at various stages of flood disaster management (Santillan et al. 2016). These tools have been observed to be cost-effective because most of the data sources and software are open source.

Santillan et al. (2016) observe that despite the power of GIS and RS as spatial decision support systems during times of disaster, the tools do not make decisions. This leads to the need to develop simplified GIS models that would not need experts to operate. The models should be simplified enough for novice GIS users to extract information from them to aid in decision-making (Twumasi et al. 2019). Information from GIS and RS must be properly communicated to non-technical persons and related organisations, otherwise the whole process becomes futile. The DRR process works well when systems have been put in place to produce output flood-risk maps understandable by the general public and decision makers (Twumasi et al. 2019). It is inescapable that floods will occur in some areas; however, the utilisation of geospatial technologies can go a long way in lessening the negative effects and in developing approaches to mitigation and preparedness (Khaing et al. 2019).

The next section is dedicated to the materials and methods used in the study. The study area and the methods used to derive the flood hazard maps are described.

14.3 Materials and Methods

The study focused mainly on the sub-catchments that are in the eastern parts of the Save Catchment as indicated in Fig. 14.1. These are the sub-catchments that are normally first hit by tropical cyclones when they reach Zimbabwe from the Mozambique Channel.

The major land use categories in the study area are forestry plantations, protected forest, national parks and large-scale commercial farms.

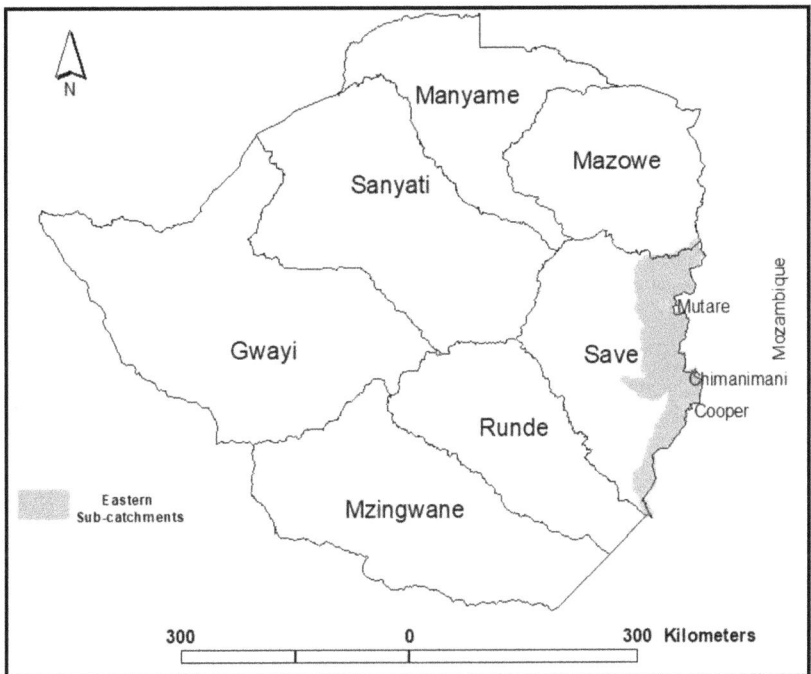

Fig. 14.1 The eastern sub-catchments of Zimbabwe. Source: Authors

The area is in Manicaland province and under the jurisdiction of the Chimanimani, Chipinge and Mutare Rural District Councils (Chikodzi et al. 2020). Water resources in the study catchments are under the management of the Zimbabwe National Water Authority's Save Catchment. The areas are subdivided into micro-catchments known as sub-zones for management purposes. The studied micro-catchments are in the E and F hydrological sub-zone regions of Zimbabwe. The main rivers in these zones are the Odzi, White Waters, Tanganda, Budzi, Rusitu, Nyanyadzi, Changadzi, Nyahodi, Umvumvumvu, Pungwe, Mutare and Zonwe rivers. The rivers are mainly perennial and flow throughout the year in an easterly and south-easterly direction into Mozambique. The geology of the eastern parts of the Save Catchment is dominated by crystalline basement rocks, which are intrusive and made of mainly granites of different ages, gneiss, dolerite and gabbro, which are largely of secondary permeability. The area is part of the Eastern Highlands of Zimbabwe, which form a series of mountain ranges and valleys extending over 250 km along the eastern border with Mozambique. The width of these mountains varies from 30 to 60 km (Chikodzi 2018). The Eastern Highlands can be divided into three peaks: Nyanga Mountains to the north where the highest peak is found, Vumba Mountains in the middle where the lowest of the three peaks is found and the Chimanimani Mountains which have the second highest peak in Zimbabwe.

Zimbabwe is classified into five distinct agro-ecological regions, and the Eastern Highlands are in region I where specialised and diversified farming can take place. The area also receives the highest amounts of rainfall in the country—more than 1000 mm/year for areas below 1700 m in altitude. For areas at higher altitude, the base rainfall is 900 mm/year, and these areas have no clear-cut dry season (Chikodzi 2018). The temperatures are comparatively low, rainfall is more effective and allows forest estates, fruit production and intensive livestock production. In places free from frost, plantations of crops such as tea, macadamia and coffee can be planted. The area stretches from Mahenye at the Save-Runde confluence to Nyanga in the north.

14.3.1 Flood-risk Maps

Two sources used to derive the flood-risk maps were the height above channel base (HACB) for the area, which determines the level of flood hazard, and also information of flood presence or absence from historical occurrence data. The HACB of the area was derived from subtracting the altitude of the nearest streams (taken at the stream base) from the altitude on the 30 × 30 m Shuttle Radar Topography Mission (STRM) digital elevation model (DEM) in a GIS. The channel bed level itself will have an HACB of 0, and on surrounding landscape the height increases relative to its height from the channel base. The HACB is a more robust hydrologically relevant parameter to use in flood modelling compared to a DEM (Quinn et al. 2019).

Once the HACB of the area was derived, the next stage was to extract in a GIS the topographical characteristics (HACB) of places that had flood presence and flood absence. A total of 30 flood present and 30 flood absence training samples were used in the spatial logistic regression. Spatial logistic regression is a type of stochastic model involving regression analysis where the dependent variable is a dummy variable (coded 0, 1). It was used to build a mathematical function in which places with a higher probability of flooding had a pixel value closer to 1 and those with the least probability of being flooded had a pixel value closer to 0. Logistic regression was chosen because it only requires the absence and presence data to derive the probability of flooding occurring (Dai and Lee 2002) and has been used in similar studies before (Pourghasemi et al. 2012; Bui et al. 2015). The binary logistic regression takes the following form:

$$P = EXP\left(a + bx / 1 + \exp\left(a + bx\right)\right)$$

where p = Probability of flooding; a = Beta value (β) for the constant; b = Beta value (β) for the HACB; x = Environmental variable (HACB).

In building the flood probability map in a GIS after running the logistic variables on the HACB map, three steps were taken:

1. With the logistic response function or logit function, the continuous predictors were mapped to a function (the logit) of the response variable, which is also continuous.
2. The logit was converted into odds. This is easy because logit is nothing but the logarithm of odds of the response variable.
3. Once the odds are known, the probability score can be calculated using the formula:

$$\text{Probability} = \text{Odds} / \left(1 + \text{Odds}\right)$$

Gumindoga et al. (2014) argue that for the purpose of communicating flood danger, maps showing flood hazards need to be understood by all stakeholders, including decision makers, hydrologists and the community. Therefore, the flood hazard map was generated through the classification of the flood probability map into three classes where 0–0.5 probability was deemed as low risk, 0.5–0.85 probability was deemed to be moderate risk and 0.85–1 was deemed to be high risk. The data processing stages explained above are summarised in Fig. 14.2.

14.3.2 Validation

Firstly, the accuracy of the flood-risk map produced was assessed through the training springs correctly predicted. The flood-risk map was checked by overlaying it with the training flood presence and flood absence map and then calculating the predictive accuracy. Secondly, of the 30 samples each for flood presence and flood absence used in the study, 25 of each were used in the flood model calibration as the training samples and the remaining 5 each were used as test samples for model validation. The final flood-risk map was checked by overlaying it with the test samples. A flood hotspot was considered a correct prediction when at least part of the test spring present sample was located within a high probability value, a probability of 0.5 being used as the cut-off value (Dai and Lee 2002). Otherwise, the prediction was considered to be wrong.

Fig. 14.2 Stages in deriving the flood hazard map. Source: Authors

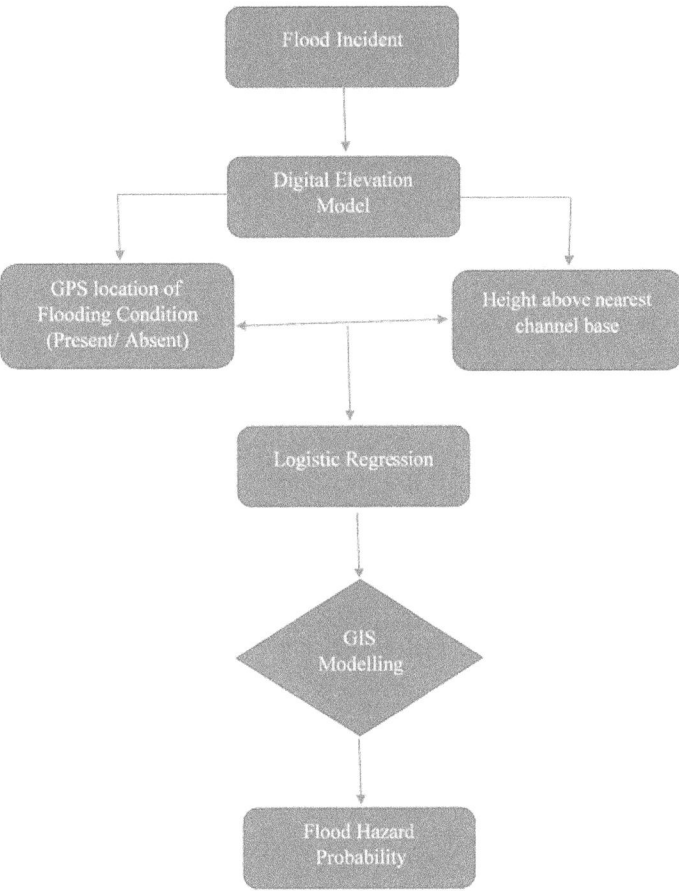

14.4 Findings and Discussions

Figure 14.3 shows the probability of floods occurring at different places of the study area as derived from the logistic regression. Note that the areas of least HACB are the most vulnerable to flooding, regardless of altitude. This means that places at high altitude are equally at risk of flooding as low-lying areas, as long as their HACB is low. During tropical cyclone Idai, for example, places such as Ngangu, the Village, Charleswood and Rathmore Estates in Chimanimani were flooded even though they are located at a relatively high altitude. In logistic regression there was a significant negative relationship between the HACB and flood occurrence probability, $B = -0.09$, $p = 0.001$. This means that as the HACB increased in the study area, the risk of a place flooding decreased. Franci et al. (2015) also observed in their study of floods that flood presence or flood absence was to a large extent determined by the height above channel base of a particular place, which is hydrologically more robust compared to a simple digital elevation model.

The next stage is to show the classified risk of flooding within the study area. The noted hotspots of flooding are not necessarily areas that have experienced widespread flooding in the past, but they show similar topographical characteristics to places that have flooded before (Fig. 14.4). This implies that the places have a high risk of flooding when subjected to similar amounts of precipitation and conditions to those that have flooded before. It is also not advisable for com-

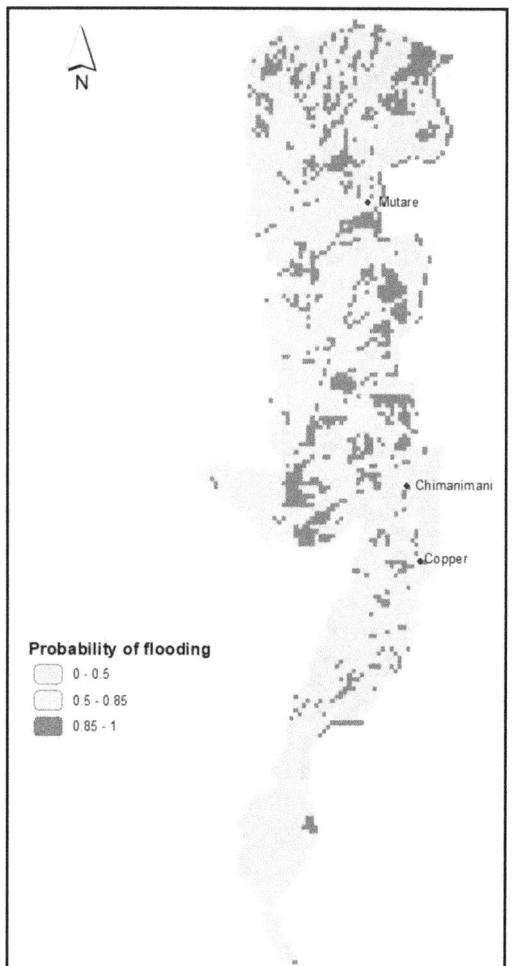

Fig. 14.3 Probability of flood occurrence in the eastern catchments of Zimbabwe. Source: Authors

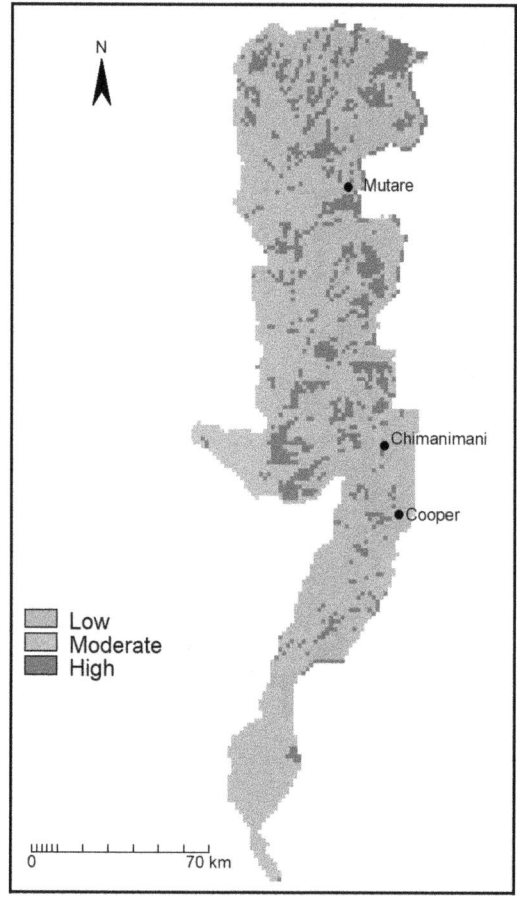

Fig. 14.4 Classes of flood risk in the study area. Source: Authors

munities to settle within these areas because they will be exposed to the risk of flooding and its negative consequences, such as loss of life, injury, damage to property and destruction of livelihoods. A field visit to some of the areas flooded during tropical cyclone Idai in Chimanimani and Chipinge showed that all the properties flooded were not well sited and were located in places with a high risk of flooding, such as waterways or valleys with very low height above channel base.

It was also observed in the study area that places of similar flood risk sometimes experienced different magnitudes and extent of damage. This is attributed mainly to differences in

land use, land cover type and the state of the environment in a particular area. It was observed that the deforested and degraded environments suffered more damage compared to healthy and pristine environments. Particular areas where bush fires were common and there was a lack of natural vegetation with deep roots and were either bare or characterised by second-generation vegetation of shrubs and grasslands were the hardest hit. Another localised factor that played a part in determining the extent of damage during the flooding is slope angle. Most areas of steep slopes above 60° experienced a combination of both flooding and landslides along natural waterways that were not so prominent before cyclone Idai. This also applied to places of recent timber harvest and where loosening of the soil was being

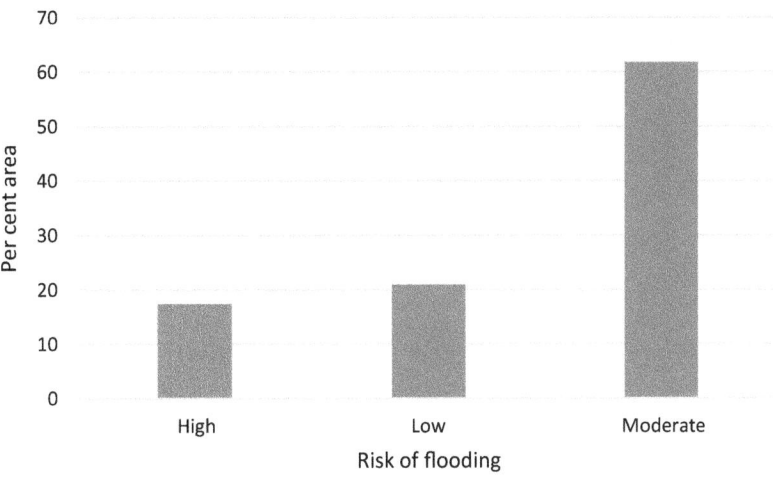

Fig. 14.5 Area of different flood risks. Source: Authors

Table 14.1 Predictive accuracy of the logistic model

Observed		Predicted		
		Flood		Percentage correct
		Absence	Presence	
	Flood absence	18	7	72.0
	Flood presence	6	22	78.6
Overall predictive percentage				75.5

Source: Authors

done due to farming on relatively shallow soil. On observation, most of the places at a high risk of flooding had high stream densities, especially in high altitude areas, or were located on flood plains or valleys.

Close to 62% of the areas in the study catchments are moderately vulnerable to flooding with a 50–85% risk of flooding (Fig. 14.5). In these zones, flooding does not happen that frequently, but may still occur when rainfall amounts reach certain thresholds or where there are localised risk factors such as land degradation and poor land use designs. This highlights the importance of carrying out systematic flood monitoring in order to issue early flood warnings in these areas, which are not all flood safe. The flood-safe zones cover an area just over 20%, with the probability of these areas flooding being extremely low, and they can be considered evacuation zones for those

that will have been affected by the floods in the high flood-risk zones. Close to 18% of the study area is zoned as being at high risk of flooding. In these areas flooding has a high risk of occurrence after even isolated heavy rains and as such must be avoided in terms of human settlements. The danger of flooding in these areas gives a compelling reason for an effective early warning system to save lives and livelihood assets.

The accuracy of the logistic model in predicting the occurrence of floods in the study catchments is presented in Table 14.1. The predictive accuracy was made using a cut value of 0.5, meaning that at least 50% or more of an area must be within a certain flood probability zone for it to be an accurate prediction. Note that 75.5% of the times, the model can predict the areas at risk of flooding and those safe from flooding in the study area using the HACB and historical occurrence; therefore it can be regarded as a reliable tool in flood management in the area. As shown in Table 14.1, the model is more robust in predicting flood present areas (78.6%) compared to flood absent areas (72%). The remaining unaccounted percentage could be due to the other factors not being analysed in the study and most likely operating at very localised scales.

Based on the household mapping done in the study area for the purposes of the 2012 population census, as shown in Fig. 14.6, the majority of housing units, i.e. over 25,000, are located in moderate flood-risk areas. In Fig. 14.6 it can also

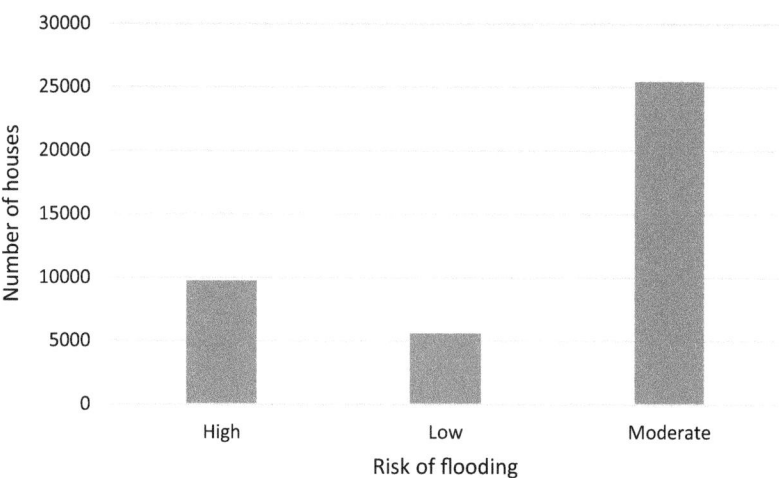

Fig. 14.6 Number of housing units and flood risk. Source: Authors

be observed that close to 10,000 houses are located in high flood-risk areas compared to just over 5000 that are in flood-safe zones. There are therefore more people at risk of flooding in the study area than those who are flood safe. Considering the fact that these areas are directly located in a cyclone zone and that there are more cyclones reaching the country than in the past, urgent steps need to be taken to reduce the vulnerability of the population in that area. Places categorised as high risk in the study also coincide with places known to be flood hotspots such as the Rusitu, Honde and Burma Valleys.

Other studies done in Zimbabwe have also confirmed that a large percentage of households at risk of flooding are located along the floodplains and waterways (Gumindoga et al. 2014). The risk is magnified by human activities such as deforestation and other forms of land degradation (Madamombe 2004). Other places have also been zoned as at high risk of flooding in Zimbabwe using different geospatial tools. These include Muzarabani, Tsholotsho, Gwayi Valley, Chikwarakwara and lower Runde (Pawaringira 2008; Gumindoga et al. 2014; Chingombe et al. 2015). The study also used presence or absence of data on previous occurrences of floods. This

data was obtained through a participatory mapping exercise that was done together with local community members. Chingombe et al. (2015) also underscore the importance of community contribution to flood hazard mapping and further argue that their participation fits into all segments of the disaster management cycle. Local community participation is best suited for use in areas that lack accurate data required for hazard modelling. This situation typifies most of the areas that were studied in this chapter. Khaing et al. (2019) note that flood hazard maps produced from GIS-based modelling or any form of earth observation are of no value if they are not simplified enough to be used by the at-risk communities and the policy makers. This bring to the fore the need to translate the produced maps into a product that is easier to understand. This may include impact-based focusing or a contingent flood hazard management plan for the noted hazard zones.

Given that a significant portion of the study area is at risk of flooding, Table 14.2 shows a contingent flood management plan formulated for the area.

Next, a summary is given of the findings from the study and recommendations are made.

Table 14.2 Flood hazard contingent plan for the studied areas

Action	Activities	Responsible institutions
Flood forecasting, run-off observation networks Information systems, impact-based forecasting	Modernise observation network and methods Regularly analyse obtained data Model for different forms of floods and causes of flooding Highlight exactly the values to be damaged Adopt impact-based forecasting.	Zimbabwe National Water Authority (ZINWA) Meteorological Services Department of Zimbabwe
Modelling, mapping and classification of flood-risk areas	Prepare micro-scale flood hazard maps Map flood-safe areas as evacuation zones	ZINWA, Ministry of Local Government, local authorities, academia, local communities
Flood-risk vulnerability assessment	Periodically revise preparedness and response plans Undertake surveys of vulnerability that cover social, economic and ecological dimensions Determine change in vulnerability and risk under various climate change scenarios Join and activate the space and major disasters charter	ZINWA, academia, local authorities, Department of Civil Protection, local NGOs and other development partners
Dissemination of warnings, data and information	Provide timely access to early warning and flood hazard risk information to communities Promote reliable platforms and ICT for data and information sharing among national, district and local-level stakeholders Provide regular updates to people in areas at risk	Community radio stations, social media, telecommunication companies, local traditional leadership, local authorities, local schools and religious leaders
Multi-purpose flood shelters	Construct/identify safe buildings and sites to serve as temporary shelters for people and livestock evacuated from at-risk places	Department of Civil Protection, local authorities, traditional leadership, provincial authorities
Flood-resilient construction	Implement flood-resistant construction, strengthening and retrofitting lifeline structures and critical infrastructure Build back better after flood disasters	Ministry of Local Government, Ministry of Roads, all line ministries, local authorities
Mock drills	Promote the planning and execution of emergency drills by all at-risk communities	Emergency services, government ministries, local communities, local and district authorities
Empowering women, marginalised and vulnerable persons	Incorporate gender-sensitive and equitable approaches in capacity development, covering all aspects of early warning systems	Woman advocacy groups, local NGOs, local authorities
Community-basedearly warning systems	Strengthen ability of communities to manage and respond to flood hazard warnings Training for youth, local community organisations on various aspects of early warning systems	Local NGOs, Ministry of Youth and Women's Affairs, local authorities, traditional leadership, Department of Civil Protection

14.5 Conclusion, Recommendations and Flood-risk Management Contingent Plan

Most of the areas at high risk of flooding have very low height above the nearest channel base values, combined with high stream density, especially on the high elevated areas, and also the presence of low-lying valleys. Close to 62% of the study area is located in places at moderate risk of flooding. Close to 20% of the study area is flood safe, with 18% of the land being at high risk of flooding. Over 25,000 housing units are settled in areas of moderate flood risk where floods occur intermittently, another 10,000 houses are located in areas of high flood risk and close to 5000 houses are within the flood-safe zones. The hotspot approach to flood modelling gives a general indication of the likelihood of floods occurring within a given area; this can facilitate informed land use planning. However, detailed studies of localised areas within areas of different flood-risk probability need to be done since factors operating at micro level are also important in influencing the flood risk of an area.

In the light of the findings, the following recommendations can be made:

Flood hazard models developed in GIS need to be simplified enough for novice users to extract information and support decision-making during flood emergencies and for planning purposes.

Areas noted to be at high risk of flooding need to be avoided in terms of human settlements, and communities at extreme high risk, such as Ngangu, Rathmore, Cooper and Machongwe, need to be relocated to safer places. There must also be a harmonisation of land and water policies since the two are inseparable and intertwined. Land cover/land use change must be monitored, and there is a need to educate communities at all levels on the link between land degradation and the risk of flooding.

All-inclusive and effective early warning system structures need to be formulated, especially in the areas at high risk of flooding. Drills on evacuation and rescue need to be done continually, even during seasons when no floods are occurring. The legislation should provide for capacity in disaster risk reduction at local community level with frequent capacity building for community structures to continuously improve knowledge, skills and competencies, to implement, monitor and coordinate flood mitigation measures and management activities in the community.

Hydro-meteorological monitoring and observation need to be strengthened by adopting new and modern technology that increases the number of observatories for monitoring meteorological and hydrological conditions. These need to be regularly maintained to obtain near real-time data to facilitate accurate forecasting. There is also a need for research on traditional early warning systems and how they can be synchronised with the scientific early warning systems. These then need to be documented and publicised to the relevant communities.

References

Albano, R., Mancusi, L., Sole, A., Adamowski, J. (2015). Collaborative strategies for sustainable EU flood risk management: FOSS and geospatial tools – challenges and opportunities for operative risk management. ISPRS International Journal of GeoInformation, 4(4), pp. 2704-2727.

Bui, D.T., Tuan, T.A., Klempe, H., Pradhan, B., Revhaug, I. (2015). Spatial prediction models for shallow landslide hazards: a comparative assessment of the efficacy of support vector machines, artificial neural networks, kernel logistic regression, and logistic model tree. Landslides. https://doi.org/10.1007/s10346-015-0557-6.

Chadburn, O., Ocharan, T.J., Kenst, K. (2010). Cost benefit analysis for community based climate and disaster risk management: synthesis report. Retrieved from: https://www.preventionweb.net/files/14851_FinalCBASynthesisReportAugust2010.pdf. (Accessed 10 September 2019).

Chikodzi, D. (2018). Unusual waterscapes and precarious rural livelihoods: Occurrence, utilisation and conservation of springs in the Save Catchment, Zimbabwe. Thesis submitted to the University of the Western Cape, Cape Town.

Chikodzi, D., Tevera D., Mazvimavi, D. (2020). SDG 15 and Socioecological Sustainability: Spring Waterscapes and Rural Livelihoods in the Save Catchment of Zimbabwe. In G. Nhamo et al. (eds.), Scaling up SDGs Implementation,

Sustainable Development Goals Series, https://doi.org/10.1007/978-3-030-33216-7_4.

Chingombe W., Pedzisai, E., Manatsa, D., Mukwada, G., Taru, P. (2015). A participatory approach in GIS data collection for flood risk management, Muzarabani district, Zimbabwe. Arab J Geosci, 8:1029–1040.https://doi.org/10.1007/s12517-014-1265-6.

Dai, F.C., Lee, C.F. (2002). Landslide characteristics and slope instability modeling using GIS, Lantau Island, Hong Kong. Geomorphology 42, 213–228.

Deo, A., Ganer, D. (2014). Tropical Cyclone Activity Over the Indian Ocean in the Warmer Climate. Monitoring and Prediction of Tropical Cyclones in the Indian Ocean and Climate Change, Springer, pp. 72–80.

FEWSNET. (2000). Assessment of the Impact of Cyclone Eline (February 2000) on the Food, Agriculture and Natural Resource Sector in Zimbabwe. Retrieved from: https://fews.net/sites/default/files/documents/reports/1000050.pdf. (Accessed 9 September 2020).

Franci, F., Mandanici, E., Bitelli, G. (2015). Remote sensing analysis for flood risk management in urban sprawl contexts. Geomatics, Natural Hazards and Risk, 6(5-7): 583-599.

Gumindoga, W., Chikodzi, D., Rwasoka, D., Mutowo, G., Togarepi, S., Dube T. (2014). The spatio-temporal variation of the 2014 Tokwe-Mukosi floods: a GIS and Remote Sensing based approach. Journal of Science, Engineering and Technology, 1(2), pp 1-10.

Jaha, A. K., Bloch, R., Lamond, J. (2012). Cities and flooding: a guide to integrated urban flood risk management for the 21st century. World Bank Publications. pp 496-573.

Keys, C., Angus, D., Benning, N. (1996). Developing our expertise in the management of flooding: some recent initiatives. Australian Journal of Emergency Management, 11 (4), 38 – 43.

Khaing, Z.M., Zhang, K., Sawano, H., Shrestha, B.B., Sayama T., Nakamura, K. (2019). Flood hazard mapping and assessment in data-scarce Nyaungdon area, Myanmar. PLoS ONE 14(11): e0224558. https://doi.org/10.1371/journal.pone.0224558.

Kwang, C., Osei, E.M. (2017). Accra flood modelling through application of geographic information systems (GIS), remote sensing techniques and analytical hierarchy process. J Remote Sensing & GIS 6: 191. https://doi.org/10.4172/2469-4134.1000191.

Madamombe, E.K., (2004). Zimbabwe: Flood Management Practices Selected Flood Prone Areas: Zambezi Basin. World Meteorological Organisation,2004. Associated Programme Flood Management Technical Document No. 1, 2nd Edition. http://www.apfm.info/pdf/case_studies/zimbabwe.pdf. (Accessed 23 November 2019).

Mahmood, A. (2017). Flood hazard mapping in integrated flood risk management: importance and problems associated to Pakistan. Retrieved from: https://ssrn.com/abstract=3057194. (Accessed 15 May 2020).

Moe, I.R., Rizaldi, A., Farid, M., Moerwanto, A.S., Kuntoro, A.A. (2018). The use of rapid assessment for flood hazard map development in upper Citarum river basin. MATEC Web of Conferences 229, https://doi.org/10.1051/matecconf/2018229.

Nelson, S.A. (2008). Flood Hazards, Prediction and Human Interventions. Tulane University.

NOAA. (2009). Flood losses: compilation of flood loss statistics. NOAA climate research centre. www.weather.gov.oh/hic/flood_stats/Flood_loss_time_series.shtml. (Assessed 20 November 2019).

Pawaringira, R. (2008). Flood hazard modelling in Tsholotsho district, Zimbabwe. Thesis submitted to the University of Zimbabwe.

Pourghasemi, H.R., Gokceoglu, C., Pradhan, B. (2012). Application of fuzzy logic and analytical hierarchy process (AHP) to landslide susceptibility mapping at araz Watershed, Iran. Natural Hazards, 63(2): 965-996.

Quinn, N., Bates, P. D., Neal, J., Smith, A., Wing, O., Sampson, C., Smith, J., Heffernan, J. (2019). The spatial dependence of flood hazard and risk in the United States. Water Resources Research, 55, 1890–1911. https://doi.org/10.1029/2018WR024205.

Santillan, J.R., Marqueso, J.T., Makinano-Santillan, M., Serviano, J.L. (2016). Beyond flood hazard maps: detailed flood characterization with remote sensing, GIS and 2D modelling. The International Archives of the Photogrammetry, Remote Sensing and Spatial Information Sciences, Volume XLII-4/W1. https://doi.org/10.5194/isprs-archives-XLII-4-W1-315-2016.

Sidek, L.M., Rostam, N.E., Hidayah, B., Roseli, Z.A., Majid, W.H.A.W.A., Zahari, N.Z., Salleh, S.H.M., Ahmad, R.D.R., Ahmad, M.N. (2016). Hydrology Analysis and Modelling for Klang River Basin Flood Hazard Map. IOP Conf. Series: Earth and Environmental Science 32, 012069 https://doi.org/10.1088/1755-1315/32/1/012069.

Twumasi, N.Y.D., Shao, Z., Orhan, A. (2019). Remote sensing and GIS methods in urban disaster monitoring and management – an overview. International Journal of Trend in Scientific Research and Development. 3(4), pp. 918-926.

UNISDR and CRED (United Nations Office for Disaster Risk Reduction; Centre for Research on the Epidemiology of Disasters). (2018). Economic losses, poverty and disasters: 1998–2017. Retrieved from: http://www.preventionweb.net/files/61119_credeconomiclosses.pdf (Accessed 10 May 2020).

United Nations (2015). United Nations. Sendai Framework for Disaster Risk Reduction 2015–2030. UNISDR. Geneva.

Part VI

Conclusion and Policy Recommendations

Tropical Cyclone Idai: Summary of Key Findings and Recommendations to Enhance the B4 Model in Zimbabwe

Abstract

This chapter presents the summary of key findings and recommendations to enhance the building and building back better (B4) model in disaster risk reduction (DRR) and management. The chapter will draw from a framework that focuses on the key thematic areas addressed in the book that include use and contestations of earth observation technologies in DRR, the naming of tropical cyclones, impacts on energy infrastructure, cyclone-induced floods in the midst of the drought and domestic and irrigation water supply systems in the aftermath of Tropical Cyclone Idai. Additional aspects covered in the chapter focus on ethical philanthropy and social responsibility, human rights dilemmas in DRR, religious engagements and the potential of dark tourism in the aftermath of natural disasters and hazards such as Tropical Cyclone Idai. The chapter can also be used as a policy brief by key stakeholders in DRR that include the government, development partners, civil society, academia, labour and industry.

Keywords

BBB · DRR · Ethics · Energy · Water and sanitation · Philanthropy · SDGs

15.1 Introduction

This chapter begins with a recap of the DRR institutional and policy framework in Zimbabwe. This will be followed by sections presenting the key findings and recommendations. The chapter draws from the DRR and management cycle, which includes early warning systems, search and rescue, relief and recovery. Perspectives on cross-cutting matters in dealing with Tropical Cyclone Idai are considered. Although not fully discussed in this book, two other cyclones hit Chimanimani in the 2020/2021 rainy season, namely Tropical Cyclone Chalane which made landfall in Mozambique and hit Zimbabwe in December 2020 and January 2021, as well as Tropical Cyclone Eloise which also made landfall in Mozambique and hit Zimbabwe in January 2021. This brought additional pain as those residing in temporary shelter, particularly the tents in Chimanimani, had to be relocated to other temporary locations twice within a period of less than 2 months. Furthermore, in the midst of Tropical Cyclone Idai, the COVID-19 pandemic also hit, with extended severe impacts into 2021 as the government had to impose two hard lockdowns, one in March/April 2020 and the other in January/February 2021 following the second wave that also took the lives of many high-profile individuals and politicians in the country. All these disasters have already retarded progress towards the 2030 Agenda for Sustainable Development and the achievement of its 17 SDGs as outlined in the book.

15.2   A Recap of the DRR Institutional Framework in Zimbabwe

During colonial times, the management of disasters in Zimbabwe was referred to as civil defence. After independence in 1980, there was a continuation of the status quo with DRR issues being administered under the provisions of the Civil Defence Act of 1982. Manyena et al. (2013) observe that the concept of civil defence had been crafted by design to serve the interests and inclinations of the minority white population during the war of liberation. The Act embraced a command and control structure derived from military systems; hence the largely top-down approach to DRR which had more clarity in terms of responsibility at national level but not as clear structurally at local levels.

DRR in Zimbabwe was then transformed from civil defence to civil protection after the enactment of the Civil Protection Act (Chapter 10:06) of 1989. The Act was later revised in 1992 and 2001 (GoZ 2001). The Department of Civil Protection (DCP) administers the Civil Protection Act Chapter 10:06 of 1989. The department falls directly under the Minister responsible for Local Government Rural and Urban Development. The key result area for the department is disaster risk management, which involves a wide spectrum of activities such as prevention/mitigation of disaster risks, preparedness planning, timely early warning and response to rehabilitate affected elements. The mandate of the DCP is overall coordination of DRR actors drawn from the public and private sectors, including development partners such as the United Nations Development Programme (UNDP), IOM, Red Cross Society and other NGOs (Mavhura 2015). The DCP had a staff complement of seven officers at the head office as of December 2019. The management structure comprises a Director, Executive Assistant, Deputy Director and four administra-

tive officers responsible for Administration and Liaison, Operations, Research and Training. Provincial and district development coordinators facilitate the work of the DCP at their respective levels. Given the scope of operations of the department, the human resources afforded to it are simply not enough to administer their mandate adequately. Most key informants observed that there was a need to increase the staff complement of the department to have direct representation at both provincial and district levels. Others advocated for a standalone institution to be established by the government to spearhead the management and coordination of disasters in the country to do away with the bureaucracy associated with government departments.

The Civil Protection Act (Chapter 10:06) of 1989 also establishes a civil protection organisation (CPO) and provides for the operation of civil protection services in times of disasters. The CPO is a national platform made up of line ministries or departments, state enterprises, relevant private sector bodies and NGOs whose regular activities are related to DRR and community development. The Act directs every province and district to have responsibility for the protection and preservation of the lives and property of their citizens (GoZ 2001).

To discharge its mandate, the DCP is supported by the National Civil Protection Coordination Committee (NCPCC), consisting primarily of the Secretary for the Ministry of Health and Childcare, Commissioner General of Police, military commanders, Zimbabwe Red Cross Society, Director of Prison Services, Director of Civil Aviation and the fire brigade. The NCPCC can appoint subcommittees, whose membership is not clearly enshrined in the Act. The subcommittee members are in most cases drawn from government line ministries, quasi-governmental bodies and NGOs. These may include the national power utility ZESA and Zimbabwe National Water Authority and United Nations agencies such as UNICEF, OCHA, ILO and UNDP (Mavhura 2015). Most key informants observed that Zimbabwe's DRR structure was outdated and out of step with modern standards as enshrined in the Sendai Framework of

DRR. The most concerning observation was that most DRR interventions still remain heavily militarised, extremely top-down, with key decisions on actions to be taken being made at the highest levels of the central government and local structures mere implementers down the chain of command. This ignores a common best practice fact which sees the most effective development and emergency plans as being best developed at grassroots level and not at national or ministerial level as spelt out in section 11(5) of the Civil Protection Act of 1989.

This may also bring paralysis in the implementation especially of the hot phase of the disaster cycle at local level. There are reported cases of military personnel refusing to take orders from anyone except the military chain of command, even when working as a team with other stakeholders. In some cases, they do not feel obliged to act with the necessary speed without authorisation by the army commanders, thus critically delaying response to disaster (Chikoto 2004). The same also applies to the release of resources when many organisations are supposed to be working together but have lines drawn between them either due to institutional rivalries or jealousies. Key informants observed that producing an effective blueprint during a DRR incident is difficult if there are many centres of power and resources are constrained.

In the event of an emergency incident, a crisis committee can be initiated at provincial or district level to handle the situation pending further instruction from the national government. This is because outside the national level, there is limited financial capacity to deal with disasters. This has huge implications for the disaster victims who must wait longer for resources to be dispatched at national level. The crisis committee can also appoint subcommittees to deal effectively with the disaster. For Tropical Cyclone Idai, the Manicaland Provincial Civil Protection Committee established 17 subcommittees to promote efficiency and effectiveness: Admin and Finance; Search and Rescue; Relief and Psycho-Social Support; Transport and Logistics; Roads; Small and Medium Enterprises; Shelter; Health; Water and Sanitation; Weather; Agriculture;

Education; Industry and Commerce; Telecommunications; Power; Information and Human Resources.

The current Act covers mainly emergency and disaster response and recovery programmes, with the education and early warning components being very limited. Community participation in DRR projects is very limited in the disaster legislation. In addition, there are no guidelines for stakeholder involvement at grassroots level, especially multi-sectoral interventions. Although local authorities are important DRR players in their jurisdictional areas (Bang 2014), this Act is silent on their role in performing this function. Furthermore, the Act does not leave room for the involvement of traditional leadership in the prevention and mitigation of disasters. Yet these institutions are important players in these respects (Manyena et al. 2013).

Due to the noted gaps in the Civil Protection Act of 1989, the Zimbabwean parliament has since 2003 been trying to repeal the Act and replace it with the Emergency Preparedness and Disaster Management Act. The Emergency Preparedness and Disaster Management Bill was still at the time of writing in 2020 at the consultation stage and has been delayed many times due to the highly polarised Zimbabwean legislature. As observed by Madamombe (2004), the new Act would allow the formation of a standalone Emergency Preparedness and Disaster Management Authority. The Authority would be assigned with key activities such as the forecasting of and planning for emergencies at different geographical levels such as from local authority, district, provincial and national levels. The above stated tiers of government will be required to produce operational disaster emergency preparedness and response plans which would be activated when a disaster hit. The new Act must also make provision for setting up emergency planning and disaster management committees which will be composed of all key stakeholders, including research institutions, private sectors and NGOs. The function of these committees would be to ensure coordinated efforts by all stakeholders in times of disaster and to improve the country's preparedness and capacity to cope with disasters and threats to communities (Mavhura 2015).

15.3   Early Warning Systems

15.3.1 Key Findings

It emerged that there is a severe challenge regarding energy supply at the Zimbabwe Meteorological Services Department (MSD). When Tropical Cyclone Idai hit, there was an electricity blackout at some point, with no diesel to power the generators. This may have disrupted updates. In the current era of extreme weather events, the MSD is as much a national key point as is defence. It lacks modern equipment, such as radars, to assist in early warning predictions. For example all models run on precipitation indicating a maximum of 100 mm of rainfall within 24 h, yet as per both the official and fieldwork figures, this rose to levels between 400 mm and 2000 mm in 24–48 h during Tropical Cyclone Idai. This was equivalent to the normal rainfall of about one season in the study area. The satellite MSD weather stations across the country were reported not to be in proper functioning condition at the time. This included human resources as most of them were managed by a single officer. Under normal circumstances, the satellite weather stations are in a position to provide a more localised forecasting. The spatial distribution of MSD weather stations is still not within the 50 km radius as per the World Meteorological Organization. Early warning during the cycle was further complicated by limited experience of the Chimanimani District Civil Protection Unit that could not deliver timely information on the pending disaster.

There was also clear evidence of widened rivers and shallow river beds across the Chimanimani. This presents a flood risk in the future. Some river channels widened by hundreds of metres. The unique behaviour of the Nyahode River was revealed as noticeable over the years. The residents of Kopa clearly described the river as washing bridges away repeatedly over the years. From our observations, the river gains velocity from the drainage from the sloping and mountainous areas, as well as the huge volumes from many tributaries upstream of Kopa.

There was also a dummy normal flow of the Nyahode River during that fateful day at Kopa. An account by one of the Agritex extension officers who survived indicates that the water subsided temporarily, and he subsequently went to sleep and no longer monitored the river. The survivor was later woken up by his son when the water was at window level and doors could not open due to mud. It became evident later that the floodwater was temporarily held by logs and other deposits at the Nyahode bridge, which later washed away, resulting in a new river channel to the left. Residents were therefore up against the challenge of historical knowledge and experiences. Tropical Cyclone Eline of 2000 remained the benchmark, yet Tropical Cyclone Idai was different.

15.3.2 Key Recommendations

There remains an urgent need to provide an alternative source of energy in the form of solar for the MSD. The relevant government arms are encouraged to consider prioritising resources for the MSD. The equipment needs to be modernised, including a dedicated radar system. Considerations to link up the MSD with the High-Performance Computer under the Ministry of Higher and Tertiary Education, Science and Technology Development would help modelling and accurate predictions.

The first phase of radar installations can target high-risk and cyclone-prone provinces such as Manicaland. The MSD and the relevant ministry should consider amending their internal policies to cover 24-hour shifts during times like Tropical Cyclone Idai. This implies putting in place the necessary human resources and support, including transportation and on-site accommodation. Satellite MSD weather stations urgently need to be upgraded in terms of equipment and functionality. Furthermore, weather forecasting should be decentralised to provinces so that forecasts for regions are done by the people on the ground. Given that extreme weather events will continue, especially as shown by Tropical Cyclones Chalane and Eloise, the MSD and its line ministry should work towards mobilising resources to revive old weather stations that used to be over 60 in number at some point and were indicated to be down to 47 by the time of fieldwork in 2019.

Authorities can also capitalise on the high levels of mobile phone penetration in the country, including Chimanimani. However, network coverage in this mountainous area needs to be improved. The Chimanimani Rural District Council and partners should audit the risk of those still located close to these rivers as their flood risk is now higher. The communities along Nyahode River need to be on high alert during rainy seasons. This can be enhanced through cyclone risk awareness and capacity development. Overall, every resident of the village, ward, district, province and country should be reached in terms of DRR awareness raising.

15.4 Search and Rescue

15.4.1 Key Findings

It emerged that the outside search and rescue teams, including the army, arrived late (roughly 4–5 days later) in Chimanimani. This certainly increased the death toll. An explanation by both the government officials and other stakeholders was that bad weather prevented aircraft rescue missions, and all the access roads had been washed away. This was confirmed by many residents during fieldwork. The local communities ended up being the ones at the frontlines, rescuing the injured, retrieving bodies and burying them.

There was also a major outcry regarding how the deceased were buried as there were instances where more than one body was placed in one grave. The explanation given by those on the ground and the government's District Development Coordinator's office was plausible, namely that the bodies were at an advanced state of decomposition and that the entire community was overwhelmed by the event. However, a genuine request after Tropical Cyclone Idai had passed was for reburials to be done as there were instances where people of different tribes and

totems were buried together, which is an abomination under local cultural beliefs.

It also emerged that when Tropical Cyclone Idai struck, there were no on-the-ground DRR and management protocols in terms of handling the situation. For example dealing with the injured, homeless, dead and general documentation was problematic. Hospitals and clinics were overwhelmed by the event, aggravated by the lack of medicines and other basics such as bandages. The injured had to be transferred to nearby district and provincial hospitals in Mutambara and Mutare hundreds of kilometres away. There was also a challenge with insufficient and functioning mortuaries to keep the dead. Furthermore, there were no suitable helicopter landing locations across the district, and this delayed the search and rescue missions. Conflict and strife also emerged as the government was uncomfortable with an influx of outside personnel, which could result in security threats. This is an element that delayed the movement of both relief goods and human resources into the epicentre of the disaster.

15.4.2 Key Recommendations

The government, especially the District Development Coordinator's Office and the Chimanimani Rural District Council, need to partner with other key stakeholders to plan future search and rescue missions. Appropriate long-term protocols for border clearances of goods and visa issuance should be put in place. The government needs to consider standby rescue helicopters and, where possible, partner with other stakeholders that may have helicopters that can fly under conditions such as those witnessed during Tropical Cyclone Idai.

The government, the district council and traditional leaders further need to consider the request by the communities to properly identify and rebury the deceased according to their customs and beliefs. Reburials are not new to the country and communities in Chimanimani since witnessing and experiencing these after the liberation war. Revised protocols on how to handle disasters of the magnitude of Tropical Cyclone Idai are needed. The documentation of survivors and the displaced requires careful attention, and, where practical, this should be done as soon as appropriate for accountability and relief purposes.

We further propose that a clear and simplified DRR and management system, particularly an early warning system, be put in place from the district, through the ward to the village levels where communities sit. Village evacuation centres must be identified, including performing early warning drills to test the new systems. Where possible, sirens of different codes/sounds could be installed, and this should be nationwide. This recommendation seemed to have been scaling up after Tropical Cyclone Idai as ward-level evacuation points were identified for Tropical Cyclone Chalane in Chipinge by the Chipinge Rural District Council (Table 15.1). Wards are

the second-lowest administrative tier in Zimbabwe, with the village unit being the lowest.

Excluded from Table 15.1 are evacuation centres for Chipinge Town area, namely Gaza Community Hall, Chipinge Country Club and Chipinge Junior Primary School. In addition, seven other centres were also identified for the repositioning of resources: Tanganda Primary School, Government Complex, Manzvire Rescue Centre, Mahenye District Development Fund Lodge, Mt Selinda Primary School, Chikore High School and Junction Gate Clinic.

There was also some talk that the government had plans to build a district hospital with the right capacity. This was confirmed by the Chimanimani Member of Parliament. This proposal needs to be followed up. In addition, basics such as bandages could be stocked across the district for strategic

Table 15.1 Identified Ward Evacuation Centres

Ward	Evacuation Centre	Ward	Evacuation Centre	Ward	Evacuation Centre
1	Chipinda Secondary School Maunganidze Information Centre	11	Madziwa Secondary School	21	Tuzuka Secondary School Rimbi High School
2	Samhutsa Secondary School Ngaone Secondary School	12	Chiriga Primary School Nyaututu Primary School	22	Manzvire Red Cross Rescue Centre Matezwa Primary School
3	Manesa Training Centre	13	Mapote Primary School Chipinge Vocational Training Centre	23	Zamuchiya Primary School
4	Tanganda Primary School Musani Secondary School	14	Mapungwana Primary School Tamandai Primary School	24	Checheche Primary School Madhuku Primary School
5	Chipangayi Secondary School	15	Muzite Primary School Chief Gwenzi Homestead	25	Rimai Primary School
6	Makandi Banana Company Sterksroom Secondary School	16	Kondo Primary School	26	Chisumbanje Primary School Green Fuel Estate
7	Clearwater Primary School	17	Munoirirwa Primary School Murenje Primary School Chinaa Secondary School	27	Vheneka Primary School
8	Simudza Primary School Mugiyo Primary School	18	Tafara Primary School Musirizwi Primary School	28	Garahwa Primary School Chinyamukwakwa Primary School
9	Ndiadzo Primary School Rusitu Valley Primary School Paidamoyo Primary School Mafumise Primary School	19	Emerald Hill Primary School Beacon Hill Primary School Mt Selinda High School	28	Maparadze Primary School
10	Chipinge Prison	20	Chibuwe Primary School Mbeure Primary School	30	Mahenye Primary School Mahenye Secondary School

Source: Authors, Fieldwork 2021

recalling in disaster situations of the magnitude of Cyclone Idai.

15.5 Response and Relief

15.5.1 Key Findings

The displaced communities complained that the wrong beneficiaries were on the lists, with those in real need being sidelined. The alleged wrong people on the lists included landlords whose houses were swept away, yet lodgers stayed in them, friends and relatives of headmen and others that did not face the challenges of Tropical Cyclone Idai. There was also evidence of expired food, some with best-before dates dating back to 2013 and 2016, with some dates written in other languages such as Chinese. It emerged that the Auditor General had dispatched a team to investigate the allegations. The leading role of NGOs became integral during the disaster. There was also major involvement of the private sector, including the housing of victims in hotels such as the Chimanimani Hotel that accommodated hundreds for about 2 months.

One of the main issues that was raised during relief was increased illegal gold mining. There was, in fact, a gold rush that also led to extensive degradation of biodiversity and the natural Chimanimani (Mawenje) Mountains aesthetics. Illegal gold miners came from far and wide, colonising the gold hotspots such as the Chimanimani Mountains, Nyabamba, Tarka Forest, Chimanimani National Park and Blocky. Water pollution, dying aquatic life, damaged cultural and traditional sites, as well as extensive deforestation was also evident. Illegal gold mining and its associated activities are a real threat to the tourism industry as many tourists hike along trails in the Chimanimani Mountains.

15.5.2 Key Recommendations

Given the situation where the community became the first port of call in terms of relief, a mechanism to replenish depleted food stocks for the community needs to be put in place for similar future events. This could entail a programme where relief in terms of food goes to all the community members, including the victims at first and then to the indirect victims subsequently. Relief lists must be captured early as this avoids clean-up to remove names of identified false beneficiaries later. This could be assisted by tapping into the rich experiences of the NGOs that were involved and continue to get involved during the disaster.

Concerning illegal mining, particularly gold mining, we recommend that the central government, in partnership with the Chimanimani Rural District Council, find lasting solutions to the problem. These solutions may include scaling-up patrols and rooting out corruption by security officials in the ZimParks. There is also a need to capacitate and educate the illegal miners to protect the environment and use a carrot and stick approach to address the situation. Matters of illegal gold mining were the same during recovery.

15.6 Recovery (Including Rehabilitation and Reformation)

15.6.1 Key Findings

At the time of fieldwork in 2019, some school roofs remained open, with some teachers staying in tents. Those displaced were also in tents as of February 2021, almost marking two full years of being there. Internally displaced people from the Ntamatanda and Garikai temporary camps and in tents were further relocated twice to Mutambara and Nyanyadzi, which are about 70 and 100 km away, respectively, during Tropical Cyclones Chalane and Eloise in 2020 and 2021, respectively. This scenario does not reflect the zeal associated with the proposed B4 Model in this book.

Other matters observed in the recovery phase include the fact that many civil servants (excluding teachers) that were involved since Tropical Cyclone Idai were still on duty as at the time of fieldwork in October 2019. These workers expressed the desire to have special leave for a considerable period granted to them. There were also noticeable transfers out by some civil servants, and this had resulted in burnout from the staff shortage. Delays in building back better housing resulted in victims remaining in tents for long periods, with the major challenge highlighted as the lack of technical capacity and/or delay in employing land surveyors for the newly identified settlement areas. There were also critical matters regarding what to do with the remaining vulnerable settlements.

A number of development partners in the affected areas were working with specific targets and timelines. This probably had unintended consequences in that some projects were bound to be left unfinished. Such projects include water and sanitation, especially. Unfortunately, the Rural District Council was also hit hard by a reduced budget due to eroded rates from paying residents. Many of the paying Ngangu and Kopa residents were displaced, yet they were among the key revenue generators for the council. The council had to, at some point, suspend revenue collection, and this was likely to continue, having a severe negative impact.

The documentation and documents of displaced individuals and households was another major issue that emerged, especially in Kopa. Overall, the following documents were needed (the displaced in Ngangu indicated that they were better covered): national IDs, death certificates, birth certificates, driver's licences, marriage certificates, all educational certificates, etc. While the ministry responsible for national documentation (Home Affairs) was said to be working on a statutory instrument to reduce the number of years legally required to declare missing persons as deceased, which is apparently 5 years, the process took longer for those affected. There were also issues of psycho-social support that was mainly short term. There were still some communities that had not received this support at the time of fieldwork, especially civil servants outside the education sector and those in commercial farming areas.

It emerged that most of the settlements severely impacted by Tropical Cyclone Idai were located in extremely vulnerable places, such as slopes, flood plains (Kopa) and waterways. In fact, residents and evidence point to the fact that old locations in the Ngangu area were not as severely damaged as new settlements. Some residents alleged corruption by the council officials that led to stands being sold in these vulnerable areas. Furthermore, some houses were poorly constructed in Kopa, with no compacted materials used in foundations and flooring. Bricks were simply laid on top of alluvial soils. Given the deep alluvial deposits in Kopa, there could be a chance that some foundations are not up to standard. However, there are some houses that have remained, and these could be good cases in terms of their resilience.

There was also extensive damage to the energy infrastructure, especially transition poles and lines. Three transformers and two small-scale hydropower stations were either completely washed away or damaged, and the third small-scale hydropower station was partially damaged. The Zimbabwe Electricity Distribution Authority was, however, praised for the repair and reconstruction work, having being done within about a month.

Roads were extensively damaged, with bridges washed away and some roads permanently damaged. At the time of fieldwork, there were visible signs of work progressing, although concern was mounting as the rains had already started. In January 2021, there was also an incident where one of the major roads repaired after Tropical Cyclone Idai just after Charter Sawmill collapsed, developing a huge sinkhole that resulted in a vehicle falling in and one life lost on 30 January 2021. Describing the situation, one key informant had this to say on a WhatsApp communication on 31 January 2021:

> [S]ince Tropical Cyclone Idai and the effected repair road works, there was no structural inspection carried out to determine whether or not the road substructure was still intact. Instead there was a rush to get timber and other produce out so there was overloading plus neglect of critical assessments. This is the same malaise we have witnessed with key infrastructure across the country ... roads, water supply, electricity and telephone network, rail transport, hospitals, schools etc...

On investigating further, it emerged that the collapsed road had been built by a known local contractor whose name was supplied. The key informant went further to paint a picture of the difficulty of recovery based on what they considered an abnormal rain season. In their view:

> The Eastern part of Chimanimani District has received over 800mm of rain since the start of the 2020/2021 rain season including rain from Topical Cyclone Chalane and Tropical Cyclone Eloise in the past three weeks. Additionally, the Intertropical Convergence Zone has been sitting over Mozambique, Malawi, Zambia and Zimbabwe depositing more rain. As such, the ground is saturated. Rivers are flooded and many water bodies, especially dams are spilling and this is January, which in the past two decades and more would be normally dry.

Agricultural land, cash crop plantations such as bananas, livestock and irrigation infrastructure were washed away. School blocks and health facilities were also extensively damaged in some instances, as were water and sanitation facilities, with many being washed away. However, the repair and reconstruction work was progressing well during the time of fieldwork. One organisation which we interviewed had worked on institutional water and sanitation projects, as well as repairing 16 boreholes, with five others still outstanding as per its targets.

Telecommunications repair and recovery was also going on well, although TelOne, a government entity, had not managed to reconnect 5 km of cable linking Ngangu Township and Chimanimani CBD. Some of the tourism, heritage and cultural sites were also severely damaged and/or washed away. These included popular resorts such as Bridal Veil and Tessa's Pool. There was also a negative impact on biodiversity and wildlife. With so much stone and rock exposed by Tropical Cyclone Idai, there was evidence of artisanal quarrying, which has the potential to lead to further land degradation and gully formation.

15.6.2 Key Recommendations

School infrastructure destroyed during Tropical Cyclone Eline had still not been repaired at the time of fieldwork. Priority must be given to build back better destroyed schools (classrooms, dormitories and teachers' accommodation). Special leave with pay should also be granted to some civil servants that were involved since Tropical Cyclone Idai hit Chimanimani in March 2019. Such special leave may be granted by the responsible authorities on a rotational basis to minimise work flow disruptions and gradually be included in human resources and other policies. Vacant posts created after Tropical Cyclone Idai need to be filled. Some vacancies that were identified include those in the Departments of Education and Agriculture.

There will be a need to climate-proof settlements that may be deemed to be relatively low risk, probably through the construction of floodwater barriers. As it was noticed that portions of roads with good stormwater drains were spared by the cyclone, a recommendation is that road repairs and new roads be supervised to ensure that stormwater drains are placed at appropriate intervals. Since the Chimanimani area is prone to tropical cyclones, the council should put in place a development partner projects auditing protocol to ensure that work is not left partially done. This will minimise the rush into potential short-term projects. Systematic handovers of unfinished projects by development partners should be a long-term protocol. This also applies to the orderly way of spreading development partners across the affected areas and avoiding duplicity and wasting of financial, human and other resources during times of disasters.

The Ministry of Home Affairs and traditional leadership should continue assisting Tropical Idai victims with the necessary documentation, including death certificates. This will allow families to access additional resources such as pensions and find closure in terms of loss. The issuing of national IDs and passports remains critical, especially given that many of the victims were cross-border traders frequenting both Mozambique and South Africa. Prioritising these victims makes sense in a country where a passport is one of the most sought-after documents. The Ministry of Primary and Secondary Education, and the Ministry of Higher and Tertiary Education, Science and Technology Development should also prioritise the issue of related qualification certificates. This can be done by drawing up a list of affected individuals and circulating it to all tertiary and vocational institutions in the country. In addition, having these as long-term protocols would also help.

A prolonged and long-term psycho-social support programme is essential to reach all that were impacted, and for future purposes. The structures should be put in place, with the potential to have development partners that commit to a permanent stay in the district. This programme should not only be targeted mainly at schools, particularly the learners, which is what happened, but it should also be aimed at teachers, other civil servants and the affected communities.

There must be a council disaster risk and vulnerability map (RVM) that determines the frequency of tropical cyclones in the area. This RVM could include the potential risk to remaining and/or destabilised mountain and hill slopes and settlements. Identified boulders could be systematically blasted or pulled down to remove the potential rock-fall danger. A good example is the balancing rocks at St Patrick Luangwa Secondary School that remained after landslides following Tropical Cyclone Idai. Furthermore, the certification of settlement in traditional areas should be revived so that the council understands what is taking place on the ground and where people are settling, especially which waterways are being systematically colonised for settlement.

Drawing from the EMA (2019) work, a few recommendations of interest are worth mentioning that support these proposals. EMA recommended zoning off all areas within 50 m of the highest Tropical Cyclone Idai level, particularly of the naturally defined flood area of Nyamatanda River. Some remaining houses within a 30-m distance of the waterways in Gangue were deemed too risky and therefore could be condemned unfit for occupation and demolished.

Some streams created during Tropical Cyclone Idai could be diverted back into main channels. This includes a stream of the Rusitu River which passed near the ZAOGA Church at Kopa Centre (EMA 2019). Environmental reconstruction and rehabilitation was another recommendation from EMA work, and we fully embrace it, given our before and after images shown in this book. There was excessive erosion in many places within Chimanimani.

As some houses that washed away at Kopa were poorly constructed, pointing to lack of oversight by the council building inspectorate, there is a need to investigate further if foundations were built properly. The remaining houses and structures could be useful in determining this. The findings will then assist the council in strengthening its building inspectorate oversight for new houses.

As for the energy programme, the ZETDC should continue rerouting their transmission lines and replacing 25 mm wires with 50 mm wires as appropriate, as they had already started this project. Furthermore, an investigation could be launched into the feasibility of scaling-up solar since the installations at the council clinic were spared. The ZETDC should also continue negotiating with plantation owners to leave adequate space along wayleaves so that plantation trees do not fall across powerlines during a disaster and other weather extremes.

Concerning road infrastructure, the council and line ministries should audit the quality of road repair and/or new road reconstruction, with a view to increasing capacity for some contractors. Roads and bridges need to be resilient to extreme weather events like Tropical Cyclone Idai. This also applies to other infrastructure such as the reconstruction and/or building of new school blocks, health, as well as water and sanitation facilities. Furthermore, long-term programmes involving communities in the rehabilitation of Chimanimani in the aftermath of Tropical Cyclone Idai or any other disaster need to be instituted by government and its development partners. These programmes could be food-for-work and other community-led initiatives. Gully and other land rehabilitation/landscaping projects remain the first in line.

15.7 Cross-cutting Matters

15.7.1 Key Findings

Repeated contact with the Department of Civil Protection in Harare confirmed that there were long-term delays in the finalisation of the relevant policies governing DRR and management in Zimbabwe. While some effort was being made with some form of draft bills in place, there was no indication as to when the new bills would be finalised.

There were also many theories regarding the cause of Tropical Cyclone Idai, science-based, spiritual and religious among others. Some believed Tropical Cyclone Idai was a cleansing from a spiritual world, with the main argument being that locals had ill-gotten wealth. Some believed that like the Biblical Sodom and Gomora, God was cleansing places filled with sin. Some believed it was climate change and resulting extreme weather events. Yet others believed there were locals that touched sacred things in the mountains, and the ancestors had to retaliate.

The main challenge with the B4 Model as experienced in the aftermath of Tropical Cyclone Idai is the slow progress. Those displaced faced three rainy seasons before the first set of houses could be occupied at Green Farm at Bumba, some 70 km from Chimanimani Town. As of the second week of October 2020, there were only three houses at roof level and five at the foundation level of the close-to 250 planned. The President of the country had also toured Green Farm to commission the project.

15.7.2 Key Recommendations

Chatiza (2020, p. 6) made six top-level cross-cutting government recommendations in response to Tropical Cyclone Idai. In his view, the government and its development partners should:

1. Set up a clear fund for disaster risk assessments and analyses;
2. Finalise the DRR and management policy, legislation and organisational structures;
3. Set up a decentralised "corps" of trained DRR and the management volunteers;
4. Invest in adaptive and resilience-building measures to protect women and other vulnerable groups;
5. Strengthen rural and urban settlement and infrastructure regulatory regimes; and
6. Set up social and child protection systems that are sensitive to disaster situations.

Other recommendations we draw based on our fieldwork relate to beliefs and knowledge systems. There is a need for all knowledge systems to be recognised and, where possible, harmonised with the main aim of strengthening early warnings and preparedness. The traditional leaders need to be assisted where possible to perform their traditional cleansing ceremonies as well as repair/rehabilitate some of their sacred shrines. Fortunately, this was done. There is also a need to protect the damaged sites through a by-law as communities started mining quarry and other commodities, an element that is likely to reduce the value of these sites for the purposes of dark tourism and other uses. For example the rescue rope that was taken by a certain NGO from Kopa could have been left on-site or secured in a museum. In fact, many more artefacts have already disappeared.

The B4 Model must include relocating victims to areas where they can continue with former livelihoods. The Green Farm was located in the transition zone between agro-ecological regions 3 and 4, yet the victims were from agro-ecological regions 1 and 2. It is therefore not surprising that even some victims that were relocated within 7 km of their original locations by the World Vison still refused to move to their new houses (https://www.newzimbabwe.com/cyclone-survivors-resist-relocation-to-new-donated-homes/). As for Green Farm, this location remains contested in terms of the continuation of livelihoods.

References

Bang, H.N. (2014). General overview of the disaster management framework in Cameroon. Disasters, 38(3):562–586.

Chatiza, K. (2020). Cyclone Idai in Zimbabwe: An analysis of policy implications for post-disaster institutional development to strengthen disaster risk management. Harare: Oxfam International.

Chikoto, G. L. (2004). Zimbabwe's Emergency Management Systems, Disaster Management Class Project. Georgia State University, Atlanta, Georgia.

EMA (Environmental Management Agency). (2019). Cyclone Idai Assessment Report in Chimanimani and Chipinge Districts. Harare: EMA.

GoZ (Government of Zimbabwe). (2001). Civil Protection Act. Parliament of Zimbabwe Document (Chapter 10.06, pp. 12). Harare, Zimbabwe.

Madamombe, E.K. (2004). "Zimbabwe: Flood Management Practices—Selected Flood Prone Areas Zambezi Basin." Unpublished Paper, WMO/GWP Associated Programme on Flood Management. http://www.apfm.info/pdf/case_studies/zimbabwe.pdf.

Manyena SB, Mavhura E, Muzenda C, Mabaso E. (2013). Disaster risk reduction legislations: Is there a move from events to processes? Glob Environ Change, 23(6):1786–1794. http://linkinghub.elsevier.com/retrieve/pii/S0959378013001337.

Mavhura, E. (2015) Disaster legislation: a critical review of the Civil Protection Act of Zimbabwe. Nat Hazards 80:605–621. https://doi.org/10.1007/s11069-015-1986-1.

Index